Econometrics in Practice

Companion files are available for downloading by writing to info@merclearning.com.

ECONOMETRICS IN PRACTICE

Paul Turner, PhD

MERCURY LEARNING AND INFORMATION
Dulles, Virginia
Boston, Massachusetts
New Delhi

Publisher: David Pallai
Mercury Learning and Information
22841 Quicksilver Drive
Dulles, VA 20166
info@merclearning.com
www.merclearning.com
800-232-0223

Paul Turner. *Econometrics in Practice.*
ISBN: 978-1-68392-660-3

Library of Congress Control Number: 2021934962

212223321 This book is printed on acid-free paper in the United States of America.

To my wife Vicki, and my daughters,
Rachel, Catherine, and Roisin

CONTENTS

PREFACE

This book has grown out of econometrics modules that I have taught over the past twenty years at various universities. During that time there have been major changes in the subject matter of the discipline but even more significant changes in the way it is taught. Both of these developments have been due to the increasing availability of cheap, fast computing power. The subject itself has changed as numerical or "Monte Carlo" methods have allowed econometricians to explore the properties of estimators through the use of artificially generated data. The teaching of the subject has changed because students can, from an early stage, get hands-on experience of the methods being taught using personal computers and widely available econometric software.

In this book, I have attempted to make econometrics accessible for students in ways that reflect both the trends described. The emphasis throughout is on econometrics as a practical discipline. Some level of statistical theory is essential, but this is accompanied at each stage by examples drawn from either real-world data sets, or artificially created data sets designed to illustrate particular points. Wherever possible, the data sets are made available for download by instructors and students, so that they can replicate the results given in the text and try out alternative specifications. Exercises are provided at the end of each chapter to reinforce understanding of the important points, with worked answers provided for odd-numbered problems. The online resources for the book consist of Excel data sets for the problems and examples given in the text.

The subject matter of this book consists of the econometric methods necessary for a practicing applied economist. Chapters 1 to 9 present the material most often taught as part of an undergraduate degree program in economics. Much of this is concerned with basic statistical theory. Chapters 1 and 2 review basic concepts in probability theory. This leads to the development of the most basic tool in the kitbag of the econometrician – the linear regression model. Chapter 3 presents the bivariate linear regression model and uses the probability theory of the earlier chapters to derive the sampling distribution of the regression parameter estimates. This gives the basic framework necessary for the estimation and interpretation of model parameters. The main estimation method discussed is that of least squares, but the idea of maximum likelihood estimation, which takes on increasing importance in later chapters, is also introduced at this stage. In Chapter 4, the linear regression model is extended to the multivariable case. This is particularly important for econometrics, since it deals with some of the problems caused by the fact that economics is typically not an experimental discipline. In particular, it permits the isolation of phenomena of interest through the inclusion of variables designed to allow for effects which are outside the control of the econometrician.

Chapters 5 to 7 present tests for the assumptions underlying the classical statistical method. In Chapter 5, the possibility that the errors of the regression model are not independent is explored. This is a particular problem for models estimated using time series data where it is known as the problem of serial correlation. Chapter 6 considers the problem of heteroscedasticity, or situations in which the variance of the errors is not constant. This is often associated with models estimated using cross-section data but can also be found in some time series applications. Chapter 7 modifies the standard regression model to allow for limited dependent variables, that is, situations in which the dependent variable is grouped according to some qualitative property. The simplest case of this is where the data are binary in nature, for example, an individual is either employed or unemployed. In cases like this, the linear regression model becomes hard to interpret and the method of maximum likelihood becomes the statistical tool of choice.

Classical statistics typically assumes that the independent variables in a regression are "fixed in repeated samples." This means that they are the outcome of an experimental process in which the investigator chooses the model inputs. In Chapter 8, we consider the implications of modifying this assumption so that the independent variables are themselves stochastic variables. That is, dependent and independent variables are generated by a joint random process with unknown parameters. This permits the analysis of the problems of "errors in variables" and "simultaneous equations" which require the introduction of new estimators such as indirect least squares and instrumental variables. It also permits discussion of the issue of the "identification" of models, that is, the extent to which the unknown structural parameters of interest can be isolated when the data is generated by a complex system of interacting variables.

Chapter 9 discusses the issue of dynamic adjustment in time series relationships. This is a central feature of most time series econometric models. Models in which dynamic relationships are present typically include lagged values of both the dependent and explanatory variables. The statistical implications of including such lags in the regression model and the methodology for choosing an appropriate model to capture the process of dynamic adjustment are discussed in the chapter along with the role of the General to Specific modeling methodology in econometrics.

Chapters 10–13 introduce topics that are often taught as part of undergraduate econometrics programs but not, generally, as part of the core econometrics module. Instead, these topics typically form the basis of more advanced option modules. The particular focus of interest is the use of time series methods in applied econometrics. Chapter 10 discusses the historical origin of this approach in the form of Box-Jenkins or ARIMA modeling of individual time series. This theme is developed further in Chapter 11, where the importance of unit root processes for both estimation and inference is discussed. Chapter 12 builds on the discussion of individual unit root processes in Chapter 11, to introduce the idea of cointegration which allows for links between multiple unit root processes. Finally, in Chapter 13, there is a brief survey of the topic of vector autoregressions. This is both a topic of interest in itself and a way of bringing together the material developed in Chapters 10 to 12.

Companion Files. *(available for downloading by writing to the publisher at info@merclearning.com)*

Microsoft Excel Data Files. Each chapter in this book comes with a set of exercises. Many of these are hands-on exercises based on data which can be accessed as part of the resources for instructors and students. There are numerous alternative econometric software packages available such as EViews, PC-Give, and Stata which can be used to reproduce the results of the text and to work through the chapter exercises. Rather than link this text to one specific package, the data is supplied in the form of Microsoft Excel data files. There are 22 data sets in total and the only software which users need to access them is Microsoft Excel or a similar program. All of the standard econometrics software packages have the ability to input data either directly from Excel files or by copying and pasting from Excel. This is a generic format which allows the user to read the data quickly into the package of choice. Most of the examples given in the text have been generated using either the EViews regression package or software written by the author.

Figures in the Text. All of the figures from the text, including those originally in full-color, appear in the companion files.

Instructor Ancillaries. The following instructor ancillaries are available to adopters of the text by writing to the publisher at info@merclearning.com:

Solutions to Even-Numbered Exercises. Worked-out solutions to even-numbered exercises in the text.

Microsoft Excel Data Files. Twenty-two data sets in total to be used with the exercises in the book.

Microsoft Power Point Slides. Each chapter includes slides with key terms, equations, figures, content, etc.

Figures in the Text. All of the figures from the text, including those originally in full-color.

ACKNOWLEDGMENTS

Thanks are due to my academic colleagues for giving me the opportunity to discuss many of the topics in this book and thereby clarify my thoughts on the issues. In particular, I would like to thank Terry Mills, Eric Pentecost, and Chris Green for continuing support. All errors and omissions remain my responsibility.

Paul Turner
April 2021

CHAPTER 1

Probability and the Statistical Foundations of Econometrics

The statistical foundations of econometric analysis lie in the theory of probability. Many of you reading this textbook will have already completed an introductory module in statistics in which you will have come across some of the important ideas of probability theory. However, it will be useful to review these ideas before we move on to more advanced topics. If you haven't been introduced to these concepts already, then this chapter will cover the essential ideas you need to study econometrics.

Before we begin our discussion of probability, we must first introduce some terminology. The most basic concept of statistical theory is the idea of a *random experiment*. This is an experiment that can be repeated a number of times, under essentially similar conditions, but whose outcome is uncertain. Consider, for example, the tossing of a coin. This can be repeated any number of times, but the outcome of any single coin toss is not known in advance. The set of possible outcomes of a random experiment is known as the *sample space*. In the case of the coin toss, the sample space consists of two possibilities – heads or tails. Finally, an *event* is a subset of the sample space which corresponds to a particular outcome, for example, heads or tails in the coin toss experiment.

The coin toss experiment we described in the previous paragraph is an example of a special kind of experiment known as a *Bernoulli trial*. In this kind of experiment the sample space can be reduced to only two possible outcomes which can be classified as "success" and "failure." For example, we might define a head as a success and a tail as a failure. However, this is essentially arbitrary because nothing would change if we were to reverse these labels. Bernoulli trials are a very important special case of a random experiment because many real-world applications can be described in these terms and a surprising amount of statistical theory can be developed using this as a basis.

Let us suppose that we conduct a total of n Bernoulli trials and that we observe k successes. We can define the *relative frequency* of successes as k / n. The *probability* of a success in an individual trial can then be defined as the value to which the relative frequency converges as the number of experiments becomes large. For example, in the case of the coin toss experiment, if the coin is unbiased, we would expect the relative frequency of success (heads) to average out at ½ as the number of experiments increases. More generally, let p denote the probability of success in an individual trial. Therefore, it follows that the probability of failure is equal to $1 - p$ because the two events (success and failure) constitute the whole of the sample space. Another way of describing this is to say that the two possible events are *exhaustive*. Note also that success and failure are *mutually exclusive events*, that is, they cannot occur simultaneously.

To illustrate some of the ideas we have introduced, let us consider another example. Suppose we have a well-shuffled pack of cards. We make a draw from the pack and inspect the card. If the card drawn is a club then we deem the experiment a success. We then return the card and repeat the experiment a large number of times. The probability of drawing a club can then be calculated as the number of successes divided by the number of experiments. Since there are 13 clubs in a pack of 52 cards, it is not hard to see that the probability of drawing a club is equal to ¼ and the probability of drawing another suit is ¾, that is,

$$\begin{aligned} p(A) &= \frac{1}{4} \\ p(B) &= 1 - p(A) = \frac{3}{4}, \end{aligned} \tag{1.1}$$

where we have defined event A as the drawing of a club and the event B as the drawing of any other suit. The two probabilities defined in equation (1.1) define the *probability distribution function* for an individual Bernoulli trial. That is, they attach a probability to all possible outcomes in the sample space.

Now let us consider another experiment, this time we make two successive draws from the pack (after replacing the card following the initial draw). The sample space now consists of four possible outcomes which can be summarized as $(AA),(AB),(BA),(BB)$. To find the probability distribution function for this experiment we note that the outcomes of each draw are *independent* of each other. That is, the probability of drawing a club on the second draw is not influenced by whether a club was drawn in the first draw. It follows that probability of two successive clubs can be calculated as $p(AA)=1/4\times1/4=1/16$, similarly, the probability of a club followed by another suit is $p(AB)=1/4\times3/4=3/16$. In this manner, we can construct the probability distribution function as

$$\begin{aligned}
p(AA)&=\frac{1}{4}\times\frac{1}{4}=\frac{1}{16}\\
p(AB)&=\frac{1}{4}\times\frac{3}{4}=\frac{3}{16}\\
p(BA)&=\frac{3}{4}\times\frac{1}{4}=\frac{3}{16}\\
p(BB)&=\frac{3}{4}\times\frac{3}{4}=\frac{9}{16}.
\end{aligned} \tag{1.2}$$

Note that, because the events listed in (1.2) are exhaustive, their probabilities sum to one. This example also illustrates that, depending on the definition of an event, there are different ways in which some events can occur. For example, suppose we are interested in evaluating the probability distribution function of the number of clubs drawn. The probability that we draw one club is equal to the sum of the probability that we observe a club on the first draw followed by another suit and the probability that we observe another suit on the first draw followed by a club on the second. Thus, if we are simply interested in the distribution of the number of clubs k, and the order in which they occur is irrelevant, then we could write the probability distribution function as

$$p(k=0)=\frac{9}{16}, p(k=1)=\frac{6}{16}, p(k=2)=\frac{1}{16} \tag{1.3}$$

Historical Note: The historical origins of the theory of probability can be found in a series of letters between Blaise Pascal (1623–1662) and Pierre de Fermat (1607–1665) in 1654. They considered the problem of how to divide up the winnings in a game of chance that was incomplete. In doing so they established the key ideas of probability and expected value. The interested reader can find an excellent discussion of this in Devlin [Devlin2008].

1.1 JOINT, CONDITIONAL, AND MARGINAL PROBABILITIES

So far, we have considered events that are mutually exclusive. However, this will not always be the case. Often, we will be interested in experiments in which there are multiple outcomes, some of which are not mutually exclusive. For example, when considering our experiment of drawing a card from a pack, let us suppose that we are also interested in whether the card drawn is a face card. Note that a card can be both a face card and a club, so the two events are not mutually exclusive. Let us define event A as the card drawn being a club, and event C as the card drawn being a face card. We can define the *joint probability* that two events occur simultaneously as $p(A \cap C)$. The joint probability can be decomposed into the product of a *conditional probability* (the probability that one event occurs given that another has occurred) and a *marginal probability* (the simple probability that one event occurs irrespective of the other event). The mathematical notation for a conditional probability is $p(A|C)$, that is, the probability event A occurs given event C, and that for a marginal probability is $p(C)$. The relationship between joint, conditional, and marginal probabilities can be written as

$$p(A \cap C) = p(A|C)p(C) = p(C|A)p(A) \tag{1.4}$$

This relationship is fundamental to probability theory and many important results derive directly from this definition.

Using we can write the conditional probability as

$$p(A|C) = \frac{p(A \cap C)}{p(C)} \tag{1.5}$$

This shows that the conditional probability is determined as the ratio of the joint probability to the marginal probability. Two other results that derive directly from the definition of conditional probability are:

1. The sum of the joint probabilities across all outcomes gives the marginal probability, that is, $p(C) = p(C \cap A) + p(C \cap B)$.
2. The conditional probability relationships are symmetric, which means that $p(A|C)p(C) = p(C|A)p(A)$.

From the definition of the joint probability and the symmetry of conditional probabilities, we have $p(A \cap C) = p(C|A)p(A)$. Substituting this allows us to write

$$p(A|C) = p(C|A)\frac{p(A)}{p(C)} \tag{1.6}$$

TABLE 1.1 Contingency Table for US Market for Potatoes.

	Price rises	**Price falls**	**Total**
Quantity rises	6	21	27
Quantity falls	11	4	15
Total	17	25	42

This form of the probability relationship is referred to as Bayes' Law or Bayes' Theorem, after the eighteenth century cleric and statistician the Reverend Thomas Bayes. It is frequently used in the derivation of conditional probabilities based on experimental data.

Now let us return to our example, in which we wish to determine the joint probability that a card drawn from a pack is both a club and a face card. There are three face cards which are also clubs. Therefore, the conditional probability that the card drawn is a face card, given that it is a club is $p(C|A)=3/13$ and we already know that the unconditional probability that the card draw is a club is $p(A)=1/4$. It follows from the definition of the conditional probability that the joint probability is $p(C\cap A)= 3/13\times 1/4=3/52$.

When dealing with more complex situations in which there are multiple events that are not mutually exclusive, it is often useful to represent the probability distribution function in tabular form. To do this we will introduce the idea of a *contingency table*. Let us begin with a simple example in which we wish to examine the relationship between changes in price and quantity for a product. The events we consider are firstly, whether prices fall or rise, and secondly, whether quantity produced falls or rises.[1] Table 1.1 summarizes the different outcomes observed for the market for potatoes based on annual United States data for the period 1976–2017.

We have argued that probabilities can be thought of as relative frequencies for large samples. This means that we can estimate the probabilities of the different events defined in the contingency table by dividing each of the cell entries by the total number of observations. The results are shown in Table 1.2.

---

[1] You may wonder how we would deal with the case where the change is identically zero. If the data are measured on a continuous basis, then the probability that this occurs is very small and can be safely ignored. However, if this is a concern, then it is always possible to define one of the events to include this possibility, for example, the price falls *or* remains constant.

TABLE 1.2 Two-Way Probability Table for US Market for Potatoes.

	Price rises	Price falls	Total
Quantity rises	0.1428	0.5000	0.6429
Quantity falls	0.2619	0.0952	0.3571
Total	0.4047	0.5952	1.0000

In this form, we can interpret the cell entries as probabilities. Where an entry shows the relationship between two events, we have a joint probability. For example, the probability that price rises *and* quantity falls, is equal to 0.2619. The row and column sums of these entries give marginal probabilities. For example, the probability that the price rises is equal to 0.4047 which is the sum of two mutually exclusive joint events, i.e., the probability that price and quantity rise simultaneously, and that price rises while quantity falls. Having calculated joint and marginal probabilities, it is now straightforward to calculate conditional probabilities. Suppose, for example, that we wish to calculate the probability that price rises given that quantity falls. This is calculated as the ratio of the joint probability that these events occur simultaneously to the marginal probability that quantity falls, that is, $0.2619/0.3571 = 0.7334$. A common mistake is to confuse joint and conditional probabilities when discussing related events but, as this example illustrates, this can be very misleading.

Example: Suppose we are interested in the behavior of the Federal Reserve when setting the Federal Funds Rate (FFR). We assume that it has three options which we will label as follows, Y_1 is the case where it cuts the FFR, Y_2 is the case where it leaves it constant and Y_3 is the case where it increases the FFR. There are also three different states of the economy which might influence this decision, X_1 is the case where inflation is below target, X_2 is the case where inflation is equal to the target (or within the target range) and X_3 is the case where inflation is above target. Now suppose we have observed how the Federal Reserve behaves over a period of time and determined the relative frequencies (or joint probabilities) for these events. These are shown in Table 1.3 as the numbers in the central rectangle. For example, the joint probability of inflation being too low and the interest rate being cut is equal to $p(X_1 \cap Y_1) = 0.1$ while the probability of inflation being too high and the interest rate being cut is $P(X_3 \cap Y_1) = 0.01$. Once we have determined these joint probabilities then it is straightforward to determine the marginal and conditional probabilities. The marginal probabilities are calculated as either the row or column sums of the joint probabilities. The conditional probabilities can then be calculated as the ratio of joint probabilities to marginal probabilities using (1.5).

TABLE 1.3 Probability Distribution for Federal Reserve Interest Rate Decision.

		Inflation too low X_1	Inflation equal to target X_2	Inflation too high X_3	
Cut interest rate	Y_1	0.10	0.10	0.01	0.21
Keep interest rate constant	Y_2	0.08	0.40	0.04	0.52
Increase interest rate	Y_3	0.02	0.10	0.15	0.27
		0.20	0.60	0.20	

For example, consider the marginal probability that the interest rate will be cut. This depends on the relationship between the actual rate of inflation and the target rate. There are three possible scenarios that need to be considered and the probability of an interest rate cut is different in each. However, because these scenarios are mutually exclusive, we can calculate the overall probability of an interest rate cut as the sum of the three joint probabilities as shown in the following equation:

$$\begin{aligned} p(Y_1) &= p(Y_1 \cap X_1) + p(Y_1 \cap X_2) + p(Y_1 \cap X_3) \\ &= 0.10 + 0.10 + 0.01 = 0.21 \end{aligned} \tag{1.7}$$

This equation defines the marginal probability of an interest rate cut. Historically, probabilities of this kind were calculated as the sum of the row (or column) entries of the contingency table and then written in its margins – hence the term marginal probability. The marginal probabilities of the interest rate being held constant and of it increasing are given by the other row sums and are equal to $p(Y_2) = 0.52$ and $p(Y_3) = 0.27$. Similarly, the column sums give the probabilities of inflation being below, equal to or above target which are, respectively, $p(X_1) = 0.20$, $p(X_2) = 0.60$, and $p(X_3) = 0.20$. Since the events associated with the marginal probabilities are both mutually exclusive and exhaustive, it follows that the marginal probabilities sum to one in both cases.

Our interest is often in the conditional probabilities rather than the probabilities that appear in the contingency table. However, these can easily be calculated using the information in the table. For example, suppose we wish to calculate the probability that the Federal Reserve will cut the interest rate if inflation is below target. Using the definition of conditional probability, we have

$$p(Y_1 | X_1) = \frac{p(Y_1 \cap X_1)}{p(X_1)} = \frac{0.10}{0.20} = 0.50 \tag{1.8}$$

that is, there is a 50% chance that the interest rate will be cut when inflation falls below target. Similarly, if we wish to calculate that probability that the interest rate will *not* change, even when inflation is above target, then we can write this as

$$p(Y_2 | X_3) = \frac{p(Y_2 \cap X_3)}{p(X_3)} = \frac{0.04}{0.60} = 0.067 \tag{1.9}$$

Calculation of the other conditional probabilities is left as an exercise for the interested reader.

1.2 THE PROBABILITY DISTRIBUTION FUNCTION

So far the random variables we have considered have been *discrete random variables*. This means that the number of possible outcomes for the random experiment is limited. The *probability distribution function* defines the probability of all the possible outcomes in the sample space. This function is important because it can be used to define the mean and the variance of the distribution in question. Suppose we have an experiment in which there are $n + 1$ possible outcomes, corresponding to $x = 0,1,...,n$. The mean, or expected value, of the random variable X can be defined as

$$\mu_X = E(X) = \sum_{x=0}^{n} p(x)x \tag{1.10}$$

while the variance, defined as the expected value of the squared deviation of the random variable from its mean, can be written

$$\sigma_X^2 = E(X-\mu)^2 = \sum_{x=0}^{n} p(x)(X-\mu)^2 \tag{1.11}$$

For example, let X be the number of successes in a set of n Bernoulli trials. This is a random variable that can take on the values $x = 0,1,...,n$. Such a random variable follows the binomial distribution and has a probability distribution function of the form

$$p(x) = \frac{n!}{(n-x)!x!} p^x (1-p)^{n-x} \tag{1.12}$$

We can show that the mean of this distribution is $\mu_X = E(X) = np$ and the variance is $\sigma_X^2 = E(X - E(X))^2 = np(1-p)$. We can also define the *cumulative probability distribution function* or CDF as the function $F(x) = p(X \leq x)$. The CDF is calculated by adding the individual probabilities over the range 0 to x. The binomial distribution is interesting in its own right but is also of historical importance because it led to the development of the *normal distribution* which we will consider in detail later.

Consider again the Bernoulli experiment of drawing a card from a pack and in which a success corresponds to drawing a club. If five draws (with replacement) are made, then the sample space consists of six alternative outcomes ranging from no successes to all five draws being clubs. We can calculate the probability distribution function of the number of clubs drawn using (1.12). This gives the results shown in Table 1.4.

The probability distribution function given in Table 1.4 can be presented as a bar chart, as shown in Figure 1.1. This shows that the probabilities are unevenly distributed. The probabilities attached to lower values of X are larger than those for higher values. Thus, the distribution appears to be asymmetric, with large probability values at the lower end and probabilities that decline gradually towards zero at the upper end.

As the number of trials increases, the shape of the PDF changes. Figure 1.2 shows PDFs for binomial distributions with $n = 10$ and $n = 30$. For $n = 10$ the shape remains basically similar to that for $n = 5$ with high probabilities for low values and a gradual decline in probability for $x > 2$. However, relative to $n = 5$, there is already a reduction in the degree of

TABLE 1.4 Probability Distribution for the Binomial Distribution with $n = 5$ and $p = ¼$.

Number of successes = x	Probability $p(X = x)$	Cumulative probability $p(X \leq x)$
0	0.23730	0.23730
1	0.39551	0.63281
2	0.26367	0.89648
3	0.08789	0.98437
4	0.01465	0.99902
5	9.77E-04	1.00000

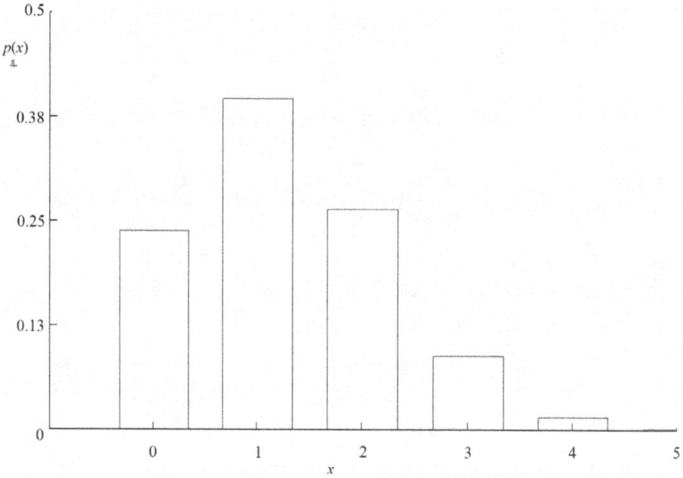

FIGURE 1.1 Probability Distribution Function for the Binomial Distribution with $n = 5$ and $p = ¼$.

asymmetry of the function. This process continues as we increase the number of trials. When $n = 30$ the asymmetry in the function is hardly visible at all. The probability of the number of successes is close to zero when $x = 0$, increases to a maximum when $x = 7$, and then declines back to 0 as $x \to 30$. In this case, however, the function is close to being symmetric around the maximum point.

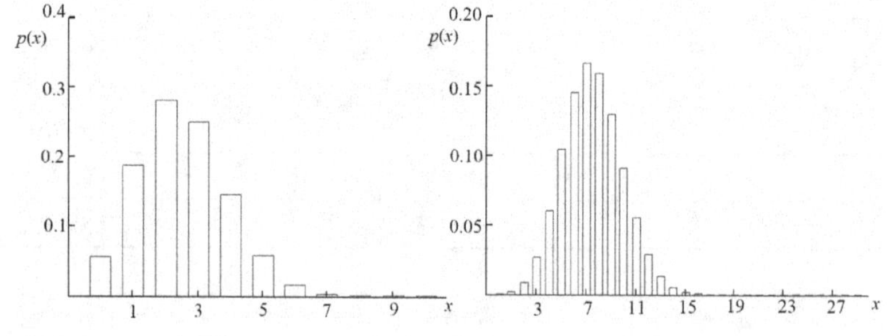

FIGURE 1.2 Probability Distribution Functions for the Binomial Distribution with $p = ¼$.

The tendency for the PDF to become more symmetric as the number of trials increases is no accident. As the number of trials increases, the shape of the PDF function can be more closely approximated by a continuous function $f(x)$ which eventually converges to the function given in equation (1.13) where $\mu = np$ and $\sigma^2 = np(1-p)$. This is the equation of the normal probability density function. The convergence of the binomial distribution is an example of a much more general phenomenon known as the *Central Limit Theorem*. This is an important theorem for statistical theory which shows that whatever the process determining the probabilities of success in an individual trial, the shape of the distribution of the number of successes will eventually converge on the normal probability density function as given in equation (1.13)

$$f(x) = \frac{1}{\sigma\sqrt{2\pi}} \exp\left(-\frac{(x-\mu)^2}{2\sigma^2}\right), \tag{1.13}$$

where μ and σ^2 are the mean and the variance of the distribution. For example, Figure 1.3 shows that the equation for the normal distribution given by provides a very close approximation to the PDF for the binomial distribution with $n = 30$. The bars in Figure 1.3 show the probabilities of the number of successes from the binomial distribution, that is, $p(X = x)$ from (1.12). The continuous line shows the curve generated by equation (1.13) with $\mu = np$ and $\sigma^2 = np(1-p)$. We can see that for any given value of X the normal curve is a good approximation to the binomial probabilities. This approximation will continue to improve as we increase the number of trials.

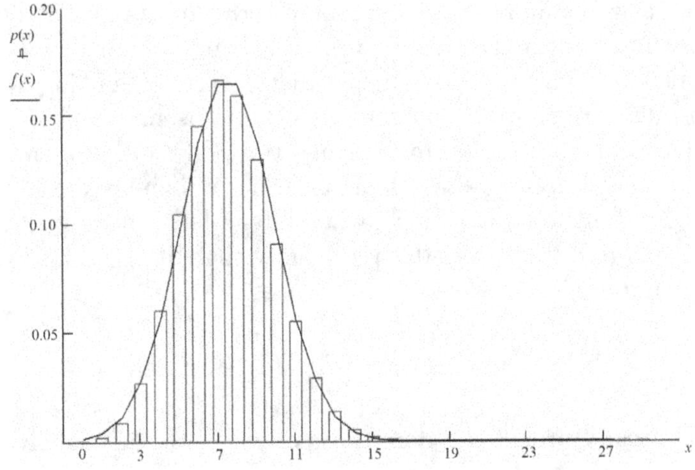

FIGURE 1.3 Normal Approximation to the Binomial PDF.

Historical Note: The binomial theorem was first set out by Jakob Bernouilli (1654–1705) in a posthumously published book in 1713 [Bernouilli1713]. Bernouilli showed that the probability of k successes in n trials is given by the coefficient of $p^k q^{n-k}$ in the expansion of $(p+q)^n$ where $q=1-p$. This generates the probabilities shown in (1.12). The special case where $p=1/2$ had previously been considered by Pascal in his correspondence with Fermat. The use of the normal function to approximate the binomial coefficients when n is large was introduced by De Moivre [DeMoivre1718] in the *Doctrine of Chances* (1718).

1.3 THE NORMAL DISTRIBUTION

We have seen that the normal curve provides a continuous curve which gives a good approximation to the binomial probabilities for a large number of trials. However, it is also an example of a general class known as *continuous distributions*. These allow us to calculate probabilities for *continuous random variables*. A continuous random variable is one that can take any real value on some interval. For example, we might wish to measure the heights of a group of people, the temperature in different locations at noon on a particular date or the distance between the place of residence and the place of work for the employees of a company. In all these cases the random variable is more naturally thought of as lying somewhere on a continuum of possible values rather than taking one of a discrete number of possibilities.

If the random variable X can take on a continuum of values along some range, it makes more sense to think in terms of the probability that X lies between two particular values within that range rather than being equal to a particular point value. This means that instead of thinking in terms of the probability distribution function, which assigns probabilities to particular point values of X, we need to think in terms of the rather more difficult concept of a *probability density function*. The probability density function, or PDF, is a function $f(x)$ which, when integrated with respect to x between two limits a and b, gives the probability that the random variable X lies between these limits, that is

$$p(a \le X \le b) = \int_a^b f(x)\,dx. \tag{1.14}$$

To be a valid probability density function $f(x)$ must satisfy two criteria. Firstly, it must be positive for all values of x, $f(x) \ge 0$ and, secondly, the area under the curve must equal one, $\int_a^b f(x)\,dx = 1$, where a and b are the limits of the range of possible values for the random variable. The normal curve satisfies both these properties. Therefore, if X is a random variable that follows a normal distribution with known mean and variance, then we can calculate the probability that X lies between any two real numbers by a process of integration using equation (1.13). Note that the PDF of the normal distribution defined in equation 1.13 is a function of only two parameters the mean μ and the variance σ^2.

The *standard normal distribution* is the normal distribution with mean zero and variance one. Any normally distributed random variable can be transformed to create another random variable with the standard normal distribution by subtracting the mean and dividing by the standard deviation (or square root of the variance). This transformation is illustrated in the following expression

$$X \sim N\left(\mu,\sigma^2\right) \Rightarrow Z = \frac{X-\mu}{\sigma} \sim N(0,1). \tag{1.15}$$

Transformation to the standard normal distribution is useful because integrals for this function are tabulated and available in books of statistical tables. This allows us to calculate critical values and confidence intervals for any arbitrary normal distribution without the computational difficulty of evaluating integrals. This is illustrated in Figure 1.4 where the shaded area is equal to 5% of the total mass of the distribution. This is the integral of the function between the limits 1.645 and ∞ which gives the probability that the random variable with this distribution lies between these limits.

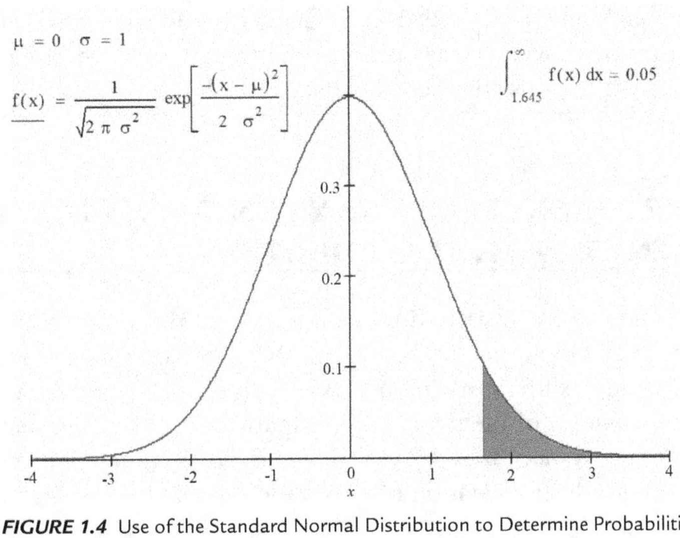

FIGURE 1.4 Use of the Standard Normal Distribution to Determine Probabilities.

An important feature of the normal distribution is that linear combinations of normally distributed random variables will themselves follow a normal distribution. For example, let $X_1 \sim N(\mu_1, \sigma_1^2)$ and $X_2 \sim N(\mu_2, \sigma_2^2)$ be independent normal random variables. If a and b are constants, then a linear combination of the variables using a and b as weights has the following normal distribution

$$aX_1 + bX_2 \sim N(a\mu_1 + b\mu_2, a^2\sigma_1^2 + b^2\sigma_2^2), \tag{1.16}$$

If X_1 and X_2 are not independent, then their covariance, that is, $E\{(X_1 - \mu_1)(X_2 - \mu_2)\}$, is not zero, and this expression becomes

$$aX_1 + bX_2 \sim N(a\mu_1 + b\mu_2, a^2\sigma_1^2 + b^2\sigma_2^2 + 2ab\sigma_{12}) \tag{1.17}$$

where σ_{12} is the covariance. The normal distribution is unique in having this property and therefore the assumption of normality is very useful in deriving the distribution of random variables which are functions of other random variables. Of course, the fact that the normal distribution has convenient properties is not a reason in itself to make the assumption but, as we saw earlier in our discussion of the central limit theorem, there are often good reasons to assume that random variables are approximately normally distributed in large enough samples.

Historical Note: Although the normal function had been introduced by Abraham De Moivre (1667–1754) as a method of approximating the coefficients of the binomial expansion, Carl Friedrich Gauss (1777–1855) was the first to interpret it as a probability density function in its own right in his book of 1809 [Gauss1809]. Gauss's particular contribution was to interpret the normal curve as capturing the distribution of measurement errors with imperfectly recorded data.

1.4 THE PROBABILITY DENSITY FUNCTION AND THE MOMENTS OF THE DISTRIBUTION

The *moments* of a distribution are defined as the expectations of integer powers of the random variable in question. For example, if X is a random variable, then its first three moments are $E(X), E(X^2)$ and $E(X^3)$. These are the *raw moments* of the distribution. Apart from the first moment, it is usually more convenient to work in terms of the *central moments* which are the expectations of the deviation of the random variable from its mean (or first moment). Thus the second central moment of the random variable X can be written as $E(X-E(X))^2 = \sigma^2$ which is the variance of X. Higher-order moments are often scaled by the standard deviation to obtain measures such as *skewness* $= E(X-E(X))^3 / \sigma^3$ and *kurtosis* $= E(X-E(X))^4 / \sigma^4$. These measures are useful in characterizing the shape of a distribution and are often referred to as the moments of the distribution even though, strictly speaking, they are transformations of the raw moments. We will adopt this convention in the rest of this chapter.

If we know the PDF of a distribution, then we can write the moments in terms of this function. For example, the mean of the distribution can be written

$$\mu = E(X) = \int_a^b x f(x)\,dx. \tag{1.18}$$

By integrating $x f(x)$ over the range of possible values for X (where b is the maximum possible value of X and a is the minimum possible value) we are effectively taking a weighted average of these possible values with the weights being the probabilities $X = x$ is observed. Similarly, the variance of X can be written

$$\sigma^2 = E(X-E(X))^2 = \int_a^b (x-\mu)^2 f(x)\,dx. \tag{1.19}$$

Higher-order moments can then be calculated by integrating a function of the form $(x-E(x))^k f(x)dx$ and then scaling by σ^k.

We have already seen that, in the case of the normal distribution, the first two moments fully characterize the shape of the PDF and are therefore the only parameters we need to know. This can be seen by the fact that the equation has only two parameters μ and σ^2. This is not the case for other distributions where higher-order moments become important. In particular, the third and fourth moments become important because they capture features such as skewness and kurtosis of the distribution.

1.5 OTHER USEFUL DISTRIBUTIONS

The normal distribution can be used to derive a number of other distributions that are important in econometric analysis. These are the *chi-squared*, F and *Student's t* distributions which will all figure prominently in econometric theory. We will consider each in turn and discuss the nature of the distribution, the sorts of data that might be characterized by such a distribution and what the theory tells us about their moments.

Let us first consider the chi-squared distribution. Suppose we have k independent random variables $Z_j : j = 1,\ldots,k$ each of which follows a normal distribution with mean zero and variance one. This is not unduly restrictive because we have already seen that any normal distribution can be written in this form with an appropriate transformation. Now let us define the following random variable

$$X = \sum_{j=1}^{k} Z_j^2. \tag{1.20}$$

The random variable defined by is said to follow a *chi-squared distribution with k degrees of freedom*. Variables with a chi-squared distribution arise naturally when we consider statistics which are defined as the sum of squared variables. This occurs frequently in econometric analysis when we consider the residual sum of squares for a regression equation. Critical values for the chi-squared distribution with different degrees of freedom are given in most standard books of statistical tables.

For values of k greater than 2, the chi-squared distribution has the characteristic shape shown in Figure 1.5 which shows the PDF for a chi-squared random variable with four degrees of freedom. The value of the PDF is zero at $x = 0$, it then increases to a peak value for some positive value of x and then declines asymptotically to zero as x becomes large. The distribution exhibits

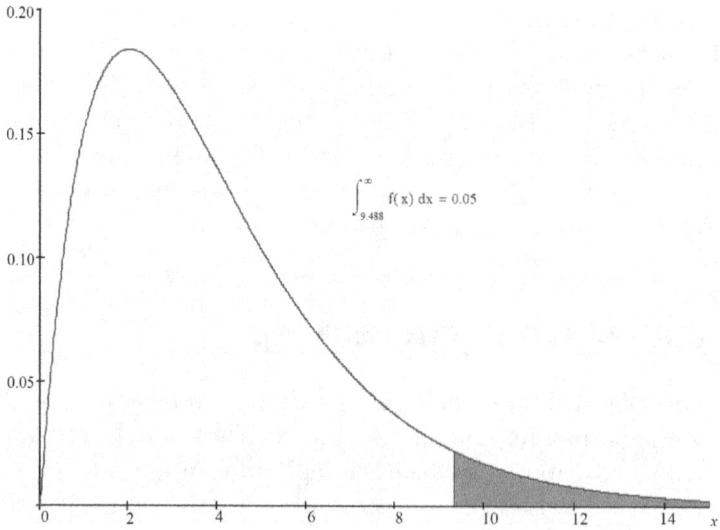

FIGURE 1.5 Probability Density Function for the Chi-Squared Distribution with Four Degrees of Freedom.

positive or *right skew* in that the right tail of the distribution is longer than the left tail. This characteristic shape is only observed for $k > 2$. If $k = 1$ or 2 then it is no longer the case that the chi-squared distribution has a PDF that takes the value 0 at $x = 0$. Instead, the value of the PDF either tends to infinity as x tends to zero when $k = 1$ or to a positive, non-zero value when $k = 2$.

The mean and the variance of a random variable that follows a chi-squared distribution with k degrees of freedom are given by the values k and $2k$, respectively. Most books of statistical tables give tables of critical values of the chi-squared distribution for different degrees of freedom. As k becomes large, the asymmetry which in the PDF of the chi-squared distribution becomes less pronounced. In the limit, for large k, the chi-squared distribution will look more and more like the normal distribution (as predicted by the central limit theorem).

Historical Note: The chi-squared distribution was first considered by Friedrich Helmert (1843–1917) in a paper published in German in 1876 [Helmert1876], but was later discovered independently by Karl Pearson (1857–1936) in his paper of 1905 [Pearson1905]. It was given the name "chi-squared" as one of the family of "skew" distributions which Pearson believed could form the basis for all continuous probability distributions.

The next distribution we will consider is the F distribution. Suppose we have two random variables each of which follows a chi-squared distribution. In particular, let us assume that $X_1 \sim \chi^2_m$ and $X_2 \sim \chi^2_n$. Now let us define the following random variable as the ratio of the two chi-squared variables each of which is divided by its degrees of freedom

$$X = \frac{X_1 / m}{X_2 / n} = \frac{X_1}{X_2}\frac{n}{m}. \tag{1.21}$$

The random variable X defined in equation 1.21 follows an *F distribution with m and n degrees of freedom*. This is written as $X \sim F_{m,n}$. The F-distribution arises naturally in econometric analysis when we consider the ratios of variables which are constructed as the sum of squared random variables. As we will see in later chapters this situation arises frequently when we perform tests of restrictions in linear regression models.

For $m \geq 3$ the F distribution has a similar shape to the *chi-squared* distribution in that its PDF takes the value 0 at its lower limit, has a single interior maximum and is right skewed. This is illustrated in Figure 1.6, which shows the probability density function for the F distribution with 10 and 10 degrees of freedom. F distributions with m less than three do not have the typical shape illustrated in Figure 1.6. For $m = 1$, $f(x) \to \infty$ as $x \to 0$, rather like we

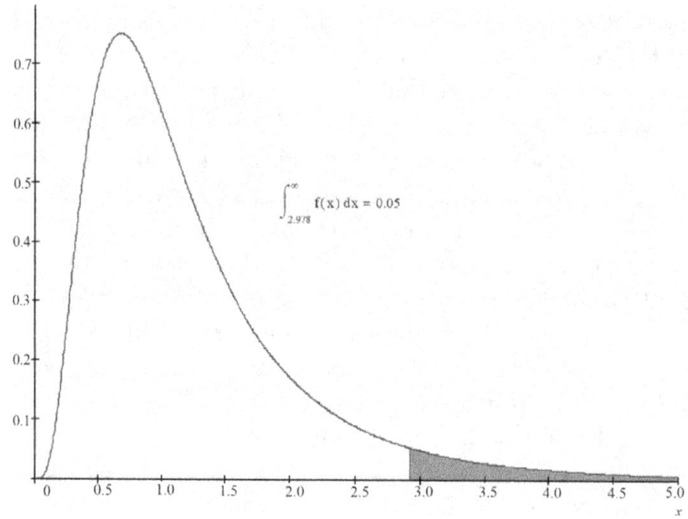

FIGURE 1.6 Probability Density Function for the *F* Distribution with 10 and 10 Degrees of Freedom.

saw for the chi-squared distribution while for $m = 2$, it will tend to a non-zero value. Another similarity with the chi-squared distribution is that as both m and n become large, the shape of the F distribution becomes symmetric and eventually converges to a normal distribution.

Historical Note: The form of the F distribution was first set out by Ronald Fisher (1890–1962) in his 1922 paper [Fisher1922]. It was later tabulated and given its name (in honor of Fisher) by George Snedecor (1881–1974) in his 1934 book *Calculation and Interpretation of Analysis of Variance and Covariance* [Snedecor1934].

Student's t distribution is defined as follows. Suppose X_1 is a random variable that follows a standard normal distribution, $X_1 \sim N(0,1)$ and X_2 is an independent random variable that follows a chi-squared distribution with k degrees of freedom, $X_2 \sim \chi_k^2$. It can be shown that the random variable T defined in equation 1.22 follows Student's t distribution with k degrees of freedom,

$$T = \frac{X_1}{\sqrt{X_2 / k}} \tag{1.22}$$

Student's t distribution is often referred to simply as the t distribution. It is useful when we wish to conduct hypothesis tests on a variable which we

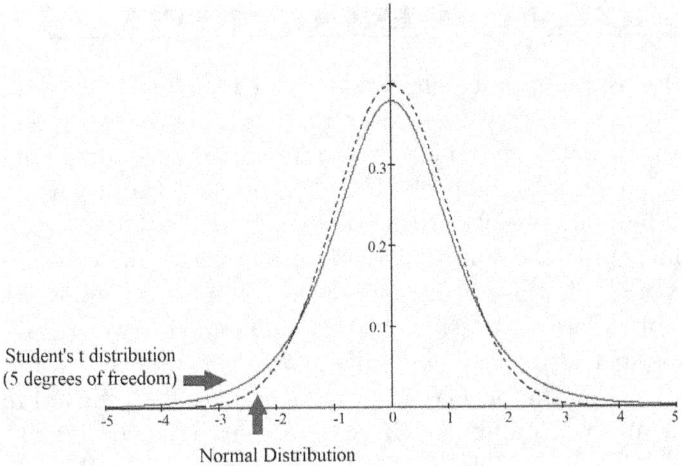

FIGURE 1.7 Relationship Between the Probability Density Functions of the t-distribution and the Standard Normal Distribution.

assume is normally distributed but for which we do not know the variance. We will see in subsequent chapters that tests for the significance of regression coefficients fall into this category. The shape of the PDF of the t distribution looks very much like that of the standard normal distribution. It is symmetric around zero and has the characteristic bell shape of the normal distribution. However, the t-distribution has "fatter tails" than those of the normal distribution. By this, we mean that more of the mass of the distribution lies in its tails than is the case of the normal. This means that extreme events (or values of the random variable that lie in the tails) are more likely for the t distribution. The difference between the two distributions is illustrated in Figure 1.7.

The t-distribution is useful when constructing tests based on small samples. As the sample size gets larger the differences between the t distribution and the normal get smaller. In the limit, as the sample size becomes arbitrarily large, the t distribution converges on the normal. In practice, for sample sizes more than 30, the difference between the t-distribution and the normal distribution is negligible.

Historical Note: The t distribution was first set out by William Sealy Gosset (1876–1937) in 1908 [Student1908]. Gosset was working for the Guinness brewing company at the time who (allegedly) forbade employees to publish research on the ground that it might be commercially sensitive. As a result, Gosset published extensively under the name of "Student."

1.6 CLASSICAL AND BAYESIAN STATISTICS

The discussion of probability and statistical distributions in this chapter has implicitly assumed that we can repeat the experiment generating the data however many times we like. For example, in generating the probability of drawing a club from a pack of cards, it is assumed that we can repeat this experiment a large enough number of times for the measured frequency to converge to the true underlying probability. This makes sense for simple examples but becomes more difficult in more complex situations where experiments are not possible. For example, suppose we are asked to state the probability that the economy will emerge from recession during the coming year. In circumstances like this, we do not have the luxury of re-running history an arbitrary number of times to measure relative frequency.

If it is not possible to repeat experiments, then the interpretation of probabilities in terms of relative frequency becomes problematic. Some statisticians argue that it is still possible to interpret probabilities in this way even when repeated experiments are not physically possible. This is the standpoint taken by the *classical* or *frequentist* school. A characteristic of the classical school is that the parameters of distributions of random variables are treated as *objective*. That is, they are fixed numbers that exist independently of the experiment being conducted or the person conducting the experiment. In contrast, the *Bayesian* school of statisticians argues that the inability to repeat experiments means that it is not possible to treat probabilities or parameters as objective. Instead, they begin with the premise that these parameters are inherently subjective. This means that they reflect the beliefs of the investigator rather than something external. In many economic examples, the Bayesian interpretation of probabilities is arguably more plausible than the classical interpretation. This is because economic situations are often non-repeatable in nature. Despite this, the statistical foundations of econometrics remain firmly rooted in the classical approach and we shall continue with this interpretation.

1.7 SUMMARY

This chapter has been concerned with the statistical foundations necessary for an understanding of econometrics. We begin with the idea of probability and the probability distribution function of a discrete random variable. This is illustrated by the binomial distribution which gives the probability of x successes in a set of n Bernoulli trials. If the number of trials is large, then we show that the binomial distribution can be approximated by the function which can also be interpreted as the probability density function of the normal distribution. The normal distribution is important because the central limit theorem shows that if we take the mean of a large number of independent random variables then this will follow a normal distribution. It is also important because it provides the basis for the development of the chi-squared, F and Student's t distributions. All of these are important for econometric analysis.

EXERCISES

EXERCISE 1.1

The following table gives population data for the United Kingdom in 2007 taken from the Office of National Statistics (ONS) database. The data are broken down into categories of employment and by gender. This can be thought of as a contingency table.

	Male	Female	Total
Employed	12,950	12,254	25,204
Self-employed	2762	1054	3816
Unemployed	944	709	1653
Not economically active	13,260	17,042	30,302
Total	29,916	31,059	60,975

UK Population in 2007 (thousands) taken from ONS database

a. Create a new table that contains the estimated joint probabilities of an individual worker falling into each of the different categories.

b. Calculate the marginal probabilities for the rows and columns and check that these add up to one (there may be a slight rounding error).

c. Calculate the conditional probability that an individual is male given that they are self-employed.

d. Calculate the conditional probability that an individual is unemployed, given that they are male.

EXERCISE 1.2

The uniform distribution for a continuous random variable X has PDF $f(x)=1/(b-a)$ where b and a are the maximum and minimum values of X. Using the definitions in terms of moments given in equations (1.18) and (1.19), show that the mean and the variance of X are given by the following expressions

$$\mu = \frac{a+b}{2}$$

$$\sigma^2 = \frac{(b-a)^2}{12}$$

EXERCISE 1.3

X and Y are independent normal random variables with distributions $X \sim N\left(\mu_X, \sigma_X^2\right)$ and $Y \sim N\left(\mu_Y, \sigma_Y^2\right)$. Calculate the distributions of $X+Y$ and $X-Y$, respectively.

REFERENCES

[Bernouilli1713] Bernoulli, J., *Ars Conjectandi, Opus Posthumum. Accedit Tractatus de Seriebus Infinitis, et Epistola Gallicé Scripta de Ludo Pilae Reticularis, (The Art of Conjecturing)*. Basel: Thurneysen Brothers, 1713.

[DeMoivre1718] DeMoivre, A., *The Doctrine of Chances: Or a Method for Calculating the Probabilities of Events in Play*. London: W. Pearson, 1718.

[Devlin2008] Devlin, K., *The Unfinished Game: Pascal, Fermat and the Seventeenth Century Letter that Made the World Modern*. New York: Basic Books, 2008.

[Fisher1922] Fisher, R. A., "The Goodness of Fit of Regression Formulae and the Distribution of Regression Coefficients." *Journal of the Royal Statistical Society*, 1922, 85: pp. 597-612.

[Gauss1809] Gauss, C. F., *Theory of the Motion of Heavenly Bodies Moving About the Sun in Conic Sections*. New York: Dover, 1809. English Translation by C.H. Davis, 1963.

[Helmert1876] Helmert, F. R., "Ueber die Wahrscheinlichkeit der Potenzsummen der Beobachtungsfehler und Über Einige Damit im Zusammenhange Stehende Fragen." *Zeitschrift für Mathematik und Physik*, 1876, 21, pp. 102–219.

[Pearson1905] Pearson, K., "On the Criterion That a Given System of Deviations From the Probable in the Case of a Correlated System of Variables Is Such That It Can Be Reasonably Supposed to Have Arisen From Random Sampling." *Philosophical Magazine*, 1905. Series 5. 50 (302): pp. 157–175.

[Snedecor1934] Snedecor, George W., *Calculation and Interpretation of Analysis of Variance and Covariance*. Ames Iowa: Collegiate Press, 1934.

[Student1908] Student (W. S. Gosset) "The Probable Error of a Mean." *Biometrika*, 1908, 6 (1): pp. 1–25.

CHAPTER 2

STATISTICAL INFERENCE

Classical statistical theory assumes that the random variables that are of interest to us are generated by a distribution which is unknown, but about which we can draw inferences based on observation. In this chapter, we discuss the theory of statistical inference. To do so, we first need to define some terms. First, suppose we have a set of N independent random variables each of which has probability density function (PDF) $f(x)$. Because the variables are independent, we can therefore write the joint PDF as the product of the individual PDFs, that is, $f(x_1)f(x_2)...f(x_N)$. A set of random variables of this type is referred to as a *random sample*. Next, we define a *statistic* as a function of one or more random variables, which does not depend on any unknown parameters. For example, given a random sample $X_1, X_2, ..., X_n$, the sample mean is defined as

$$\overline{X} = \frac{1}{N}\sum_{i=1}^{N} X_i. \tag{2.1}$$

This satisfies the condition for a statistic because it does not depend on any unknown population parameters. Similarly, the sample variance is defined as

$$\hat{\sigma}_X^2 = \frac{1}{N-1}\sum_{i=1}^{N}\left(X_i - \overline{X}\right)^2, \tag{2.2}$$

and again, this qualifies as a statistic because it does not depend on any unknown parameters. Note, however, that in both cases, the *distribution* of these statistics will depend on the unknown parameters that describe the distribution of the random variables X_i.

These are examples of *estimators* of unknown population parameters based on sample data. In both cases, we can show that these estimators are *unbiased*. That is, if the population mean of the X variable is $E(X_i) = \mu$ for all values of i, and its variance is $E(X_i - \mu)^2 = \sigma_X^2$, then $E(\overline{X}) = \mu$, and $E(\hat{\sigma}_X^2) = \sigma_X^2$.

To get an unbiased estimator of the population variance, we divide the sum of the squared deviations of the X variables from the sample mean by the number of observations minus one. This is because the use of the sample mean, rather than the population mean in (2.2), implies that there are only $N - 1$ independent squared deviations in this expression. Thus, the *degrees of freedom* for the sum of squared deviations around the sample mean is equal to the number of observations minus one. This correction follows naturally from the construction of the estimator of sample variance. To demonstrate this, note that we can write

$$\begin{aligned}\sum_{i=1}^{N}(X_i-\bar{X})^2 &= \sum_{i=1}^{N}\{(X_i-\mu)-(\bar{X}-\mu)\}^2 \\ &= \sum_{i=1}^{N}(X_i-\mu)^2+\sum_{i=1}^{N}(\bar{X}-\mu)^2-2\sum_{i=1}^{N}(X_i-\mu)(\bar{X}-\mu).\end{aligned} \tag{2.3}$$

To find the expected value of this expression, we take each of its elements in turn. First, since $E(X_i-\mu)^2=\sigma_X^2$, we have $E\left[\sum_{i=1}^{N}(X_i-\mu)^2\right]=N\sigma_X^2$. Next, by definition of the sample mean, we have $\bar{X}-\mu=1/N\sum_{i=1}^{N}(X_i-\mu)$ and therefore

$$E\left[\sum_{i=1}^{N}(\bar{X}-\mu)^2\right]=N\times\frac{1}{N}E(X_i-\mu)^2=\sigma_X^2. \tag{2.4}$$

Finally, since we have $\sum_{i=1}^{N}(X_i-\mu)(\bar{X}-\mu)=1/N\sum_{i=1}^{N}(X_i-\mu)^2$, taking expectations yields

$$E\left[\sum_{i=1}^{N}(X_i-\mu)(\bar{X}-\mu)\right]=\sigma_X^2. \tag{2.5}$$

Combining these expectations yields

$$E\left[\sum_{i=1}^{N}(X_i-\bar{X})^2\right]=N\sigma_X^2+\sigma_X^2-2\times\sigma_X^2=(N-1)\sigma_X^2. \tag{2.6}$$

Therefore, to obtain an unbiased estimator of the sample variance, we must divide by $N - 1$ rather than N. As the sample size N increases, then the bias resulting from dividing by N rather than $N - 1$ will become arbitrarily small.

Estimation is defined as the process of using sample data to construct estimates of unknown population parameters. In doing so, we often have different estimators available to us and must choose between them.[1] Consider an estimator $\hat{\theta}$ of an unknown population parameter θ. The criteria we use to assess whether this is a good estimator (and whether it is better than alternatives) can be summarized as follows:

1. An estimator is *unbiased* if its expectation is equal to the population value, that is, $E(\hat{\theta}) = \theta$. For unbiasedness, this property must be true whatever the sample size.
2. An estimator is *consistent* if it converges in probability on the population value, that is, $\lim_{N\to\infty} \hat{\theta}_N = \theta$, where $\hat{\theta}_N$ is an estimator based on N observations. This is a large sample property that is often used when it is not possible to prove unbiasedness. Note that unbiasedness and consistency are different properties. Estimators that are biased in small samples can often be consistent. Although it is less common, it is also possible to find unbiased estimators that are not consistent.
3. An estimator is said to be *efficient* if it has lower variance than other possible estimators. Unlike the first two criteria, efficiency is defined as a comparison between alternative estimators rather than an intrinsic property of an individual estimator.

In addition to estimation of parameters, we also often wish to make *inferences* about them. That is, we wish to test hypotheses about population parameters. Inference is also concerned with judgments concerning the range of possible values within which parameters might lie. The topics of estimation and inference will form the major part of this chapter.

The statistical methodology for estimating parameters and making inferences based on these estimates is well established. However, most of this analysis assumes that the data we examine are generated experimentally and are, therefore, under the control of the investigator. The main practical problems for the econometrician arise because of the nature of the data we work with. In most cases, economic data are nonexperimental and are passively observed by the investigator. The implications of this are far-reaching and the role of econometrics as a discipline is to analyze the implications of data generated in this way and to suggest methods for dealing with the problems it creates.

---

[1] An *estimate* is a particular numerical value based on data. An *estimator* is a method, or algorithm, by which data is processed to form estimates.

2.1 SAMPLING

A sample is said to be random if there is an equal probability of selecting any member of the population as part of the sample to be examined. In classical statistical theory, this is often motivated by stylized examples such as the drawing of different colored balls from an urn. A standard scenario is one in which we have an urn containing both black balls and white balls and we wish to draw a random sample to test the hypothesis that there is an equal number of each color in the urn. For a controlled experiment of this type, it is easy to construct a random sample – we simply make sure that the experimenter cannot see the color of the balls prior to making the draw. An artificial scenario of this kind helps identify the strict criteria under which a sample can be said to be random.

In the more complex situations, we encounter in the real world, it may be more difficult to ensure a random sample. For example, suppose we wish to generate a random sample of households to investigate expenditure on a consumer product. We might dial random telephone numbers from the directory and interview the person answering. Although this sounds like a reasonable procedure, it is anything but random. First, this procedure automatically eliminates from the sample all those households that do not have a listed number. This may be because they do not have a telephone or because they choose not to be listed. In either case, a group of households, who are likely to have somewhat different characteristics from the rest of the population, are excluded. Second, only those calls that are answered will be considered. This will bias the sample according to the time of day at which the calls are made. If the calls were made during working hours, then the sample will tend to overrepresent households in which there is at least one member who is not currently employed. In general, sampling procedures, which look random at first sight, may be subject to subtle forms of sample selection bias when we think about them more carefully.

Since it is often very difficult to obtain a truly random sample, statisticians often use a system of *stratified sampling* to obtain a sample which is genuinely representative of the population as a whole. Usually, this will involve sampling different subgroups in numbers that reflect their share in the overall population. For example, we might divide the sample up into different age bands and ensure that the numbers we interview in each reflects the proportions that they make up of the total population. Although procedures like this may look nonrandom, they are nevertheless more likely to produce samples that are representative of the population than less-structured approaches.

One way of thinking of the sampling problem is to think of each observation as an experiment. To be a genuine experiment, the results must be independent of the experiments that have gone before. When using cross-section data, this seems to be a reasonable assumption. We can think of the process of generating a new observation as akin to conducting a new experiment and enlarging the sample. However, when dealing with time-series data, the analogy begins to break down. In what sense does a new time-series observation constitute an independent observation? The answer is that in many cases, it does not. For example, when new Gross Domestic Product (GDP) estimates are reported each quarter, the figures released do not constitute a random drawing from the population of possible outcomes. Instead, they depend heavily on the recent history of the economy and on the behavior of GDP over the recent past. To justify the use of classical statistical methods with time-series data, it is necessary to make strong assumptions about the distribution of the variables in questions. In particular, we need to make the assumption that the series in question is *stationary*. That is, we assume that its moments are independent of time. We will discuss this issue in greater detail in subsequent chapters but, for the moment, we will simply assume that the necessary conditions hold and that we can treat time-series data in the same way as we treat experimental or survey data.

Taking all these considerations into account, and assuming that we can generate a true random sample, then we can define the *sampling distribution* of a statistic as the probability distribution based on a random sample of size N. Note that the sampling distribution does not refer to a particular sample of data. Instead, it is the distribution of all possible samples of a given size. The sampling distribution is determined by the underlying distribution of the population generating each observation, the statistic concerned, and the sample size. The sampling distribution of a statistic is distinct from its *asymptotic distribution* which is the limit of the sampling distribution as the sample size becomes large, that is, as $N \to \infty$.

For example, let us consider the derivation of the sampling distribution of the arithmetic mean from a random sample $X_i : i = 1,\ldots,N$ when each individual observation is assumed to follow a normal distribution with mean μ and variance σ_X^2, that is, $X_i \sim N(\mu, \sigma_X^2)$. The sample mean is a linear combination of normal random variables and will therefore itself follow a normal distribution. Consider the first moment or expectation

$$\begin{aligned} E(\bar{X}) &= E\left(\frac{1}{N}\sum_{i=1}^{N} X_i\right) = \frac{1}{N}E(X_1 + X_2 + \ldots + X_N) \\ &= \frac{1}{N}\left(E(X_1) + E(X_2) + \ldots + E(X_N)\right). \end{aligned} \tag{2.7}$$

Now, by assumption $E(X_i)=\mu$ for all values of i, which means that the expected value of the arithmetic mean is equal to the population mean. The sample mean is therefore an unbiased estimator of the unknown population mean. Next, consider the variance of the sample mean which is defined as

$$E\left(\bar{X}-E\left(\bar{X}\right)\right)^2, \tag{2.8}$$

and we have already shown that $E\left(\bar{X}\right)=\mu$, so this can be written as $E\left(\bar{X}-\mu\right)^2$. Expanding this expression yields

$$\begin{aligned} E\left(\bar{X}-\mu\right)^2 &= E\left(\frac{1}{N}\sum_{i=1}^{N}X_i-\mu\right)^2=\frac{1}{N^2}E\left(\sum_{i=1}^{N}X_i-N\mu\right)^2 \\ &= \frac{1}{N^2}E\left(\sum_{i=1}^{N}(X_i-\mu)^2\right). \end{aligned} \tag{2.9}$$

Since $E\left(X_i-\mu\right)^2=\sigma_X^2$ for all values of i, we have

$$\sigma_{\bar{X}}^2=\frac{N\sigma_X^2}{N^2}=\frac{\sigma_X^2}{N}. \tag{2.10}$$

Combining these results shows that, under our assumptions, the sample mean will follow a normal distribution with mean equal to the population mean of the distribution for the underlying random variable. The variance of the sample mean is equal to the variance of the underlying distribution divided by the number of observations, that is,

$$\bar{X}\sim N\left(\mu,\frac{\sigma_X^2}{N}\right). \tag{2.11}$$

One implication of this is that the variance of the sample mean will fall as the number of observations increases and, in the limit, will go to zero as $N\to\infty$.

Now that we have derived the sampling distribution of the arithmetic mean, we can use this to derive test statistics for the purposes of statistical inference. Using the transformation for the standard normal distribution, we have

$$\frac{\bar{X}-\mu}{\sigma_X/\sqrt{N}}\sim N(0,1). \tag{2.12}$$

If σ_X is known, then this would be a valid test statistic. However, σ_X is not known in most circumstances. In order to derive a valid test statistic, we must substitute the sample variance for the unknown population variance. This, in turn, requires us to derive the distribution of the sample variance. This is not easy, and we will offer only an heuristic derivation here. Consider the expression $\sum_{i=1}^{N}(X_i - \overline{X})^2$. From the definition of the sample mean, this consists of the sum of $N - 1$ independent sums of squares, each of which has expected value σ_X^2. Dividing by σ_X^2 means that we have an expression of the form $\sum_{i=1}^{N}((X_i - \overline{X})/\sigma_X)^2$ which consists of the sum of $N - 1$ squared standard normal random variables and hence has a chi-squared distribution with $N - 1$ degrees of freedom. Now consider again the definition of the sample variance given in equation (2.2). Multiplying both sides by $N - 1$ and dividing by σ_X^2 yields

$$(N-1)\frac{\hat{\sigma}_X^2}{\sigma_X^2} = \sum_{i=1}^{N}\left(\frac{X_i - \overline{X}}{\sigma_X}\right)^2. \tag{2.13}$$

We have established that the right-hand side of this expression has a χ_{N-1}^2 distribution. Hence, it follows that $(N-1)\hat{\sigma}_X^2/\sigma_X^2 \sim \chi_{N-1}^2$. In the next section, we will make use of this result to derive a valid test statistic for the purposes of inference.

2.2 HYPOTHESIS TESTING

Hypothesis testing can be thought of as the process of using a sample of data to draw inferences about population parameters. A hypothesis test requires the following elements:

1. A hypothesis to be tested (usually described as the *null hypothesis*) and another hypothesis against which it will be examined (the *alternative hypothesis*).
2. A *test statistic* whose distribution is known under the assumption that the null hypothesis is true.
3. A *decision rule* that determines the circumstances under which the null hypothesis will be rejected.

The first of these elements is normally determined by economic theory. However, the second two elements depend more on statistical theory.

The test statistic we use will depend on the assumptions we make about the statistical distribution of the variables we examine, whereas the decision rule will depend on the costs of making either Type I or Type II errors. A Type I error is the case where we reject the null hypothesis when it is true, whereas a Type II error is the case where we fail to reject the null hypothesis when it is false.

> **Historical Note:** The first use of the term *null hypothesis* comes in Ronald Fisher's 1935 book *The Design of Experiments* [Fisher1935]. The term *alternative hypothesis* had been used earlier by Jerzy Neyman and Egon Pearson [Neyman1928] who developed much of the methodology and terminology discussed this chapter.

Usually, the decision rule involves fixing the *size* of the test or the probability that we make a Type I error. The size of the test gives the proportion of experiments that would be expected to incorrectly reject the null hypothesis, that is, to generate a *false positive* result. It is usually expressed as a percentage. For example, a 5% size implies that we would be willing to accept five false positive results in every 100 experiments. Test size is usually set at a low level so that we reduce the probability of a false positive result in any individual experiment. However, common test sizes such as 5% and 1% are arbitrary choices which are often used because they are conventional rather than because of any conscious choice by the researcher. The choice of test size reflects the researcher's view of the costs of a false positive result. In addition to the cases in which the test produces errors, we are also interested in the extent to which it gets the answer correct. Ideally, we would like tests to have both high degrees of *specificity* and *sensitivity*. Specificity is the ability of the test to correctly identify the null when this is correct (a true negative), whereas sensitivity is the ability to correctly identify the alternative when this is correct (a true positive).

As an example of the testing process, let us consider a situation in which we wish to use a sample of data to test a hypothesis about the population mean. For simplicity, we will assume that each observation is a random experiment in which the outcome follows a normal distribution. The first stage is to specify the hypothesis we wish to text. Suppose, for example, that we wish to test the null hypothesis $H_0 : \mu = \bar{\mu}$ against the alternative $H_1 : \mu \neq \bar{\mu}$. From the previous section, we have shown that if the population variance is known, then a possible test statistic would take the following form

$$\frac{\bar{X} - \bar{\mu}}{\sigma_X / \sqrt{N}} \sim N(0,1). \tag{2.14}$$

In this case, the statistic will follow a standard normal distribution. If the population variance is unknown, then we must substitute an estimate for σ_X in (2.14) in order to construct an operational test statistic. By doing this, however, we will change its distribution. Consider, for example, the effects of replacing σ_X by its estimate $\hat{\sigma}_X$. From the previous section, we have $(N-1)\hat{\sigma}_X^2/\sigma_X^2 \sim \chi_{N-1}^2$. Recall that the t distribution is defined as the distribution of the random variable defined as the ratio of standard normal random variable to the square root of a chi-squared random variable divided by its degrees of freedom. Therefore, if we divide the random variable $\sqrt{N}\left(\bar{X}-\bar{\mu}\right)/\sigma_X$ by $\hat{\sigma}_X/\sigma_X$, then the result will be a random variable that follows a t distribution with N – 1 degrees of freedom. This gives us a valid test statistic of the form

$$t = \frac{\bar{X}-\bar{\mu}}{\hat{\sigma}_X / \sqrt{N}} \sim t_{N-1}. \tag{2.15}$$

Unlike the previous expression (2.14), this does not contain any unknown parameters and therefore constitutes a usable test statistic.

Next, we need to determine a critical value as a basis for comparison with the test statistic. The critical value is normally chosen so that it fixes the size of the test or the probability of making a Type I error. The value chosen will depend on the nature of the alternative hypothesis. If the alternative to H_0 is $H_1 : \mu \neq \bar{\mu}$, then we have a *two-sided alternative*, that is, we are equally concerned about positive and negative deviations of the estimate from the hypothesized value. However, if the alternative takes the form $H_1 : \mu > \bar{\mu}$, then we have a *one-sided alternative* in which only positive deviations are of interest.

Let us first consider the case of two-sided alternative. The decision rule will involve choosing a critical value t_{crit} such that, if $|t| > t_{crit}$, we reject the null. t_{crit} is set so that $p(\text{Reject } H_0 \mid H_0 \text{ true}) = a$, where a is the size of the test. This is illustrated in Figure 2.1 for a t-test in which we have 20 observations and a 5% significance level. We must find t_{crit} so that 95% of the mass of the distribution lies between $-t_{crit}$ and t_{crit}. Alternatively, because the distribution is symmetric, we need to find t_{crit} so that 2.5% of the area under the curve lies to the right of this value. In our case, this gives a value $t_{crit} = 2.093$. Tests of this kind, using a two-sided alternative, are often referred to as *two-tailed tests* because the critical value is determined by the area under both tails of the PDF.

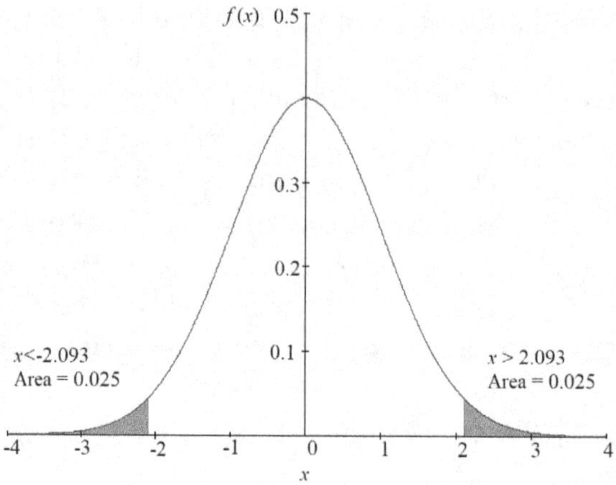

FIGURE 2.1 Determination of *t* Critical Values for a Two-Sided Alternative.

Critical values are often written to indicate the size of the test. For example, for a 5% critical value and a two-tailed based on the Student's t distribution, we could write the critical values as $\pm t_c^{0.025}$ or $\pm t_c^{2.5\%}$. For a one-tailed test, the notation is somewhat simpler because there is only one critical value that we would write as $t_c^{0.05}$ or $t_c^{5\%}$. At the risk of being pedantic, when we have a two-tailed test, we must find a pair of critical values. For example, let Z be a statistic with a known distribution. For a test of size a, we need to find critical values such that $p\left(Z < z_c^L\right) = a\ /\ 2$ and $p\left(Z > z_c^U\right) = a\ /\ 2$. Taken together, these critical values give the correct size for the test because $p\left(z_c^L < z < z_c^U\right) = 1 - a$. In the case of symmetric distributions, such as Student's t or the normal, we have $z_c^L = -z_c^U$, and therefore, we choose z_c such that $p\left(|z| > z_c\right) = a$. Unfortunately this shortcut cannot be applied for nonsymmetric distributions, such as the chi-squared or F distributions. In such cases, we need to find a distinct pair of critical values to conduct a two-tailed test.

Consider now the case in which there is a one-sided alternative, for example, $H_1 : \mu > \bar{\mu}$. In this case, we are only interested in cases in which the test statistic exceeds its expected value under the null. This means that the critical value is determined only by the right tail of the distribution as illustrated in Figure 2.2. The critical value in this case is given by $t_c^{0.05} = 1.729$. Tests of this kind are referred to a *one-tailed tests* for obvious reasons.

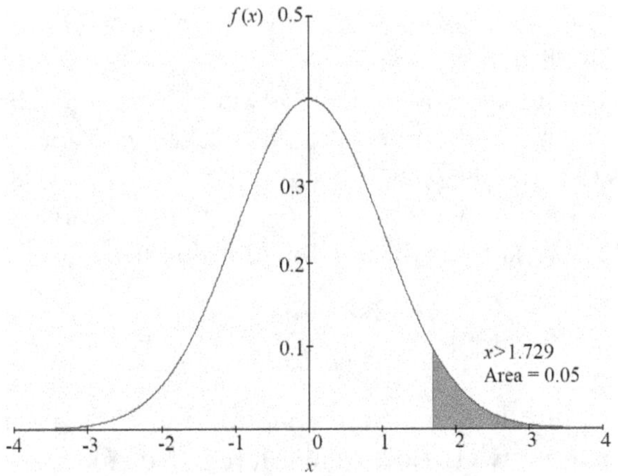

FIGURE 2.2 Determination of t Critical Value for a One-Sided Alternative.

Example: The average annual growth rate for US consumption expenditures between 1970 and 2019 is calculated as 2.9569 with standard deviation $\hat{\sigma}_X = 1.7495$. Can we reject the null hypothesis that the underlying growth rate is equal to 3% per annum?

Assuming that the underlying growth rate is normally distributed, we first need to set out the null and alternative hypotheses. The form of the question implies a two-sided alternative. Therefore, we will test $H_0 : \mu = 3$ against $H_1 : \mu \neq 3$. The test statistic we will use is written as

$$t = \frac{2.9569 - 3}{1.7495 / \sqrt{49}} = -0.1724.$$

From the t tables, we obtain critical value(s) ± 2.011 for a t-distribution with 48 degrees of freedom. The t-ratio lies between the critical bounds and therefore we cannot reject the null hypothesis in this case.

In many situations, we do not want to test a null hypothesis that specifies a specific value for the unknown parameter. We may simply be interested in testing whether a parameter is greater or less than some specific value. In such cases, it is more natural to use a one-sided alternative and a one-tailed test. For example, we might wish to test $H_0 : \mu \leq \bar{\mu}$, in which case, it is natural to specify the alternative hypothesis as $H_1 : \mu > \bar{\mu}$. The following example may help to make this clearer.

Example: Let us assume that the growth rate of the Canadian economy is a normally distributed random variable X with mean μ and variance σ_X^2. The average annual growth rate between 1962 and 2019 is calculated as $\bar{X} = 3.2221$ with standard deviation $\hat{\sigma}_X = 2.2658$. Should we reject the null hypothesis that $\mu \leq 3$ against the alternative that it is >3?

In this case, the wording of the question indicates a one-sided alternative. The test statistic can be written as follows:

$$t = \frac{3.2221 - 3}{2.2658 / \sqrt{58}} = 0.7465.$$

The 5% critical value for a t-distribution with 57 degrees of freedom with a one-sided alternative is 1.672. Therefore, we cannot reject the null hypothesis in favor of the alternative at the 5% level.

When we fail to reject the null, does this mean that we implicitly "accept" it? Strictly speaking, the answer is no. Failure to reject the null means precisely that–there is no implication that the null is accepted, simply that there is not enough evidence to reject it. However, it is not unusual to see failure to reject as being described as acceptance of the null hypothesis.

2.3 CONFIDENCE INTERVALS

Hypothesis testing is a useful tool but can sometimes lead to a very black and white approach to statistical inference. We are only allowed two possible choices in a hypothesis test – either we accept (or fail to reject!) the null hypothesis or we reject it. In many cases, a more interesting, and possibly a more honest, approach is to express our results in such a way as to indicate our degree of uncertainty about the parameter, or hypothesis, in question. One approach of this kind is to express the results in terms of a *confidence interval*. This consists of an upper or lower bound for the parameter in question that defines a $100(1 - \alpha)\%$ degree of confidence about the value of the unknown parameter, where a reflects an acceptable probability of making a Type I error. For example, if we set $a = 0.05$, then this would be consistent with a 95% confidence interval. We can interpret such an interval as stating that there is a $100(1 - \alpha)\%$ chance that the range quoted contains the true unknown value of the parameter in question.

When we define the confidence interval, it is important to note that it is the confidence interval itself which is treated as a random variable. A statement of the form "there is a 95% probability that the population mean lies between these limits" is not valid in classical statistical theory. In the classical framework, the population mean is not a random number and therefore we cannot make probabilistic statements about it. A more accurate, though not so intuitive, statement would be to say that, if the experiment used to construct the confidence interval was to be carried out 100 times, then 95 of the intervals obtained would be expected to contain the population mean. Thus, the probabilistic statement we make refers to the interval not to the population parameter. A Bayesian statistician, however, would have no such qualms about making probabilistic statements about the population mean. This is because, within the Bayesian framework, there is no assumption that population parameters are fixed numbers that are independent of the investigator. Instead, it is assumed that they are subjective parameters that reflect the investigator's beliefs. It is therefore perfectly valid within a Bayesian framework to refer to "probability intervals" (or more usually "credible intervals") for the parameters. While it would certainly be interesting to discuss this further, most econometrics is conducted within the classical, or frequentist framework, and we will adopt this terminology throughout our book.

Historical Note: The idea of a confidence interval was first introduced by Jerzy Neyman in his 1937 paper in the *Philosophical Transactions of the Royal Society* [Neyman1937]. It was highly controversial at the time in that statisticians from the frequentist tradition regarded it as being dangerously close to Bayesian methodology.

For any statistical distribution, it is possible to find lower and upper bounds that define a $100(1-\alpha)\%$ confidence interval. This is particularly easy for symmetric distributions centered on zero such as the standard normal or t-distribution. For example, let us consider the case of generating a confidence interval for the population mean under the assumption that our data are generated by a normal distribution of the form $X_i \sim N\left(\mu, \sigma_X^2\right);\ i=1,\ldots,N$. We can show that $\sqrt{N}\left(\bar{X}-\mu\right)/\hat{\sigma}_X \sim t_{N-1}$, where $\bar{X}$ and $\hat{\sigma}_X$ are the usual estimates of the mean and standard deviation. Let $t_{N-1}^{0.025}$ be that number such that 2.5% of the mass of the t distribution with N – 1 degrees of the freedom lies to the right of this value (it immediately follows that 2.5% of the mass of the distribution lies to the left of $-t_{N-1}^{0.025}$). From the results in the previous section, we can therefore write

$$p\left(-t_{N-1}^{0.025} < \frac{\sqrt{N}\left(\bar{X}-\mu\right)}{\hat{\sigma}_X} < t_{N-1}^{0.025}\right) = 0.95. \tag{2.16}$$

We can transform the inequality on the right and write this as

$$p\left(\bar{X}-t_{N-1}^{0.025}\frac{\hat{\sigma}_X}{\sqrt{N}}<\mu<\bar{X}+t_{N-1}^{0.025}\frac{\hat{\sigma}_X}{\sqrt{N}}\right)=0.95. \tag{2.17}$$

The pair of numbers $\left\{L_c^{0.025}=\bar{X}-t_{N-1}^{0.025}\hat{\sigma}_X/\sqrt{N},\ U_c^{0.025}=\bar{X}+t_{N-1}^{0.025}\hat{\sigma}_X/\sqrt{N}\right\}$ give the 95% confidence interval for the unknown population mean.

The construction of a confidence interval becomes a little more complicated for nonsymmetric distributions. Let us consider, for example, the population variance. We have already seen that based on a random sample of data $X_1, X_2, \ldots, X_N$, where $X_i \sim N\left(\mu, \sigma_X^2\right)$, we have $(N-1)\hat{\sigma}_X^2/\sigma_X^2 \sim \chi_{N-1}^2$. It is straightforward to determine lower and upper bounds for the chi-squared distribution such that

$$p\left(L_c^{a/2}<(N-1)\frac{\hat{\sigma}_X^2}{\sigma_X^2}<U_c^{a/2}\right)=a. \tag{2.18}$$

From this, we obtain

$$p\left(\frac{N-1}{L_c^{a/2}}\hat{\sigma}_X^2>\sigma_X^2>\frac{N-1}{U_c^{a/2}}\hat{\sigma}_X^2\right)=a, \tag{2.19}$$

which gives the $100(1-\alpha)\%$ confidence interval for the population variance.

Example: Suppose we wish to construct a 95% confidence interval for the standard deviation of the growth rate of US consumption expenditures. We saw earlier that the sample standard deviation based on 49 annual observations from 1970 to 2019 was 1.7495. From the chi-squared tables, we have $L_c^{0.025}=30.755$ and $U_c^{0.025}=69.023$. Therefore, the lower and upper bounds of the confidence interval for the population variance can be calculated as

$$\frac{48}{69.023}\times 1.7495^2=2.1285 \quad \text{and} \quad \frac{48}{30.755}\times 1.7495^2=4.7770.$$

Taking square roots gives the lower and upper bounds of the 95% confidence interval for the standard deviation as 1.4589 and 2.1856.

2.4 *P* VALUES

Another method of dealing with the "all or nothing" nature of classical hypothesis testing is to quote the p value of a hypothesis test rather than a simple accept/reject decision. Consider a random variable that follows a standard normal distribution under the null hypothesis. The 5% critical values for a two-tailed test are ±1.96. Therefore, using a 5% significance level, we would fail to reject the null if the test statistic is 1.95 but reject if it is 1.97. Any reasonable investigator would, however, realize that there was virtually no difference between these two cases. The p value is a function of the test statistic that helps avoid this problem. What it involves is evaluating the cumulative density function for the observed value of the test statistic. Instead of deciding on a critical value, and then basing an accept/reject decision on this one value, the p value approach asks the question at what level of significance would our test statistic lead us to reject the null?

Historical Note: The idea of p values has been present for many years. However, the term itself was first used by Karl Pearson [Pearson1900] in the context of a discussion of the chi-square test. This concept underlies much of the statistical methodology of Ronald Fisher.

Figure 2.3 illustrates the determination of the p value. Suppose we wish to test the null hypothesis that a parameter is equal to zero and the test statistic follows a standard normal distribution under the null. The function $F(x)$ is the cumulative normal density function. Next, we assume that we obtain a test statistic equal to one. We have $F(1) = 0.8413$ and this tells us that the probability of a standard normal random variable taking a value of one or less is equal to 0.8413. The p value is defined as $1 - F(1) = 0.1587$ which gives us the significance level at which we would reject the null hypothesis that the random variable has a mean of zero on the basis of a one-tailed test. Thus, the p value gives us a more flexible way of assessing a test statistic. Rather than allowing only an accept/reject decision, it allows us to assess the strength of the evidence for rejection of the null hypothesis.

Let us consider another example of a case in which the p value might prove useful. Suppose we have two random samples of data $X_1, X_2, \ldots, X_{N_1}$ and $Y_1, Y_2, \ldots, Y_{N_2}$. In each case, we assume that the observations are generated as independent drawings from normal distributions of the form $X_i \sim N(\mu_X, \sigma_X^2)$ and $Y_i \sim N(\mu_Y, \sigma_Y^2)$. Now suppose we wish to test the null hypothesis that the population variances are the same, that is, $H_0: \sigma_X^2 = \sigma_Y^2 = \sigma^2$.

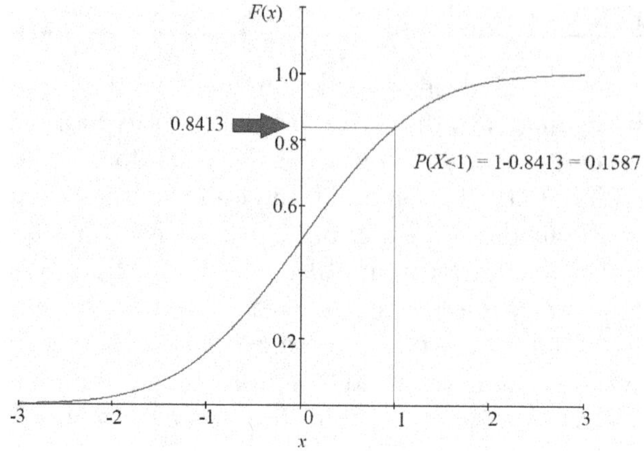

FIGURE 2.3 Determination of the *P* Value.

From results already established, we have $(N_1 - 1)\hat{\sigma}_X^2 / \sigma^2 \sim \chi^2_{N_1 - 1}$ and $(N_2 - 1)\hat{\sigma}_Y^2 / \sigma^2 \sim \chi^2_{N_2 - 1}$ under the null hypothesis. Therefore, dividing each of these expressions by the degrees of freedom and taking the ratio will give us a random variable that follows an *F* distribution, that is,

$$\frac{\hat{\sigma}_X^2}{\hat{\sigma}_Y^2} \sim F_{N_1 - 1, N_2 - 1}. \tag{2.20}$$

The *p* value for this test statistic gives us the probability that we will make a Type I error if we reject the null hypothesis.

Example: Suppose we wish to test whether GDP growth is equally variable in the United States and the United Kingdom. Annual data for the United States for 1949–2019 give an estimate of the standard deviation of 2.3139, whereas that for the United Kingdom for 1949–2019 is equal to 2.0008. To test the hypothesis that growth is equally as variable for the two economies, we construct the following test statistic:

$$\left(\frac{2.3139}{2.0008}\right)^2 = 1.3375. \tag{2.21}$$

Under the null hypothesis that the growth rates are equally variable, this is distributed as F with 70 and 70 degrees of freedom. The F tables do not give enough fine detail to determine the p value for these degrees of freedom, but it can easily be determined using modern statistical software. The value we obtain is 0.113 which indicates that we would not reject the null at the 5%, or even the 10%, level of significance.

2.5 HIGHER-ORDER MOMENTS

One of the advantages of the normal distribution is that we only need to know its first two moments (the mean and the variance) to know everything about it. When we consider other distributions, we need to consider higher-order moments such as *skewness* and *kurtosis*. Skewness measures the extent to which the mass of the distribution (the area under the PDF) is unevenly distributed to the left and right of the mean. Kurtosis is a measure of the "peakedness" of the distribution, that is, the frequency of extreme deviations from the mean – usually measured relative to the normal distribution.

The skewness of a random variable X is defined as

$$\gamma_1 = E\left[\left(\frac{X-\mu}{\sigma_X}\right)^3\right]. \tag{2.22}$$

This is usually estimated using the formula given in equation (2.23) even though this will be biased in small samples. However, as the sample size gets larger, this will converge on the true value

$$\hat{\gamma}_1 = \frac{1}{N}\sum_{i=1}^{N}\left(\frac{X_i-\overline{X}}{\hat{\sigma}_X}\right)^3. \tag{2.23}$$

The skewness coefficient measures the degree of asymmetry of the sampling distribution in that it measures the extent to which the mass of the distribution lies to the right or the left of the sample mean. For a normally distributed variable, we would expect to observe a skewness coefficient close to zero. This is because observations should be evenly distributed around the mean and, because we are raising deviations to an odd power in (2.23), the effects of positive and negative deviations should approximately cancel out. If $\hat{\gamma}_1 > 0$, then this indicates positive (or right) skew in the PDF and the mass of the distribution is concentrated to the left. An example of this is the chi-squared distribution with degrees of freedom >2.

Kurtosis is based on the fourth moment of the distribution. The theoretical kurtosis coefficient is defined as

$$\gamma_2 = E\left[\left(\frac{X-\mu}{\sigma_X}\right)^4\right]. \tag{2.24}$$

For the normal distribution, we have $\gamma_2 = 3$. Because we are often interested in comparing distributions with the normal, kurtosis is sometimes expressed as $(\gamma_2 - 3)$ or *excess kurtosis*. The kurtosis coefficient can be estimated using the formula given in equation (2.25). Again, this will be biased in small samples, but the bias will go to zero as the sample size gets larger.

$$\hat{\gamma}_2 = \frac{1}{N}\sum_{i=1}^{N}\left(\frac{X_i - \overline{X}}{\hat{\sigma}_X}\right)^4. \tag{2.25}$$

If kurtosis is <3, then the distribution is said to be *platykurtic* or "flatter" than the normal distribution. An extreme example of a platykurtic distribution is the uniform distribution, which is effectively perfectly flat. In contrast, if kurtosis is >3, then the distribution is said to be *leptokurtic* or "more peaked" than the normal distribution. In cases like this, more of the mass of the distribution will be found in the tails than is the case for the normal distribution. A good example of a leptokurtic distribution is the *t* distribution. Examples of platykurtic and leptokurtic distributions are given in Figure 2.4. In each case, the PDF of the distribution is shown relative to that of the normal distribution.

> **Historical Note:** The four moments (mean, variance, skewness, and kurtosis) came to prominence in statistical theory because of the work of Karl Pearson [Pearson1895]. Pearson introduced a "family" of distribution curves based on these moments as parameters. His intention was that these would form the basis for a complete analysis of all continuous probability distributions. The Pearson family of distributions is not used in modern statistical analysis but the moments themselves have proved to be an important way of visualizing and understanding probability density functions. In recent years, the higher-order moments, skewness, and kurtosis have proved to be particularly important in understanding and interpreting the distribution of high frequency financial data.

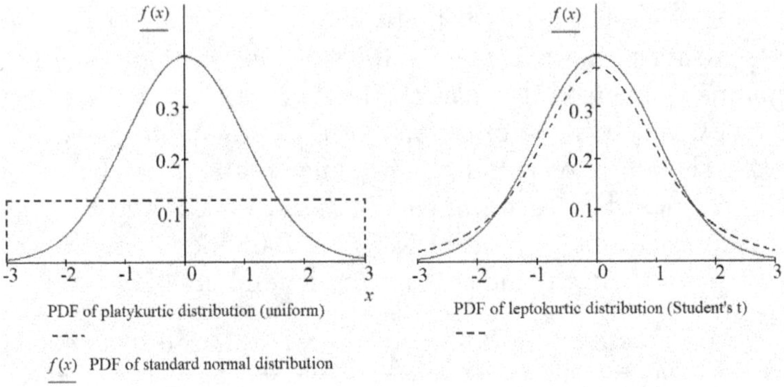

FIGURE 2.4 Different Forms of Kurtosis (Relative to the Normal Distribution).

The reason why higher-order moments are important is that many of our statistical testing procedures are based on the idea of a normal distribution. This is true even we consider tests based around Student's t, the chi-squared or the F distribution. In each of these cases, the test statistic is ultimately based on the assumption of normality. For example, in the case of the Student's t distribution, we assume a normally distributed variable with unknown variance. It is the fact that we must replace the unknown population standard deviation with its sample equivalent which means that we must use the t distribution rather than the normal. It therefore becomes extremely difficult to derive the sampling distributions of the test statistics if the underlying data are not normally distributed. Testing for normality of a random variable is therefore an important part of the econometrician's toolkit.

In order to test whether a random variable is normally distributed, we make use of the *Jarque-Bera test statistic*. This is defined in terms of the sample skewness and kurtosis coefficients as shown in equation (2.26)

$$JB = \frac{N}{6}\left[\hat{\gamma}_1^2 + \frac{1}{4}(\hat{\gamma}_2 - 3)^2\right]. \tag{2.26}$$

Under the null hypothesis that the variable in question follows a normal distribution, it can be shown that this statistic is distributed as chi-squared with two degrees of freedom. A test based on this statistic is often applied to assess if deviations from the normal distribution are severe.

Example: Consider the rate of return on a diversified portfolio of stocks. We can approximate this using one of the stock market indices used to measure overall movements in the market. In this case, we use the Financial Times 100 index, which is one of the most frequently quoted indices for the UK market. The return on holding a diversified portfolio can be measured as the first difference of the logarithm of the index. Table 2.1 gives sample statistics for the average return on the UK FTSE 100 index using this method. The sample period is from January 2015 to September 2020.

TABLE 2.1 Returns on the UK Stock Market 1/1/2015ñ12/9/2020.

Variable	DLOG(FTSE)
Mean	-0.177686 E-6
Maximum	0.086668
Minimum	-0.115124
Standard Deviation	0.010865
Skewness	-0.970049
Kurtosis	17.230814
Jarque-Bera	13262.089000

Suppose we wish to test the null hypothesis that returns are normally distributed. Under the null, the Jarque-Bera statistic follows a chi-squared distribution with two degrees of freedom and therefore the 5% critical value is 5.99. Given a test statistic of 13,262, we reject the null in favor of the alternative. From the other statistics presented in the table, we see that an important factor leading to our rejection of the null is the excess kurtosis indicated by a kurtosis coefficient of 17.23. This indicates a highly leptokurtic distribution, that is, one in which many more observations lie in the tails of the distribution than would be expected with a normal distribution. Therefore, if we had assumed a normal distribution, we would considerably underestimate the probability of extreme observations in stock market returns. Features such as this are of obvious interest to stock market traders who wish to estimate the chances of being caught out by a sudden crash in the market.

2.6 NONPARAMETRIC TESTS

The tests we have discussed so far have all been parametric in nature. That is, they are concerned with testing hypotheses about the parameters of an unknown distribution. There are, however, tests that do not rely on this process but are instead concerned with features of the data which do not depend on parameters. An example here is the use of tests based on contingency tables for the independence of two or more variables or events. Let us consider a simple example in which there are two events of interest which we label A and B. In the example we used in Chapter 1, event A was an increase in price for a good and event B was a fall in quantity. Each event has a complementary event, say A^- and B^- such that, between them, the event and the complementary event are exhaustive. Using this framework, we can draw up a 2 × 2 contingency table of the form.

TABLE 2.2 Hypothetical 2 × 2 Contingency Table.

	A	A^-	
B	A	b	$a+b$
B^-	C	d	$c+d$
	$a+c$	$b+d$	$a+b+c+d=N$

Using this framework, we can construct tests for the independence of the two events. In the example of Chapter 1, this would be equivalent to testing the null hypothesis that the direction of changes in prices is unrelated to that of the change in quantity. More formally, it amounts to a test of the hypothesis that the joint probability that events A and B occur simultaneously is equal to the product of the marginal events, that is, $P(A \cap B) = P(A)P(B)$.

The most intuitive test here is Pearson's chi-square test. This involves comparing the observed values in each cell of the contingency table with their expected values under the assumption of independence. The expected values can be calculated as $E_{ij} = p_i p_j N$; $i = A, A^-$ and $j = B, B^-$, where the superscript – indicates a complementary event, that is, if p_A is the probability that event A occurs, then p_{A^-} is the probability that event A does not occur. The marginal probabilities can be calculated as $p_A = (a+c)/N$, $p_{A^{-1}} = (b+d)/N$, $p_B = (a+b)/N$ and $p_{B^-} = (c+d)/N$. The test statistic is

$$\sum_{i=A,A^-} \sum_{j=B,B^-} \frac{\left(O_{ij} - E_{ij}\right)^2}{E_{ij}}, \tag{2.27}$$

which is distributed asymptotically as chi-squared with one degree of freedom. More generally, if the contingency table contains r rows and s columns, then the test statistic will be distributed asymptotically as $\chi^2_{(r-1)(s-1)}$.

To illustrate the use of the Pearson chi-square test, let us consider a real-world example. In Chapter 1, we looked at the relationship between current price and quantity changes in the market for potatoes and argued that this reflected movements along a demand curve because quantity was fixed in the short term. If we wish to examine the supply relationship, then we need to consider the fact that supply responds to lagged, rather than current, values of price. Therefore, let us consider the following contingency table that relates changes in quantity to the 1-year lag in price changes. If our hypothesis is correct, then we should still observe a significant relationship between these variables, but one in which a positive change in price results in a positive change in quantity which is delayed by 1 year. The contingency table we observe is given in Table 2.3.

TABLE 2.3 Contingency Table for Relationship Between Quantity Changes and Lagged Price Changes.

	Price rises (lagged 1 year)	Price falls (lagged 1 year)	Total
Quantity rises	17	9	26
Quantity falls	0	15	15
Total	17	24	41

From this table, we can first calculate the marginal probabilities of each event and then calculate the expected values of each joint event under the assumption of independence. The results of this are given in Table 2.4.

TABLE 2.4 Expected Values of Quantity Changes and Lagged Price Changes under the Assumption of Independence.

	Price rises (lagged 1 year)	Price falls (lagged 1 year)	Total
Quantity rises	10.7788	15.2193	26
Quantity falls	6.2198	8.7821	15
Total	17	24	41

If we compare Tables 2.3 and 2.4, we note that the row and column sums are the same (subject to rounding errors in the calculations). The cell entries for individual events are, however, very different. For example, under the assumption of independence, we would expect just over six cases in which quantity falls after a lagged price increase. In practice, however, we observe no such cases. Similarly, we would expect between eight and nine cases in which quantity falls after a fall in price, but we observe nearly double that number at 15. The question is, however, whether these differences are statistically significant. To test this, we calculate the Pearson chi-square statistic given in equation (2.27) that gives a value of 16.75. The 5% critical value for a chi-square test with one degree of freedom is 3.84, and therefore, we reject the null hypothesis at this level.

The Pearson test is most appropriate in large samples of data because the distribution of the test statistic is only known asymptotically. For small samples, there is an exact test provided by Fisher who shows that the distribution of the values in a 2 × 2 contingency table follows a hypergeometric distribution under the null of independence. The p value for observing the sample of values shown in Table 2.2 is given by

$$\varphi = \frac{(a+b)!(c+d)!(a+c)!(b+d)!}{a!b!c!d!N!}. \tag{2.28}$$

The value of this statistic for the data shown in Table 2.3 is 0.00002. This confirms the conclusion from Pearson's test that the events defined in this table are not independent. There is a significant relationship between current changes in quantity and lagged changes in price.

EXERCISES

Variable	Ratio
Mean	60.925342
Maximum	106.116351
Minimum	10.000000
Standard Deviation	20.922939
Skewness	-0.027204
Kurtosis	2.394301

The table above gives summary statistics for the ratio of Consumption Expenditures to GDP for 50 economies taken from the United Nations online database.

EXERCISE 2.1

Test the null hypothesis that the population mean is >55%.

EXERCISE 2.2

Test the null hypothesis that the population mean is equal to 65%.

EXERCISE 2.3

Calculate the Jarque-Bera test statistic and test the null hypothesis that this ratio follows a normal distribution.

REFERENCES

[Fisher1935] Fisher, R. A. *The Design of Experiments*, 1935. Edinburgh: Macmillan.

[Neyman1928] Neyman, J. and Pearson, E. "On the Use and Interpretation of Certain Test Criteria for Purposes of Statistical Inference, Part I and Part II." *Biometrika*, 1928, 20, pp. 175–294.

[Neyman1937] Neyman, J. "Outline of a Theory of Statistical Estimation Based on the Classical Theory of Probability." *Philosophical Transactions of the Royal Society of London. Series A, Mathematical and Physical Sciences*, 1937, 236(767), pp. 333–380.

[Pearson1895] Pearson, K. "Contributions to the Mathematical Theory of Evolution, II: Skew Variation in Homogeneous Material." *Philosophical Transactions of the Royal Society*, 1895, 186, pp. 343–414.

[Pearson1900] Pearson, K. "On the Criterion that a Given System of Deviations From the Probable in the Case of a Correlated System of Variables Is Such That It Can Be Reasonably Supposed to Have Arisen From Random Sampling." *Philosophical Magazine*, 1900, Series 5. 50(302), pp. 157–175.

CHAPTER 3

THE BIVARIATE REGRESSION MODEL

Regression analysis is the most important tool that economists use to quantify their models. Economic theory provides explanations of linkages between variables of interest, for example, the relationship between consumption expenditures and disposable income. However, theory rarely gives precise values for the size of the response of one variable to another. For this, we must turn to econometrics and, in particular, to regression analysis. The regression model provides a mechanism by which the response of one variable to another can be quantified and evaluated from a statistical perspective. It therefore acts as one of the key items in the toolkit of the applied social scientist, and the objective of this chapter is to discuss how it can be used sensibly in the investigation of economic relationships.

We will begin with a discussion of the simplest possible case – the bivariate linear regression model. This consists of a single endogenous variable Y linked to a single exogenous variable X by a linear relationship. The parameters of interest in this model are the intercept α and the slope coefficient β as shown in equation (3.1)

$$Y_i = \alpha + \beta X_i + u_i, \tag{3.1}$$

where u_i is a random error that introduces a stochastic element into the relationship. In practice, it is very rare that the applied econometrician will be interested in a relationship as simple as (3.1). Most of the time we deal with complex relationships in which there are several right-hand side variables and where the equation of interest may be one of a system of simultaneous equations. Nevertheless, the analysis of a simple equation like this gives us the opportunity to develop an understanding of the regression model that will be of value when it comes to dealing with more complex relationships.

Therefore, in this chapter, we will present a thorough review of the bivariate regression model that will cover estimation, statistical inference, and prediction.

3.1 DERIVATION OF THE OLS ESTIMATOR

The problem facing the econometrician is how best to use the data available $\{(X_i, Y_i); i = 1,...,N\}$ to estimate the unknown parameters of equation (3.1). Ordinary least squares (OLS) provide a simple method for the generation of such estimates which, under certain conditions, can be shown to have the properties that the estimates are both unbiased and efficient (in the sense that they have the lowest possible variances in the class of unbiased estimators). The method of OLS is to choose parameter estimates $\hat{\alpha}$ and $\hat{\beta}$ which minimize the sum of the squared deviations of the actual values of Y_i from the fitted values $\hat{\alpha} + \hat{\beta}X_i$. In mathematical notation, we can write the problem as

$$\min_{\hat{\alpha},\hat{\beta}} RSS = \sum_{i=1}^{N}\left(Y_i - \hat{\alpha} - \hat{\beta}X_i\right)^2. \tag{3.2}$$

This is a relatively straightforward problem in calculus since the loss function is quadratic in the variables of interest. Differentiation with respect to $\hat{\alpha}$ and $\hat{\beta}$ yields the following pair of first-order conditions for a minimum,

$$\frac{\partial RSS}{\partial \hat{\alpha}} = -2\sum_{i=1}^{N}\left(Y_i - \hat{\alpha} - \hat{\beta}X_i\right) = 0 \tag{3.3}$$

$$\frac{\partial RSS}{\partial \hat{\beta}} = -2\sum_{i=1}^{N}X_i\left(Y_i - \hat{\alpha} - \hat{\beta}X_i\right) = 0. \tag{3.4}$$

Equations (3.3) and (3.4) in turn can be used to derive the following pair of simultaneous equations in $\hat{\alpha}$ and $\hat{\beta}$ which are known as *the least-squares normal equations*. Since all summations here are over the full sample of data $i = 1,\ldots, N$, we will omit the limits of the summation operator in future expressions to simplify the notation,

$$\hat{\alpha}N + \hat{\beta}\sum X_i = \sum Y_i \tag{3.5}$$

$$\hat{\alpha}\sum X_i + \hat{\beta}\sum X_i^2 = \sum X_i Y_i. \tag{3.6}$$

The solution of these equations is interesting because it demonstrates that the OLS parameter estimates are functions of the sample moments of the data. For example, dividing equation (3.5) by N immediately yields the result that the regression line passes through the sample means of the data, that is,

$$\bar{Y} = \hat{\alpha} + \hat{\beta}\bar{X}. \tag{3.7}$$

Substituting $\hat{\alpha} = \bar{Y} - \hat{\beta}\bar{X}$ into (3.6) and rearranging yields

$$\hat{\beta} = \frac{\sum X_i Y_i - \bar{Y}\sum X_i}{\sum X_i^2 - \bar{X}\sum X_i} = \frac{\sum X_i Y_i - N\bar{Y}\bar{X}}{\sum X_i^2 - N\bar{X}^2}. \tag{3.8}$$

Given that $\sum X_i Y_i - N\bar{Y}\bar{X} = \sum (X_i - \bar{X})(Y_i - \bar{Y})$ and that $\sum X_i^2 - N\bar{X}^2$ $= \sum (X_i - \bar{X})^2$, we can write (3.8) as

$$\hat{\beta} = \frac{\sum (X_i - \bar{X})(Y_i - \bar{Y})}{\sum (X_i - \bar{X})^2}. \tag{3.9}$$

Equation (3.9) enables an intuitive interpretation of the OLS slope coefficient in terms of the sample moments of the Y and X variables. Dividing numerator and denominator by $N - 1$ enables (3.9) to be written as

$$\hat{\beta} = \frac{\sum (X_i - \bar{X})(Y_i - \bar{Y}) / (N-1)}{\sum (X_i - \bar{X})^2 / (N-1)} = \frac{\hat{\sigma}_{XY}}{\sigma_X^2}. \tag{3.10}$$

The numerator of this expression is an unbiased estimator of the population covariance and the denominator is an unbiased estimator of the variance of X. Thus, the slope coefficient for the bivariate regression model is equal to the ratio of the sample covariance of X and Y and the sample variance of X. We have therefore established that both the intercept and the slope coefficient estimates for the OLS model can be written in terms of the first and second sample moments of the data. Note that, in our estimates of the covariance of X and Y and the variance of X, we divide by $N - 1$ rather than by the number of observation to allow for the loss of degrees of freedom incurred when estimating sample means. In large samples, this makes relatively little difference to the calculation of the sample moments and, because in this case we are taking the ratio of two sample moments, the estimate of the slope coefficient is unaffected when we use the same divisor for *both* sample moments. However, we need to be careful if we use this method to calculate the regression slope coefficient because some statistical packages and spreadsheets will use $N - 1$ as the divisor for the variance of X and N as the divisor for the covariance of X and Y.

Historical Note: Adrien Marie Legendre (1752–1833) was the first mathematician to set out the least squares method in print in his book (*New Methods for the Determination of the Orbits of Comets*) published in 1805, [Legendre1805]. However, Carl Friedrich Gauss (1777–1855), in his 1809 book (*Theory of the Motion of the Heavenly Bodies Moving about the Sun in Conic Sections* [Gauss1809]), later claimed to have been using the method since 1795.

Example: An econometrician wishes to estimate the parameters of the demand curve for potatoes in the United States. A preliminary investigation indicates a negative relationship between these variables as shown in Figure 3.1.

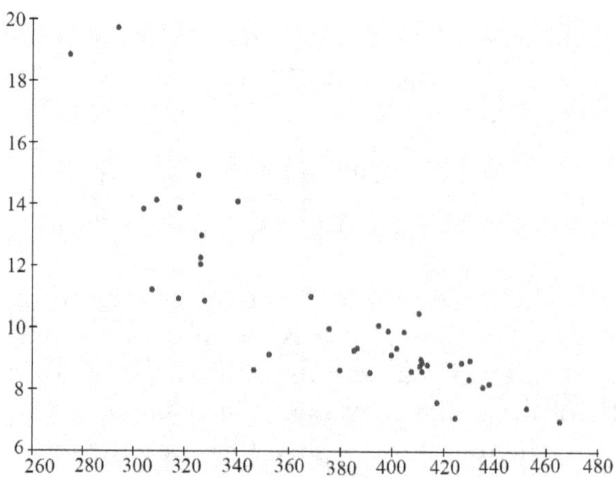

FIGURE 3.1 Scatter Diagram of the Price of Potatoes against the Quantity Sold.

The data here are taken from the *National Potato Council Yearbook 2019*. Price is measured as \$ per hundredweight (CWT) and quantity is measured as millions of CWT. The price series has been deflated by the consumer price index where the 2015 value is equal to 100. We have 43 annual observations for the period 1975–2017. From Figure 3.1, it certainly appears that there is a strong negative relationship between these series.

The equation to be estimated takes the following form: $p_t = a + \beta q_t + u_t$.[1] To calculate the least squares estimates of the regression parameters, we first calculate the sample moments of the data. This yields the following results:

$$\begin{aligned} &\overline{p} = 10.3762 \qquad \overline{q} = 378.4450 \\ &\hat{\sigma}_p = 2.8417 \qquad \hat{\sigma}_q = 48.9175 \\ &\hat{\rho}_{pq} = -0.8437. \end{aligned} \tag{3.11}$$

This contains all the information necessary to calculate the least squares parameter estimates. First, we calculate the slope coefficient estimate as the ratio of the sample covariance of p and q to the sample variance of q. The sample covariance is calculated as $\hat{\sigma}_{pq} = \hat{\rho}_{pq}\hat{\sigma}_p\hat{\sigma}_q$, where $\hat{\rho}_{pq}$ is the sample correlation coefficient. Using this, we have

$$\hat{\beta} = \frac{\hat{\sigma}_{pq}}{\hat{\sigma}_q^2} = \hat{\rho}_{pq}\frac{\hat{\sigma}_p}{\hat{\sigma}_q} = -0.0490. \tag{3.12}$$

We can then calculate the estimate of the intercept by using the property that the regression line passes through the sample means of the data. This yields

$$\hat{\alpha} = \overline{p} - \hat{\beta}\overline{q} = 28.9203. \tag{3.13}$$

The estimates are therefore consistent with the hypothesis of a downward sloping demand curve that takes the form $p_t = 28.9203 - 0.049q_t$.

3.2 INTERPRETING THE REGRESSION LINE – MARGINAL EFFECTS AND ELASTICITIES

The slope coefficient of the regression line gives us an estimate of the *marginal effect* of the variable X on the variable Y, that is, we can think of $\hat{\beta}$ as an estimate of dY/dX. This means that the units of measurement for the slope

[1] Note that we use the subscript t for this example to indicate that this is time-series data.

coefficient depend on the units of the X and Y variables. In our example, the slope coefficient of −0.049 indicates that an increase of one million hundredweight of potatoes implies a fall in price of 4.9 cents per hundredweight. The assumption of a constant marginal effect is a very strong assumption which can be misleading if taken too literally. A more reasonable assumption is that the marginal effect is approximately constant within the range of the sample data for X. If we try to use the estimated regression model to predict the value of Y using values of X which lie a long way outside the range of the data used to estimate the model, then it is likely that the predictions will prove unreliable.

The marginal effect of X on Y is not always the most interesting statistic for the investigator. In many cases, a more interesting quantity is the *elasticity*. This measures the proportional response of Y to a given proportional change in X. The elasticity can be written in mathematical terms as $\eta = (dY / Y) \div (dX / X) = (dY / dX) \times (X / Y)$. Along a linear regression line, the elasticity will change because dY / dX is constant, but X / Y changes if the intercept is nonzero. It is possible to obtain an elasticity estimate for a point on the regression line: for example, we can evaluate the elasticity at the sample means of the data $\hat{\eta} = \hat{\beta} \times \bar{X} / \bar{Y}$. In our example of the demand curve for potatoes, we obtain an estimate of the elasticity at the means of the variables as $\hat{\eta} = -1.7871$. Note that this is the elasticity of price with respect to quantity *not* the price elasticity of demand as defined in most economics textbooks. We can obtain an estimate of the price elasticity of demand using the following transformation: $\hat{\eta}_D = -(1 / \hat{\eta}) = 0.56$. This indicates that the demand for potatoes is price inelastic, that is, a change in price is associated with a less than proportionate change in quantity. One implication of this is that a cut in price will reduce total sales revenue for this product.

It is often useful to obtain a more direct estimate of the elasticity through a modification of the regression equation itself. Consider an alternative specification of the regression equation which is expressed in logarithms of the variables Y and X,

$$\ln(Y_i) = \alpha + \beta \ln(X_i) + u_i. \tag{3.14}$$

This is referred to as a *log-linear regression model* for obvious reasons. We can estimate this by OLS to obtain the slope coefficient $\hat{\beta}$, which can be interpreted as an estimate of the marginal effect $d\ln Y / d\ln X$. Now, the first-order differential of the logarithmic function can be written as $d\ln z = dz/z$ and, for small increments, the ratio of the first-order differentials of $\ln Y$ and $\ln X$ will give the derivative of $\ln Y$ with respect to $\ln X$, that is, $d\ln Y/d\ln X = dY/dX \times X/Y$. This means that the marginal effect from a log-linear model gives the elasticity of Y with respect to X. Because of this, the log-linear specification is very convenient in many econometric applications and is frequently chosen in preference to the simple linear specification.

Example: If we apply the log-linear specification to our example data for the potato market, then we obtain the following estimated demand curve

$$\ln(p_t) = 11.7061 - 1.5854 \ln(q_t) + \hat{u}_t. \tag{3.15}$$

The slope coefficient here gives us an estimate of the elasticity of price with respect to quantity. That is, it measures the percentage response of price to a 1% increase in quantity. Note that the estimate of the slope coefficient for this equation is quite close to the estimate of the elasticity of price with respect to quantity which we calculated for the linear equation at the means of the variables. In order to compare the linear and log-linear specification, we can write equation (3.15) in terms of the levels of the variables as

$$p_t = \exp(11.7061) q_t^{-1.5854} \exp(\hat{u}_t). \tag{3.16}$$

We can now compare the fit of these alternative specifications against the actual data in Figure 3.2. This shows that, although the mathematical forms of the two equations appears to be very different, they both provide reasonable fits to the data. The log-linear specification has the property that it approaches the axes asymptotically as quantity either approaches zero or tends to infinity. This property is desirable for a demand curve as it avoids predictions of negative price or quantity in extreme circumstances. In contrast, the linear specification predicts negative quantity when $p > 28.9203$ and negative price when $q > 590.2$.

Historical Note: Alfred Marshall (1842–1924) is credited as the first to use the concept of elasticity in the context of economics in his book *Principles of Economics* first published in 1890 [Marshall1890].

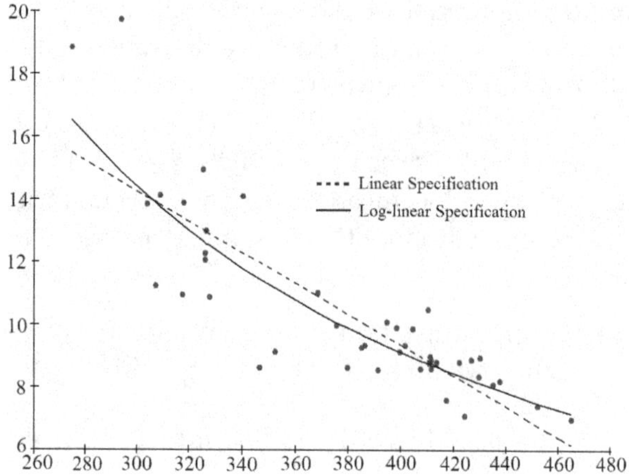

FIGURE 3.2 Linear and Log-Linear Demand Curve Estimates.

3.3 THE REVERSE REGRESSION

In some cases, the direction of causation for an economic model is obvious. However, in others, it may be less so, and there may be a sensible interpretation of the model in which Y causes X rather than vice versa. That is, instead of thinking of X as the exogenous variable and Y as the endogenous variable, we might think of Y as the variable causing changes in X. For example, in our demand curve estimates, we have chosen price as the dependent variable and quantity as the independent variable. There is a strong case for doing this when modeling agricultural markets, because it is difficult to adjust quantity in the short run while price is free to adjust immediately in response to external shocks. However, there are many markets in which this is not the case and, in which, it may make more sense to think of quantity adjusting in the short run while price remains relatively sticky.

Consider the regression equation $Y_i = \alpha + \beta X_i + u_i$. It may be tempting to assume that we could estimate the marginal effect of Y on X by estimating this equation by least squares to obtain $\hat{Y}_i = \hat{\alpha} + \hat{\beta} X_i$ and then "solving" this equation to obtain $X_i = -\hat{\alpha} / \hat{\beta} + \hat{Y}_i / \hat{\beta}$. This would yield an estimate of the marginal effect of Y on X which is equal to the reciprocal of the OLS slope coefficient from the original regression equation. Unfortunately, this procedure is quite incorrect. To see this, consider the reverse regression equation $X_i = \gamma + \delta Y_i + v_i$, where γ and δ are the intercept and slope parameters, and v is a random error. It is easy to see that the estimate of the slope coefficient from this regression will take the form

$$\hat{\delta} = \frac{\sum (X_i - \bar{X})(Y_i - \bar{Y})}{\sum (Y_i - \bar{Y})^2}. \tag{3.17}$$

This is clearly not equal to the reciprocal of the slope coefficient from a regression of Y on X. However, there is an interesting relationship between the least squares estimates of the slope coefficients of the original regression and the reverse regression. If we multiply these estimates together, then we obtain the following result:

$$\hat{\beta} \times \hat{\delta} = \frac{\left(\sum (X_i - \bar{X})(Y_i - \bar{Y})\right)^2}{\sum (X_i - \bar{X})^2 \sum (Y_i - \bar{Y})^2} = \hat{\rho}^2_{XY}. \tag{3.18}$$

This shows that the product of the slope coefficient from our original regression and the reverse regression is the square of the sample correlation coefficient of Y and X. This establishes a link between the three possible measures of association between a pair of variables Y and X that we have considered.

Example: Estimation of the reverse regression for our model of the demand for potatoes yields

$$q_t = 529.1471 - 14.5239 p_t + \hat{u}_t. \tag{3.19}$$

The product of the slope coefficients from the original and reverse regressions is equal to $-0.049 \times -14.5239 = 0.7167$ which is approximately equal to the squared value of the correlation coefficient between these two variables. The fact that this is not an exact relationship because of rounding errors in the process of the calculation.

We therefore have three measures of the association between two variables in the form of the correlation coefficient and the slope coefficients from the original and the reverse regressions. We can think of these as being the result of alternative methods of specifying a best-fit line through the scatter of points, which characterizes these variables. The simple regression is constructed by minimizing the sum of the squared vertical distances of the scatter of points from the line. The reverse regression is constructed by minimizing the sum of the squared horizontal distances, and finally, the correlation line can be thought of as minimizing the sum of the squared perpendicular distances of the scatter from the line. The method we choose to use to fit a line to the data will depend on our views of the causal relationships between the variables and what we wish to do with the model once it has been constructed.

3.4 ASSUMPTIONS OF THE CLASSICAL LINEAR REGRESSION MODEL

So far, we have concentrated on the mechanics of the regression model. However, if we wish to go further and discuss the statistical properties of the estimator, then we need to make further assumptions about the nature of the data and the properties of the random error term. We will begin with the standard set of assumptions listed in Table 3.1.

TABLE 3.1 Assumptions of the Classical Linear Regression Model (CLRM).

1. The error term has zero mean $E(u_i) = 0; i = 1,...,N$.
2. The covariance of the error term indexed i and that indexed $j \neq i$ is zero $E(u_i u_j) = 0; i \neq j$.
3. The variance of the error term is constant $E(u_i^2) = \sigma_u^2;\ i = 1,...,N$.
4. Exogeneity of regressors Strong form: The X variable is nonstochastic (fixed in repeated samples). Weak form: The covariance of the X variable and the error is zero $E(X_i u_i) = 0$.
5. The errors follow a normal distribution.

Assumption 1 is not controversial. Providing our regression equation contains a constant, that is, the intercept is nonzero, then the error term will always have expectation zero by construction. However, Assumptions 2–5 place somewhat stronger requirements on our model. Assumption 2 requires the errors to be independent. This is frequently problematic when dealing with time series data, where the assumption is described as the assumption of *serial independence*. Time series models often have errors that are related to errors in the immediate past, for example, $E(u_t u_{t-1}) \neq 0$. In such circumstances, the error is said to be serially correlated and we need to take account of this when assessing the properties of the OLS estimator. Assumption 3 is described as the assumption of *homoscedasticity*. Models that violate this condition are most often found when we are dealing with cross-section data, where the size of the variance of the error term is related to the value of the exogenous variable. For example, we might have $\sigma_{u_i}^2 = \sigma^2 X_i^2$. Again, this has implications for the properties of the least squares estimator which must be taken into account when regression results are evaluated. Assumption 4 is that the X variable should be regarded as exogenous, that is, independent of any random disturbances to the relationship. This assumption is problematic when the equation is one of a system of equations that describe the joint behavior of a set of variables of interest. Finally, Assumption 5 states that the errors should ideally follow a normal distribution. This assumption is frequently made so that the distribution of estimators can be derived easily. However, it is not necessary in all circumstances.

Assumptions 1–4 are the *Gauss–Markov assumptions*. Under these conditions, the OLS estimator has lowest variance in the class of linear unbiased estimators. Alternatively, the OLS estimator is said to be Best Linear Unbiased Estimator (BLUE). The assumption that the errors follow a normal distribution is not necessary for the OLS estimator to be BLUE, but it is included in the list of Classical Linear Regression assumptions because it proves useful in deriving the distribution of the estimator. Note that, if we replace the strong form of Assumption 4 with the weaker version given in 4(b), then proof of the Gauss–Markov theorem becomes very difficult. However, it is possible to derive equivalent large sample properties. In particular, we can show that the OLS estimator is consistent (converges in probability on the population value in large samples) and that it converges faster than other consistent estimators.

In the discussion which follows, we will maintain the strong form of Assumption 4. This is not realistic for most econometric models because it assumes the ability of the investigator to replicate the input data (X values) by experimental means. It implies that the only source of random variation in the sample data is the random error term u. While such an assumption is appropriate for experimental sciences, it is unrealistic for most economic applications. However, it will make it possible to derive distributional results for the OLS parameter estimates that would not be possible if we were to use the weaker form. We will therefore maintain this assumption for the moment and consider the effects of relaxing it later.

3.5 DISTRIBUTION OF THE OLS ESTIMATOR

Consider the OLS estimator of the slope coefficient given in equation (3.9). From the original model we have $Y_i - \bar{Y} = \beta\left(X_i - \bar{X}\right) + u_i - \bar{u}$, substituting into (3.9) and noting that $\bar{u}\sum\left(X_i - \bar{X}\right) = 0$ by definition of the arithmetic mean of x, yields

$$\hat{\beta} = \beta + \frac{\sum\left(X_i - \bar{X}\right)u_i}{\sum\left(X_i - \bar{X}\right)^2}. \tag{3.20}$$

Now, if we maintain the strong version of Assumption 4, then taking expectations yields

$$E\left(\hat{\beta}\right) = \beta + \frac{\sum\left(X_i - \bar{X}\right)E\left(u_i\right)}{\sum\left(X_i - \bar{X}\right)^2}. \tag{3.21}$$

From Assumption 1, we have that $E\left(u_i\right) = 0;\ i = 1,\ldots,N$, and therefore equation (3.21) shows $E\left(\hat{\beta}\right) = \beta$ under these assumptions, that is, the OLS estimator of the slope coefficient is unbiased. Note the crucial role of the strong version of Assumption 4 here. Without this assumption, we would have to treat the X variables as random quantities, and it would be extremely difficult to prove unbiasedness in this way. Instead, we would have to rely on the large sample concept of consistency in which, under certain assumptions, the estimator $\hat{\beta}$ can be shown to "converge" on the true value if the sample size is sufficiently large. Note also that we only require Assumptions 1 and 4 for the OLS estimator to be unbiased. Failure of either Assumption 2 or Assumption 3 (or both) does not, in itself, mean that the OLS estimator will be biased.

Next, consider the variance of the OLS estimator. From the results derived so far, we have that $\hat{\beta} - E\left(\hat{\beta}\right) = \frac{\sum\left(X_i - \bar{X}\right)E\left(u_i\right)}{\sum\left(X_i - \bar{X}\right)^2}$. Therefore, the variance of $\hat{\beta}$ is given by the expression in equation (3.22)

$$V\left(\hat{\beta}\right) = E\left(\hat{\beta} - E\left(\hat{\beta}\right)\right)^2 = E\left(\frac{\sum\left(X_i - \bar{X}\right)u_i}{\sum\left(X_i - \bar{X}\right)^2}\right)^2. \tag{3.22}$$

From Assumptions 2 and 3 of the Classical Linear Regression Model (CLRM), we have that $E(u_i u_j) = 0; i \neq j$ and $E(u_i^2) = \sigma_u^2;\ i = 1,\ldots,N$. Therefore, taking expectations of the right-hand side of equation (3.22) yields

$$V(\hat{\beta}) = \frac{\sigma_u^2}{\sum(X_i - \bar{X})^2}. \tag{3.23}$$

Finally, from Assumption 5 of the CLRM, we have that the errors are normally distributed and, from equation (3.20), we have the results that the OLS estimator is a linear combination of the errors. Since any linear combination of normally distributed variables itself follows a normal distribution, we therefore can show that under the CLRM assumptions, the OLS estimator follows a normal distribution as shown in expression (3.24)

$$\hat{\beta} \sim N\left(\beta, \frac{\sigma_u^2}{\sum(X_i - \bar{X})^2}\right). \tag{3.24}$$

Equation (3.24) illustrates an interesting feature of the regression model. Let us consider the denominator of the variance expression. As the sample size increases, then this must also increase because the summation always involves the addition of positive numbers. Therefore, since σ_u^2 is constant, it follows that the variance of the OLS estimator tends to zero as the sample size become large, that is, the distribution of the OLS estimator is degenerate. Figure 3.3 illustrates the behavior of the probability density function (PDF) of the OLS estimator as the sample size changes for the case $\beta = 1, \sigma_u^2 = 1$, and $\sigma_X^2 = 1$. As the sample size increases, the variance of the OLS estimator falls, reducing the spread of the PDF. In the limit, as the sample size becomes infinite, the PDF of the OLS estimator collapses onto a vertical line going through the population value of the parameter β.

Next, we consider the distribution of the intercept estimator in the OLS regression model. From the least squares normal equations, we have

$$\hat{\alpha} = \bar{Y} - \hat{\beta}\bar{X} = \alpha + \bar{X}(\beta - \hat{\beta}) + \bar{u}. \tag{3.25}$$

If Assumptions 1 and 4 hold, then $E(\hat{\beta}) = \beta$ and $E(\bar{u}) = 0$, and it immediately follows that $E(\hat{\alpha}) = \alpha$. Therefore, the OLS estimator of the intercept is unbiased under the same assumptions that ensure that the slope coefficient is unbiased. Since the OLS estimator of α is a linear combination of

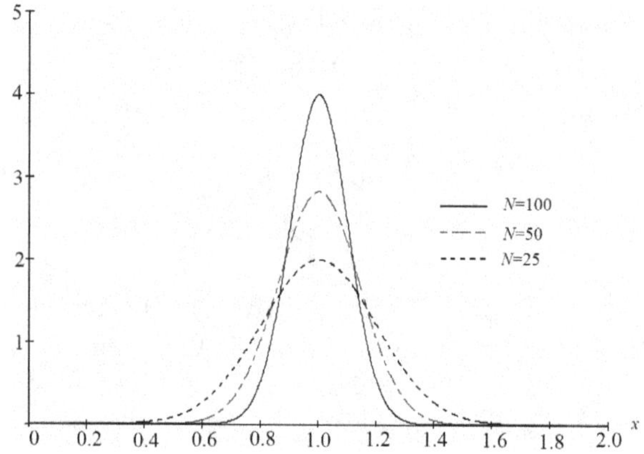

FIGURE 3.3 Effects of Increasing Sample Size on the PDF of the Least Squares Estimator.

normally distributed variables, it follows that it too is normally distributed. Substituting (3.20) for $\hat{\beta}$ in (3.25) yields

$$\hat{\alpha} = \alpha - \frac{\bar{X}\sum(X_i - \bar{X})u_i}{\sum(X_i - \bar{X})^2} + \bar{u}. \tag{3.26}$$

Therefore, the variance can be derived as

$$V(\hat{\alpha}) = E(\hat{\alpha} - E(\hat{\alpha}))^2 = E\left(-\frac{\bar{X}\sum(X_i - \bar{X})u_i}{\sum(X_i - \bar{X})^2} + \bar{u}\right)^2. \tag{3.27}$$

Expanding the term in brackets and taking expectations[2] yields

$$V(\hat{\alpha}) = \sigma_u^2\left(\frac{1}{N} + \frac{\bar{X}^2}{\sum(X_i - \bar{X})^2}\right). \tag{3.28}$$

---

[2] Note that this derivation makes use of the CLRM assumptions in the same way as the derivation of the variance of the slope coefficient estimator. Also note that the crossproduct terms are eliminated because $\sum(X_i - \bar{X}) = 0$ by construction.

For completeness, we can also derive the covariance of the slope and intercept estimates as

$$\operatorname{cov}\left(\hat{\alpha},\hat{\beta}\right)=E\left(\hat{\alpha}-E\left(\hat{\alpha}\right)\right)\left(\hat{\beta}-E\left(\hat{\beta}\right)\right)$$
$$=E\left(-\frac{\bar{X}\sum\left(X_i-\bar{X}\right)u_i}{\sum\left(X_i-\bar{X}\right)^2}+\bar{u}\right)\left(\frac{\sum\left(X_i-\bar{X}\right)u_i}{\sum\left(X_i-\bar{X}\right)^2}\right). \tag{3.29}$$

Multiplying out the parentheses and taking expectations yields

$$\operatorname{cov}\left(\hat{\alpha},\hat{\beta}\right)=-\frac{\bar{X}}{\sum\left(X_i-\bar{X}\right)^2}\sigma_u^2. \tag{3.30}$$

Therefore, the intercept and slope estimates for the OLS model can be shown to follow a joint normal distribution of the form

$$\begin{pmatrix}\hat{\alpha}\\ \hat{\beta}\end{pmatrix}\sim N\left(\begin{pmatrix}\alpha\\ \beta\end{pmatrix},\sigma_u^2\begin{pmatrix}\frac{1}{N}+\frac{\bar{X}^2}{\sum\left(X_i-\bar{X}\right)^2} & -\frac{\bar{X}}{\sum\left(X_i-\bar{X}\right)^2}\\ -\frac{\bar{X}}{\sum\left(X_i-\bar{X}\right)^2} & \frac{1}{\sum\left(X_i-\bar{X}\right)^2}\end{pmatrix}\right) \tag{3.31}$$

3.6 STATISTICAL INFERENCE WITH THE OLS ESTIMATOR

The main reason why we are interested in the distribution of the OLS estimator is that we wish to use this for the purposes of statistical inference. That is, we wish to be able to conduct hypothesis tests on the coefficients of the model and to construct confidence intervals for the unknown model parameters. First, note that the distribution of the OLS estimator of the slope coefficient can be transformed to the standard normal distribution as shown in expression (3.32).

$$\frac{\hat{\beta}-\beta}{\sigma_u/\sqrt{\sum\left(X_i-\bar{X}\right)^2}}\sim N(0,1). \tag{3.32}$$

If we knew the error variance, then statistical inference would be relatively simple. For example, suppose we wished to test $H_0:\beta=\beta_0$ against $H_1:\beta\neq\beta_0$. The test statistic for this test would be

$$\frac{\left(\hat{\beta}-\beta_0\right)}{\sigma_u / \sqrt{\sum\left(X_i-\bar{X}\right)^2}}, \tag{3.33}$$

and we could compare this with the appropriate critical value from the standard normal tables. Similarly, we could construct $a\%$ confidence intervals using the formula $\beta \pm z_c^{a/2}\sigma_u / \sqrt{\sum\left(X_i-\bar{X}\right)^2}$, where z is the critical value for a two-tailed test, again taken from the standard normal tables. However, we do not typically know the error variance and therefore we must use an estimate. Substitution of an estimated value for σ_u^2 in (3.33) means that the resulting statistic no longer follows the normal distribution. Instead, the test statistic can be shown to follow Student's t distribution. Although the t distribution is in many ways similar to the normal distribution, in that it is symmetric and has the characteristic "bell shape," it differs in that relatively more of the mass of the t distribution lies in the tails. However, this difference declines as the sample size gets larger and, in large enough samples, the t and normal distributions become indistinguishable. Nevertheless, the convention is to use the t distribution when conducting hypothesis tests on the coefficients of linear regression model because this will be valid in both large and small samples.

Example: Consider the following regression equation that relates the growth rate of household consumption expenditure for the US to the growth rate of real personal disposable income. Growth rates are calculated as the first difference of the logarithm of each variable. The data are annual from 1971 to 2019 and are taken from the Federal Reserve Economic Database (FRED). Standard errors are given in parentheses below parameter estimates and are calculated using the formulae given in equation (3.31).

$$\Delta \ln C_t = \underset{(0.0033)}{0.0052} + \underset{(0.1001)}{0.8284}\Delta \ln Y_t^D + \hat{u}_t. \tag{3.34}$$

Suppose we wish to test the null hypotheses that the coefficients are zero against the alternative that they are nonzero. Given this alternative hypothesis, a two-tailed test is appropriate and therefore the 5% critical value for a t test with 47 degrees of freedom is 2.01. The test statistic[3] for the

[3] The values of the test statistics here are calculated using the rounded values reported in the regression equation. Those calculated by the regression package will be slightly different as they use unrounded values. However, the difference does not change the conclusions of the testing procedure in either case.

intercept is $0.0052/0.0033 = 1.58$, whereas that for the slope coefficient is $0.8284/0.1001 = 8.26$. Therefore, in the case of the slope coefficient, we reject the null hypothesis that the slope parameter is zero in favor of the alternative that it is not zero. In the case of intercept, however, we cannot reject the null hypothesis that the intercept is equal to zero.

Tests of the null hypothesis that the coefficients are zero form a standard part of econometric procedure. However, this null hypothesis is not always the most interesting from the point of view of economic theory. For example, in the case of the consumption–income relationship, we might be more interested in testing the null hypothesis that the slope coefficient is equal to one against the alternative that it is less than one. In this case, the null is that there the elasticity of consumption with respect to income is equal to one which implies that consumption and income are proportionally related. In this case, the alternative hypothesis is that the elasticity is less than one so that a 1% change in income produces a less than proportionate response in consumption. The one-sided nature of the alternative hypothesis means that a one-tailed test is appropriate in this case, and so the 5% critical value in this case is 1.68. The test statistic is $(0.8284-1)/0.1001 = -1.71$, and therefore, we reject the null that the parameter is equal to one in favor of the alternative that it is less than one in this case.

The regression results given in equation (3.34) also allow us to construct confidence intervals for the model parameters. Suppose, for example, that we wish to calculate the 95% confidence interval for the slope. This will be given by the following expression:

$$\hat{\beta}_1 - 2.01 \times SE\left(\hat{\beta}_1\right) < \beta_1 < \hat{\beta}_1 + 2.01 \times SE\left(\hat{\beta}_1\right),$$

where $\hat{\beta}_1$ is the OLS estimate of the slope coefficient and $SE\left(\hat{\beta}_1\right)$ is its standard error. Using the values reported in (3.34), this gives

$$0.8284 - 2.01 \times 0.1001 < \beta_1 < 0.8284 + 2.01 \times 0.1001$$
$$0.6272 < \beta_1 < 1.0296.$$

Note that this confidence interval includes the value one. This is consistent with our earlier rejection of the null hypothesis that the parameter was equal to one because that test was conducted on the basis of a one-sided alternative. The confidence interval above has been calculated using the critical value for a two-sided alternative. Therefore, the fact that the confidence interval includes the value one indicates that we would not reject the null hypothesis that the parameter is equal to one at the 5% level using a two-tailed test and we know, from earlier examples, that the critical values for a two-tailed test are larger, in absolute value, than those for a one-tailed test of the same size.

Econometricians tend to place most emphasis on the size of the tests they conduct rather than their power. The size of the test is the probability of rejecting the null hypothesis when it is true. This can always be set by the investigator through the choice of an appropriate critical value. We can think of the size of the test as the probability of generating a false positive result, that is, a Type I error. The power of a test is defined as the probability of rejecting the null hypothesis when the alternative is true. Alternatively, we can think of this as one minus the probability of generating a false-negative result. A false negative corresponds to a Type II error, where we accept the null even though the alternative is true. For a variety of reasons, it is much more difficult to determine the power of test than it is to fix its size. However, this does not mean that power is unimportant, and it needs to be considered whenever we implement a testing procedure.

To consider the relationship between size and power, consider the case illustrated in Figure 3.4. This corresponds to a situation in which both the null and the alternative hypotheses involve specific values of an unknown

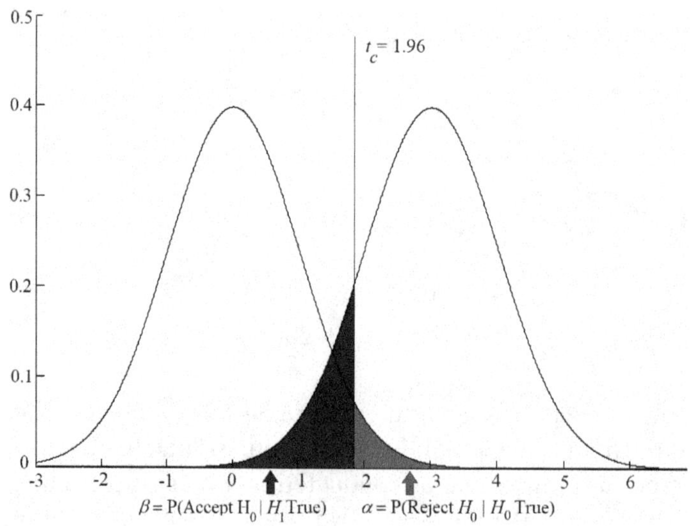

FIGURE 3.4 PDFs for Null and Alternative Hypotheses.

parameter θ. For example, we might have a test of the form $H_0 : \theta = \theta_A$ against $H_1 : \theta = \theta_B$. Figure 3.4 shows the PDFs for H_0 when $\theta \sim N(0,1)$ and for H_1 when $\theta \sim N(3,1)$.

The size of the test is illustrated by the vertical line corresponding to $t_c = 1.96$. For a standard normal distribution and a one-tailed test, this indicates a significance level of 2.5%. The shaded region labeled *a* shows 2.5% of the area under the PDF for the null lying to the right of t_c. We would reject the null hypothesis at the 2.5% level if the test statistic is >1.96. If the null hypothesis is correct, then this gives the probability of a Type I error. Next, consider the implications if the alternative hypothesis is true. The probability that we fail to reject the null is given by the area under the PDF for the alternative hypothesis to the left of the line $t_c = 1.96$. This is shown by the shaded region in Figure 3.4 labeled β and, if the alternative hypothesis is true, this gives the probability of a Type II error. The power of the test is equal to $1 - \beta$. Therefore, the choice of the critical value determines both the size of the test and its power. If we increase the critical value by increasing the critical value, then the shaded area *a* falls which lowers the probability of making a Type I error, but we simultaneously increase the shaded area β which lowers the power of the test (increases the probability of making a Type II error).

In the example given in Figure 3.4, we can write down an expression for the power of a test as $1 - \beta = 1 - \int_{-\infty}^{t_c} f\left(\hat{\theta} - \theta_B\right) d\hat{\theta}$, where f is the PDF of the random variable $\hat{\theta}$ (the estimator) and θ_B is the hypothesized value under the alternative hypothesis. Consider the case in which we are interested in estimating the slope coefficient for a least squares regression. We know that the distribution of such an estimator is degenerate, that is, its variance falls to zero as the sample size gets large. It therefore follows that $\int_{-\infty}^{t_c} f\left(\hat{\theta} - \theta_B\right) d\hat{\theta} \to 0$ as $N \to \infty$ or the power of the test approaches one as the sample size becomes large. Another implication of this is that, for any given size of such a test, we can determine the power of the test *providing that we have enough observations*. Of course, this last point is the tricky one for econometricians, who rarely have the opportunity to generate data experimentally, and are forced to take the number of observations as a given. This may help us understand the lack of discussion of power in many econometrics textbooks. In an experimental science, the investigator can control the power of a test by replicating the experiment an appropriate number of times. Since econometricians do not have such control the issue of power is often mentioned and then promptly ignored.

The example in Figure 3.4 indicates several other problems for the determination of the power of a statistical test. First, in order to determine the probability of making a Type II error, we need to assume that the alternative hypothesis takes a specific form, $H_1 : \theta = \theta_B$. In the more general cases, $H_1 : \theta < \theta_B$ and $H_1 : \theta \neq \theta_B$, we cannot draw a unique PDF for the alternative hypothesis and therefore we cannot identify the power of the test with a specific number. Second, to determine the PDF of the estimator, we need to know the parameters of its distribution such as the variance and possibly higher-order moments. These are rarely known in advance and we must usually make use of estimates that complicate the distribution and make it harder to determine both size and power for any given test. However, the general points illustrated by the diagram and the discussion remain true for more complex cases. If we increase the size of the test, taking the number of observations as fixed, then we reduce its power. The only way to increase both the size and the power of a test simultaneously is to generate more data.

3.7 PROOF OF THE GAUSS–MARKOV THEOREM

One of the reasons why the least squares estimator is given such prominence in the statistics literature is that it can be shown to be the estimator with the lowest variance in the class of linear unbiased estimators. This is often summarized by the description that the OLS estimator is BLUE. In this section, we provide a proof of this property. For simplicity, we consider the case of an equation without an intercept, that is, $Y_i = \beta X_i + u_i$. This means that we can concentrate on the estimation of a single parameter. However, the proof generalizes easily to more complex cases in which the equation contains an intercept and more than one independent variable. Our objective is to show that under the Gauss–Markov assumptions, the least squares estimator is unbiased and has the lowest variance in the class of linear unbiased estimators.

To demonstrate the Gauss–Markov theorem, we write the OLS estimator as

$$\hat{\beta} = \sum_{i=1}^{N} a_i Y_i \qquad \text{where } a_i = \frac{X_i}{\sum_{i=1}^{N} X_i^2}. \tag{3.35}$$

That is, the least squares estimator is a weighted average of the Y observations, where the weights are functions of the X observations. Now, let us consider any other linear combination of the Y variables of the form

$$\tilde{\beta} = \sum_{i=1}^{N} g_i Y_i, \quad \text{where } g_i = a_i + h_i, \tag{3.36}$$

and at least one $h_i \neq 0$. It is assumed that $\tilde{\beta}$ is also an unbiased estimator. We will now show that the variance of $\tilde{\beta}$ is necessarily greater than the variance of $\hat{\beta}$. First, note that

$$\tilde{\beta} = \sum_{i=1}^{N} g_i (\beta X_i + u_i) = \beta \sum_{i=1}^{N} g_i X_i + \sum_{i=1}^{N} g_i u_i. \tag{3.37}$$

Since $\tilde{\beta}$ is unbiased, it follows that $\sum_{i=1}^{N} g_i X_i = 1$, and therefore, it also follows that $\sum_{i=1}^{N} (a_i + h_i) X_i = 1$. Since $\hat{\beta}$ is also unbiased, we have $\sum_{i=1}^{N} a_i X_i = 1$ and therefore, $\sum_{i=1}^{N} h_i X_i = 0$. Now, consider the variance of $\tilde{\beta}$, we have

$$\begin{aligned} V(\tilde{\beta}) &= E\left(\tilde{\beta} - E(\tilde{\beta})\right)^2 = E\left(\tilde{\beta} - \beta\right)^2 = E\left(\sum_{i=1}^{N} g_i u_i\right)^2 \\ &= E\left(\sum_{i=1}^{N} (a_i + h_i) u_i\right)^2. \end{aligned} \tag{3.38}$$

Assuming that the Gauss–Markov assumptions hold we have $E(u_i u_j) = 0$ for all $i \neq j$ and $E(u_i^2) = \sigma^2$ for all i. Therefore,

$$\begin{aligned} V(\tilde{\beta}) &= \sum_{i=1}^{N} (a_i + h_i)^2 \sigma_u^2 \\ &= \sigma_u^2 \left[\sum_{i=1}^{N} a_i^2 + \sum_{i=1}^{N} h_i^2 + 2\sum_{i=1}^{N} a_i h_i \right]. \end{aligned} \tag{3.39}$$

Now, consider $\sum_{i=1}^{N} a_i h_i = \left(\dfrac{1}{\sum_{i=1}^{N} X_i^2} \right) \sum_{i=1}^{N} X_i h_i$ which is equal to zero by the assumption of unbiasedness. It therefore follows that we can write the variance of this estimator as

$$V(\tilde{\beta}) = \sigma_u^2 \sum_{i=1}^{N} a_i^2 + \sigma_u^2 \sum_{i=1}^{N} h_i^2, \tag{3.40}$$

and since

$$\sigma_u^2 \sum_{i=1}^{N} a_i^2 = \sigma_u^2 \sum_{i=1}^{N} X_i^2 / \left(\sum_{i=1}^{N} X_i^2 \right)^2 = \sigma_u^2 / \sum_{i=1}^{N} X_i^2 = V\left(\hat{\beta}\right), \tag{3.41}$$

we have

$$V\left(\tilde{\beta}\right) = V\left(\hat{\beta}\right) + \sigma_u^2 \sum_{i=1}^{N} h_i^2 . \tag{3.42}$$

The second term in this expression is a sum of squares and is therefore always positive. It follows that any linear unbiased estimator has variance which differs from the least squares estimator by a positive number. This establishes the Gauss–Markov result and demonstrates why the least squares estimator is the most efficient estimator in the class of linear unbiased estimators.

3.8 THE METHOD OF MAXIMUM LIKELIHOOD

So far, we have concentrated on the method of least squares for the estimation of model parameters. Although least squares does have good properties as shown in the previous section, there are circumstances in which alternative methods become useful. In particular, there is an alternative method known as the method of *maximum likelihood* which is applicable in a wide range of statistical frameworks. In the case of the linear regression model, this produces very similar results to the method of least squares. However, it has wider applicability, and it is useful to introduce it at this stage. The method of maximum likelihood begins by making specific assumptions about the distribution of the errors in the model and then deriving an estimator based on choosing parameter values that maximize the joint "probability" of observing the sample data. Note that we have placed the work probability in inverted commas here because there is a subtlety in the use of the concept which we will explain as we progress.

Let us consider the regression model (3.1) and assume that the errors $u_i; i = 1,\ldots,N$ are independent random variables that follow a normal distribution with mean zero and variance σ_u^2. We can write the joint probability of observing a particular sample of data as

$$P\left(u_i; i = 1,\ldots,N \middle| \alpha, \beta, \sigma_u^2\right), \tag{3.43}$$

where a, β and σ_u^2 are the parameters of interest. For the purposes of maximum likelihood estimation, we now reverse this function so that the parameters of interest become a function of the sample data. We refer to this as the *likelihood function* and it written as shown in equation (3.44)

$$L\left(a, \beta, \sigma_u^2 \middle| u_i; i = 1, \ldots, N\right). \tag{3.44}$$

The problem here is that we cannot interpret (3.44) as a joint probability because the parameters of interest are not random variables. Hence, we redefine (3.44) as a *likelihood* function. The maximum likelihood function is defined by taking the values of the parameters of interest which maximize this function for a particular sample of data.

Maintaining the assumption that the errors are independent random variables which follow a normal distribution with mean zero and variance σ_u^2 means that we can write the likelihood function as

$$L\left(a, \beta, \sigma_u^2\right) = \prod_{i=1}^{N} \frac{1}{\sigma_u \sqrt{2\pi}} \exp\left(-\frac{1}{2}\left(\frac{Y_i - a - \beta X_i}{\sigma_u}\right)^2\right). \tag{3.45}$$

The log function is monotonic so it is easier to take the log transformation of this function that defines the *log likelihood function* as shown in equation (3.46)

$$LL\left(a, \beta, \sigma_u^2\right) = -N \ln \sqrt{2\pi} - N \ln \sigma_u - \frac{1}{2\sigma_u^2} \sum_{i=1}^{N} (Y_i - a - \beta X_i)^2. \tag{3.46}$$

The first-order conditions for a maximum of this function can be written as

$$\begin{aligned}
\frac{\partial LL}{\partial a} &= \frac{1}{\sigma_u^2} \sum_{i=1}^{N} (Y_i - a - \beta X_i) = 0 \\
\frac{\partial LL}{\partial \beta} &= \frac{1}{\sigma_u^2} \sum_{i=1}^{N} X_i (Y_i - a - \beta X_i) = 0 \\
\frac{\partial LL}{\partial \sigma_u^2} &= -\frac{N}{\sigma_u} + \frac{1}{\sigma_u^3} \sum_{i=1}^{N} (Y_i - a - \beta X_i)^2 = 0.
\end{aligned} \tag{3.47}$$

Thus, we can solve for the maximum likelihood estimators of a and β from the following pair of simultaneous equations

$$\sum_{i=1}^{N}\left(Y_i - \hat{\alpha}_{ML} - \hat{\beta}_{ML}X_i\right) = 0$$
$$\sum_{i=1}^{N}X_i\left(Y_i - \hat{\alpha}_{ML} - \hat{\beta}_{ML}X_i\right) = 0. \quad (3.48)$$

These equations are identical to the least squares normal equations, and therefore, for the linear regression model under Gauss–Markov assumptions, least squares and maximum likelihood procedures yield identical estimates. Solving the third first-order condition gives us the maximum likelihood estimator of the error variance as shown in equation (3.49)

$$\hat{\sigma}^2_{u,ML} = \frac{1}{N}\sum_{i=1}^{N}\left(Y_i - \hat{\alpha}_{ML} - \hat{\beta}_{ML}X_i\right)^2. \quad (3.49)$$

From our discussion of the least squares procedure, we note that this is a biased (but consistent) estimator because it does not incorporate the degrees of freedom correction when taking the average of the residual sum of squares.

Historical Note: Carl Friedrich Gauss was the first to set out the method of maximum likelihood in his book of 1809 "Theory of the Motion of the Heavenly Bodies Moving about the Sun in Conic Sections" [Gauss1809] (the same book in which he claimed to have been using least squares since 1795).

3.9 PREDICTION WITH THE OLS ESTIMATOR

When discussing prediction, the first thing to note is that prediction and forecasting are different activities. Prediction involves the generation of a value of y given a particular value of x. For example, if we have estimated a regression equation of the form $\hat{Y}_i = \hat{\alpha} + \hat{\beta}X_i$, then the predicted value of y for $X = x_0$ is $\hat{y}_0 = \hat{\alpha} + \hat{\beta}x_0$. Forecasting generally requires us to predict values for X as well as those for Y. The main practical difference is that forecasts can be in error because we use the "wrong" value of x, whereas this is not a consideration when it comes to prediction.

We will first consider the topic of prediction. Suppose we wish to predict Y_{N+1} for a given value of $X = x_{N+1}$, having estimated the parameters of the model using data for $i = 1,\ldots,N$. The prediction error from the OLS estimator can be written as

$$Y_{N+1} - \hat{Y}_{N+1} = (\alpha - \hat{\alpha}) + (\beta - \hat{\beta})x_{N+1} + u_{N+1}. \tag{3.50}$$

Taking expectations through (3.50) gives $E(Y_{N+1} - \hat{Y}_{N+1}) = 0$ since the OLS estimates of the intercept and the slope are unbiased and the expected value of the error is zero. Next, we consider the variance of the prediction error. The prediction error variance is defined as

$$E(Y_{N+1} - \hat{Y}_{N+1})^2 = E(\alpha + \beta x_{N+1} + u_i - \hat{\alpha} - \hat{\beta}x_{N+1})^2. \tag{3.51}$$

The right-hand side of this expression can be written as

$$E((\alpha - \hat{\alpha}) + (\beta - \hat{\beta})x_{N+1} + u_{N+1})^2. \tag{3.52}$$

Expanding this expression yields

$$E\begin{pmatrix} (\alpha - \hat{\alpha})^2 + (\beta - \hat{\beta})^2 x_{N+1}^2 + u_{N+1}^2 + 2u_{N+1}(\alpha - \hat{\alpha}) + \\ 2u_{N+1}(\beta - \hat{\beta})x_{N+1} + 2(\alpha - \hat{\alpha})(\beta - \hat{\beta})x_{N+1} \end{pmatrix}. \tag{3.53}$$

Since $\operatorname{cov}(u_i u_{N+1}) = 0;\ i = 1,\ldots,N$, we have $E(u_{N+1}(\alpha - \hat{\alpha})) = E(u_{N+1}(\beta - \hat{\beta}))$ $= 0$ and therefore taking expectations through (3.51) yields

$$\begin{aligned} E(Y_{N+1} - \hat{Y}_{N+1})^2 &= \sigma_u^2\left[\frac{1}{N} + \frac{\bar{x}^2}{\sum(x_i - \bar{x})^2}\right] + x_{N+1}^2\frac{\sigma_u^2}{\sum(x_i - \bar{x})^2} + \sigma_u^2 \\ &\quad - \bar{x}x_{N+1}\frac{\sigma_u^2}{\sum(x_i - \bar{x})^2} \\ &= \sigma_u^2\left[1 + \frac{1}{N} + \frac{(x_{N+1} - \bar{x})^2}{\sum(x_i - \bar{x})^2}\right]. \end{aligned} \tag{3.54}$$

Thus, the forecast error variance can be decomposed into a part due to random errors to the underlying relationship, σ_u^2, and a part due to the variance of the parameter estimates, $\sigma_u^2\left[1/N+\left(x_{N+1}-\overline{x}\right)^2/\sum\left(x_i-\overline{x}\right)^2\right]$. The prediction error variance changes according to the deviation of the right-hand side variable from its sample mean. The greater the discrepancy between the value of x used to construct the prediction and the sample mean of x, then the larger is the prediction error variance. What this means is that we are more likely to get accurate predictions when the value of the x variable used is a "typical" value in the sense that it is close to the sample mean. The more extreme the value of x we use, that is, the further from the sample mean, then the less reliable will be the prediction. The importance of this effect will vary depending on the nature of the data used.

3.10 SUMMARY

In this chapter, we have introduced one of the basic statistical tools of econometrics in the form of the linear regression model. We have shown that the least squares estimator can be derived straightforwardly from a simple calculus problem. In addition, we have shown that under certain assumptions, this estimator has desirable properties in that it is unbiased and has the smallest variance in the class of unbiased estimators. These assumptions, known as the *Classical Linear Regression Model assumptions* or alternatively the *Gauss–Markov assumptions*, also allow us to determine the distribution of the least squares estimator and therefore, to conduct hypothesis tests concerning the unknown parameters of the relationship between the Y and X variables. However, it should be emphasised that the Gauss–Markov assumptions rarely hold in practice when estimating econometric models based on real-world data. We therefore need to investigate further the implications of these assumptions and to develop techniques for dealing with situations in which they fail. Finally, we have considered the method of maximum likelihood as an alternative way of constructing an estimator for the parameters of interest. In the case of the linear regression model, with normally distributed errors, the least squares and maximum likelihood methods yield identical parameter estimates for the slope and intercept. However, the maximum likelihood method allows us to consider more general cases in which the assumption of normality fails, and it is therefore useful to introduce it at this stage.

EXERCISES

EXERCISE 3.1

For the least-squares regression model $Y_i = a + \beta X_i + u_i;\ i = 1,\ldots,N$, show that the residual sum of squares is equal to $(N-1)\{\hat{\sigma}_Y^2 - \hat{\sigma}_{XY}^2 / \hat{\sigma}_X^2\}$, where $\hat{\sigma}_Y^2$ and $\hat{\sigma}_X^2$ are the sample variances of Y and X, and $\hat{\sigma}_{XY}$ is the sample covariance of X and Y.

EXERCISE 3.2

For the least-squares regression model $Y_i = a + \beta X_i + u_i;\ i = 1,\ldots,N$, show that the regression residuals are uncorrelated with the X variable.

EXERCISE 3.3

The following sample moments have been calculated for observations on the price and quantity produced of oranges in the United States. The data are annual from 1981 to 2016 and are calculated as percentage changes relative to the previous year:

```
Mean of price changes                          0.441150
Mean of quantity changes                      -2.034617
Variance of price changes                      211.5094
Variance of quantity changes                   246.5624
Covariance of price and quantity changes      -145.3071
```

a. Calculate the correlation coefficient for price and quantity changes.

b. Calculate the slope and intercept coefficients for a regression of price changes on quantity changes.

c. Calculate the slope and intercept coefficients for a regression of quantity changes on price changes.

d. Explain why the correlation coefficient in part a) and the regression slope coefficients in parts b) and c) are different and set out the relationship between them.

Exercises 3.4–3.6 make use of the data contained in the workfile SHARES. XLSX. This contains daily data for the share prices of a number of leading UK companies as well as the FTSE 100 share price index for the period January 2003 to May 2008. The aim is to estimate the *market model* which relates the daily return on a particular share to the return on the market as a whole. Daily returns are defined as the percentage change in the value of the share or the overall market index. Thus, the model we will estimate takes the form

$$R_t^i = a + \beta R_t^M + u_t^i,$$

where $R_t^i = 100 \times \left(P_t^i - P_{t-1}^i\right) / P_{t-1}^i$ and $R_t^M = 100 \times \left(P_t^M - P_{t-1}^M\right) / P_{t-1}^M$. P_t^i is the price of share i at date t and P_t^M is the value of the market index at date t.

EXERCISE 3.4

If we estimate the market model for AstraZeneca shares, then we obtain the following results:

```
Ordinary Least-squares Regression Results
Sample period: 2 to 1359
Dependent Variable DASTRA
Sample Size 1358
```

Variable	Coefficient	Std Err	T-Ratio
C	-0.008039	0.034037	-0.236183
DFTSE	0.891929	0.035108	25.405183

R-squared	0.3224	F-statistic	645.4233
SEE	1.253479	RSS	2130.563220
Durbin-Watson	1.7797	LogL	-2232.722030
ARCH(1) Test	5.4683	AIC	3.291195
Jarque-Bera	1519.7986	SIC	3.298874

a. Test the null hypothesis $H_0 : \beta = 0$ against the alternative $H_1 : \beta \neq 0$ using a 5% level of significance.

b. Test the null hypothesis $H_0 : \beta = 1$ against the alternative $H_1 : \beta < 1$ using a 5% level of significance.

c. Explain why the critical value you use in these two tests is different.

EXERCISE 3.5

An econometrician has estimated the following market model for British Airways shares

```
Ordinary Least-squares Regression Results
Sample period: 2 to 1359
Dependent Variable DBA
Sample Size 1358
```

Variable	Coefficient	Std Err	T-Ratio
C	2.742235 E-3	0.052921	0.051818
DFTSE	1.523805	0.054585	27.915937

R-squared	0.3650	F-statistic	779.2996
SEE	1.948886	RSS	5150.302785
Durbin-Watson	2.0365	LogL	-2832.054370
ARCH(1) Test	16.3298	AIC	4.173865
Jarque-Bera	457.4844	SIC	4.181544

In addition, we are given that the average value of DM is 0.035445 and $\sum\left(DM-\bar{DM}\right)^2 = 1274.731713$.

a. Calculate the predicted return on BA shares at the sample mean of the market return and calculate a 95% confidence interval for the prediction.

b. The maximum and minimum values for the daily change in the market are 6.081533 and –5.481471, respectively. Calculate the central predicted values for the change in BA shares as well as 95% confidence intervals based on these values. Are the confidence intervals noticeably wider than that calculated for the mean?

EXERCISE 3.6

Estimate the market model for Vodafone shares.

a. Test the null hypothesis $H_0 : \beta = 0$ against the alternative $H_1 : \beta \neq 0$ using a 5% level of significance.

b. Test the null hypothesis $H_0 : \beta = 1$. Is it sensible to use $H_1 : \beta < 1$. Does the model support this hypothesis? If not, suggest another alternative and use this for your test.

REFERENCES

[Gauss1809] Gauss, C. F. *Theory of the Motion of Heavenly Bodies Moving about the Sun in Conic Sections*, 1809, English Translation by C. H. Davis 1963, New York: Dover.

[Legendre1805] Legendre, A.-M. *Nouvelles méthodes pour la détermination des orbites des comètes* (New Methods for the Determination of the Orbits of Comets), 1805, Paris: F. Didot.

[Marshall1890] Marshall, A. *Principles of Economics*, 1890, first edition, London: Macmillan.

CHAPTER 4

THE MULTIVARIATE REGRESSION MODEL

In Chapter 3, we set out the bivariate regression model and discussed its properties. This model is most suited to the experimental sciences, where the data are generated by the investigator, and the errors are essentially errors of measurement or purely random effects outside the control of the investigator. When we apply regression models to economic data, however, this is rarely appropriate. In most situations, economic data are generated by a complex, multivariable process in which there are numerous interacting variables, none of which is controllable by the modeler. It therefore becomes necessary to develop multivariable approaches that allow for the interaction of groups of variables.

Multivariate regression analysis extends the model discussed in Chapter 3 to the case where there are potentially many variables on the right-hand side of the model. In this way, we can start to develop methods for dealing with the complex world of economic data. Suppose we wish to determine the relationship between two variables of interest. Experimental sciences can focus on the relationship between these variables by conducting experiments in which all other factors are held constant. The multivariable regression model can be used to achieve a similar effect by controlling for the influence of other factors through the inclusion of variables in the regression equation which capture their effects on the variable of interest.

Let us begin with a model of the form

$$Y_i = \beta_1 + \beta_2 X_{2i} + \beta_3 X_{3i} + \cdots + \beta_k X_{ki} + u_i, \tag{4.1}$$

in which we have data for each of the variables for a sample $i = 1,...,N$. Now suppose we have estimates for each of the model parameters $\hat{\beta}_1, \hat{\beta}_2, \ldots, \hat{\beta}_k$. The sum of the squared deviations of the Y variable from the fitted values can be written as

$$RSS = \sum_{i=1}^{N} \left(Y_i - \hat{\beta}_1 - \hat{\beta}_2 X_{2i} - \hat{\beta}_3 X_{3i} - \cdots - \hat{\beta}_k X_{ki} \right)^2. \tag{4.2}$$

The OLS estimator is defined by taking the partial derivatives of (4.2) and setting them equal to zero. This creates a system of k equations in the k unknown parameters that can be written as follows[1]:

$$\begin{aligned} &N\hat{\beta}_1 + \sum X_{2i}\hat{\beta}_2 + \cdots + \sum X_{ki}\hat{\beta}_k = \sum Y_i \\ &\sum X_{2i}\hat{\beta}_1 + \sum X_{2i}^2\hat{\beta}_2 + \cdots + \sum X_{2i}X_{ki}\hat{\beta}_k = \sum X_{2i}Y_i \\ &\cdots\cdots\cdots\cdots\cdots\cdots\cdots\cdots\cdots\cdots\cdots\cdots \\ &\sum X_{ki}\hat{\beta}_1 + \sum X_{ki}X_{2i}\hat{\beta}_2 + \cdots + \sum X_{ki}^2\hat{\beta}_k = \sum X_{ki}Y_i. \end{aligned} \tag{4.3}$$

These are the least squares normal equations for the multivariable model. They are analogous to the pair of normal equations derived for the bivariate regression model in Chapter 3 in that they define a system of k linear equations in k unknown variables, that is, the regression parameters. If these equations are not collinear, then they can be solved to yield the OLS estimates of the regression parameters in equation (4.1). For a solution to exist, these equations must be linearly independent, that is, it must not be possible to write any one of the equations in (4.3) as a linear combination of other equations in the system. Note that, as in the bivariate case, the first of the normal equations establishes the property that the regression passes through the sample means of the data. This property can be used to write the model in mean deviation form for the purposes of solving for the slope coefficient estimates.

A derivation of the OLS estimator with simpler notation can be obtained by rewriting the model in matrix form. For example, we can write (4.1) in matrix form as

$$y = X\beta + u, \tag{4.4}$$

---

[1] Since the limits are 1 and N for all summation operations, we omit them to simplify the notation.

where $\boldsymbol{y}$ is an $N \times 1$ vector of observations for the endogenous variable, $\mathbf{X}$ is an $N \times k$ matrix whose first column consists of ones and whose other columns are the observations for each of the exogenous variables in turn, $\boldsymbol{\beta}$ is a $k \times 1$ vector of parameters, and $\boldsymbol{u}$ is an $N \times 1$ vector of random errors. To derive the OLS estimator, we first specify a loss function consisting of the following quadratic form:

$$RSS = \left(\boldsymbol{y} - \mathbf{X}\hat{\boldsymbol{\beta}}\right)' \left(\boldsymbol{y} - \mathbf{X}\hat{\boldsymbol{\beta}}\right) = \boldsymbol{y}'\boldsymbol{y} + \hat{\boldsymbol{\beta}}'\mathbf{X}'\mathbf{X}\hat{\boldsymbol{\beta}} - 2\hat{\boldsymbol{\beta}}'\,\mathbf{X}'\boldsymbol{y}. \tag{4.5}$$

Differentiating with respect to $\hat{\boldsymbol{\beta}}$ and setting the derivative equal to zero yields

$$(\mathbf{X}'\mathbf{X})\hat{\boldsymbol{\beta}} = \mathbf{X}'\boldsymbol{y}. \tag{4.6}$$

Equation (4.6) is the matrix form of the least-squares normal equations. A comparison of the simplicity and elegance of equation (4.6) with the equivalent scalar expression (4.3) illustrates the value of using the matrix form of the model. Another advantage of this form of the model is that it makes the conditions for the existence of a solution to the normal equations transparent. For a solution to exist, we require $(X'X)$ to be invertible. This in turn requires the matrix $\mathbf{X}$ to have rank k, that is, there must be no linear dependent relationships between the columns of $\mathbf{X}$. If we assume that a solution to (4.6) exists, then it takes the form

$$\hat{\boldsymbol{\beta}} = (\mathbf{X}'\mathbf{X})^{-1}\,\mathbf{X}'\boldsymbol{y}. \tag{4.7}$$

This form of the solution is considerably more elegant than that obtained through scalar algebra, again illustrating the value of the matrix approach.

Example: As with the bivariate regression model, the least square parameter estimates for the multivariable model are functions of the sample moments of the variables in the model. This can be illustrated using the following numerical example, in which we estimate a demand curve for gasoline for the US economy. The data are annual observations from the period 1950 to 2016 and our estimating equation takes the form

$$\Delta g_t = \beta_1 + \beta_2 \Delta y_{2t} + \beta_3 \Delta p_{3t} + u_t. \tag{4.8}$$

Δg is the first difference of the logarithm of consumption of gasoline measured in millions of barrels, Δy is the first difference of the logarithm of gross domestic product (GDP) in \$bn at 2000 prices, Δp is the first difference of the logarithm of the ratio of the price index for gasoline and the GDP deflator, and u is a random error. From the data, we obtain the means of the data and the sums of squares and crossproducts given in (4.9)

$$\begin{aligned}
&\overline{\Delta y} = 0.0316 \qquad \overline{\Delta p} = 2.57\times 10^{-4} \qquad \overline{\Delta g} = 0.02022\\
&\sum\left(\Delta y_t - \overline{\Delta y}\right)^2 = 0.03437\\
&\sum\left(\Delta p_t - \overline{\Delta p}\right)^2 = 0.83355\\
&\sum\left(\Delta y_t - \overline{\Delta y}\right)\left(\Delta p_t - \overline{\Delta p}\right) = 4.055\times 10^{-4}\\
&\sum\left(\Delta y_t - \overline{\Delta y}\right)\left(\Delta g_t - \overline{\Delta g}\right) = 0.02628\\
&\sum\left(\Delta p_t - \overline{\Delta p}\right)\left(\Delta g_t - \overline{\Delta g}\right) = -0.08357.
\end{aligned} \tag{4.9}$$

The slope coefficients for this regression equation can therefore be calculated as

$$\begin{aligned}
\begin{pmatrix}\hat{\beta}_2\\ \hat{\beta}_3\end{pmatrix} &= \begin{pmatrix}0.03437 & 4.055\times 10^{-4}\\ 4.055\times 10^{-4} & 0.83355\end{pmatrix}^{-1}\begin{pmatrix}0.02628\\ -0.08357\end{pmatrix}\\
&= \begin{pmatrix}0.7657\\ -0.1006\end{pmatrix}.
\end{aligned} \tag{4.10}$$

An estimate of the intercept can then be obtained by using the condition that the regression line must pass through the sample means of the data. This yields

$$\begin{aligned}
\hat{\beta}_1 &= 0.02022 - 0.7657\times 0.03162 + 0.1006\times 2.57\times 10^{-4}\\
&= -0.003966.
\end{aligned} \tag{4.11}$$

4.1 DERIVATION OF THE DISTRIBUTION OF THE OLS ESTIMATOR

TABLE 4.1 Assumptions of the Classical Linear Regression Model (CLRM)

Assumption 1:	The errors have zero mean: $E(\boldsymbol{u}) = \mathbf{0}$.
Assumptions 2 and 3:	The errors are serially independent and have constant variance: $E(\boldsymbol{uu}') = \sigma_u^2 \boldsymbol{I}_N$
Assumption 4:	The right-hand side variables are exogenous (a) Strong form - $\mathbf{X}$ is fixed in repeated samples (b) Weak form - $E(\mathbf{X}'\boldsymbol{u}) = \mathbf{0}$
Assumption 5:	The errors follow a normal distribution: $\boldsymbol{u} \sim N(0, \sigma_u^2 \boldsymbol{I}_N)$

To derive the distribution of the multivariate OLS estimator, we establish the matrix equivalents of the CLRM assumptions that we discussed in Chapter 3. These are listed in Table 4.1. Given these assumptions, we can demonstrate that the result that OLS estimator is the Best Linear Unbiased Estimator (BLUE) continues to hold in the multivariable case. We can also derive the distribution of the OLS estimator using the same methods as for the bivariate model in Chapter 3.

First, we will show that the OLS estimator for the multivariable model is unbiased. We have $\hat{\boldsymbol{\beta}} = (\mathbf{X}'\mathbf{X})^{-1}\mathbf{X}'\boldsymbol{y}$ and substituting for $\boldsymbol{y}$ yields $\hat{\boldsymbol{\beta}} = \boldsymbol{\beta} + (\mathbf{X}'\mathbf{X})^{-1}\mathbf{X}'\boldsymbol{u}$. From Assumption 4(a), we have that the only stochastic element in this expression is the vector of random errors $\boldsymbol{u}$. Taking expectations yields $E(\hat{\boldsymbol{\beta}}) = \boldsymbol{\beta} + (\mathbf{X}'\mathbf{X})^{-1}\mathbf{X}'E(\boldsymbol{u})$ and, by Assumption 1, we have $E(\boldsymbol{u}) = \mathbf{0}$ which ensures that $E(\hat{\boldsymbol{\beta}}) = \boldsymbol{\beta}$. Therefore, under the Gauss–Markov assumptions, the OLS estimator is unbiased. Moreover, if Assumption 5 holds, then the OLS estimator is a linear combination of normally distributed random variables and is therefore itself normally distributed.

To derive the variance of the OLS estimator, we first note that the variance in the multivariate case will consist of a $k \times k$ symmetric matrix with variances of the individual OLS coefficient estimates on the diagonal and their covariances off the diagonal. This is referred to as the *variance–covariance matrix of the regression parameters*. As we have demonstrated unbiasedness, we can write

$$\begin{aligned}\operatorname{var}\left(\hat{\boldsymbol{\beta}}\right) &= E\left(\hat{\boldsymbol{\beta}}-\boldsymbol{\beta}\right)\left(\hat{\boldsymbol{\beta}}-\boldsymbol{\beta}\right)' \\ &= E(\mathbf{X}'\mathbf{X})^{-1}\mathbf{X}'\boldsymbol{uu}'\mathbf{X}(\mathbf{X}'\mathbf{X})^{-1}.\end{aligned} \tag{4.12}$$

Assumption 4(a) allows us to write the right-hand side of this expression as $(\mathbf{X}'\mathbf{X})^{-1}\mathbf{X}'E(\boldsymbol{uu}')\mathbf{X}(\mathbf{X}'\mathbf{X})^{-1}$. From Assumptions 2 and 3 of the CLRM, we have that $E(\boldsymbol{uu}')=\sigma_u^2\boldsymbol{I}_N$ and therefore with some minor algebra this yields

$$\operatorname{var}\left(\hat{\boldsymbol{\beta}}\right)=\sigma_u^2(\mathbf{X}'\mathbf{X})^{-1}. \tag{4.13}$$

Again, note that the use of matrix algebra permits a considerable improvement in terms of the elegance of the notation and the ease of the derivation. Using only a few lines of algebra, we have been able to show that under the CLRM assumptions,

$$\hat{\boldsymbol{\beta}} \sim N\left(\boldsymbol{\beta},\sigma_u^2(\mathbf{X}'\mathbf{X})^{-1}\right). \tag{4.14}$$

The derivation of the distribution of the OLS estimator would have been considerably more difficult, and would have involved far more complex expressions, if we had retained the use of scalar notation. Thus, the initial costs of writing the model in matrix form are more than justified in terms of the subsequent ease with which we can derive important results for the OLS estimator.

Historical Note: The first clear statement of the distribution of regression coefficients in the multivariable model comes in a paper by Fisher (*Journal of the Royal Statistical Society*, 1922) [Fisher1922]. However, he credits "Student" (W.S. Gosset) as having developed the theory.

4.2 PRINCIPLES OF TESTING

One purpose of estimating multivariate regression models is the testing of hypotheses derived from economic theory. We have already discussed the construction and implementation of testing procedures in the context of the bivariate regression model in Chapter 3. When we move to a multivariate framework, the nature of the hypotheses to be tested changes as it becomes possible to test hypotheses relating different parameters as well as those relevant to a single parameter. We will therefore spend some time discussing testing in a multivariate framework. Before we begin this, however, we will expand a little on the different approaches to testing which form the basis of statistical inference in multivariate econometrics.

There are three main approaches to testing which form the basis of most of the tests used in econometrics. These are the Wald, Lagrange Multiplier, and Likelihood Ratio approaches. Although all of these can be applied in the context of least squares estimation, it is easier to explain them in terms of the maximum likelihood approach. In this section, therefore, we discuss testing approaches using maximum likelihood examples, on the understanding that this generalizes easily to the least squares framework.

Consider an investigator who estimates a model using the maximum likelihood approach. That is, he or she seeks to maximize a log-likelihood function of the form $LL(\boldsymbol{\theta}|\Omega)$, where $\boldsymbol{\theta}$ is vector of parameters and Ω is the information set available including the data. Any restriction on the parameter vector will result in a value of the likelihood which is lower than the maximum given a free choice of parameters. The question is whether this reduction is significant or not and we can approach testing this in three different ways:

1. The Wald approach examines the difference between the values of the parameters at the restricted and unrestricted solutions to the problem, that is, $\hat{\boldsymbol{\theta}}_U - \hat{\boldsymbol{\theta}}_R$.

2. The Lagrange multiplier approach examines the derivative of the (log) likelihood function at the restricted solution, that is, $\partial LL(\boldsymbol{\theta}) / \partial\boldsymbol{\theta}\big|\hat{\boldsymbol{\theta}}_R$. If the restrictions are valid, then this should be close to zero.

3. The likelihood ratio approach examines the difference between the values of the likelihood function when evaluated at the maximum and at the restricted values of the parameters.

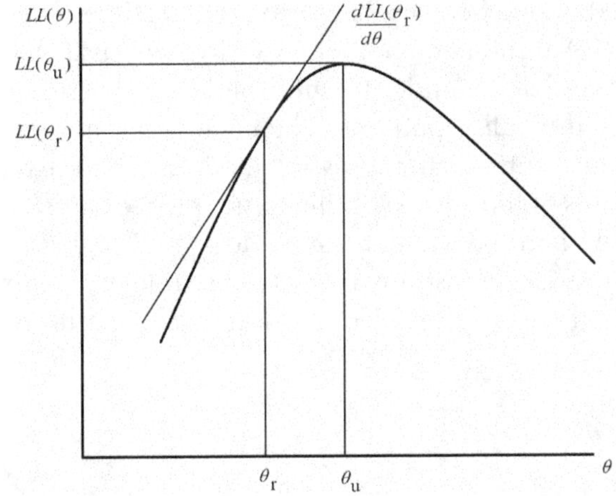

FIGURE 4.1 Wald, Lagrange Multiplier, and Likelihood Ratio Approaches to Testing.

The three approaches to testing are illustrated in Figure 4.1 for the example of a model with a single parameter. The Wald approach is based on the distance $\theta_u - \theta_r$, the Lagrange multiplier approach is based on $\partial LL(\theta)/\partial\theta$, and the likelihood ratio approach is based on $LL(\theta_u) - LL(\theta_r)$. All three tests lead to test statistics with the same asymptotic distribution. However, there are differences in small sample properties. The tests also differ in how easily they can be set up and implemented in different circumstances. For example, in many circumstances in econometrics, it is easy to estimate an unrestricted model and generate a test statistic based on these estimates. This means that the Wald testing approach is most appropriate, and this leads naturally to the t tests we frequently apply to individual coefficient estimates. In contrast, there are some circumstances in which it will be much easier to estimate a restricted rather than an unrestricted model. This will be the case in subsequent chapters when we wish to test for misspecification, where it is natural to define a restricted model that can then be relaxed in several different ways. This leads naturally to the Lagrange multiplier approach. The likelihood ratio approach is most appropriate when both the unrestricted and restricted versions of the model are straightforward to estimate. This is not uncommon in econometrics and again, there are many circumstances in which this will be the natural testing approach.

4.3 HYPOTHESIS TESTING IN THE MULTIVARIATE REGRESSION MODEL

In the previous section, we discussed testing approaches in very general terms. In this section, we will look at the details of how tests are constructed and applied in practice. We have seen that testing a hypothesis requires the following: (1) a null and an alternative hypothesis, (2) a test statistic whose distribution is known under the null, and (3) a decision rule for acceptance/rejection of the null hypothesis. The main difference between testing in the bivariate and multivariate regression models is that we have a greater variety of hypotheses of interest. We will consider three cases of interest: the first is where we wish to test a hypothesis relating to a single coefficient, the second is where we wish to test a hypothesis that relates two or more coefficients, and the third is where we wish to test several hypotheses simultaneously.

4.3.1 Testing a Hypothesis Relating to a Single Coefficient

Suppose we wish to test $H_0 : \beta_j = \bar{\beta}_j$ against the alternative $H_1 : \beta_j \neq \bar{\beta}_j$. From (4.14), we have that $\hat{\beta}_j \sim N\left(\bar{\beta}_j, \sigma_u^2 \varsigma_{jj}\right)$ under the null – where ς_{jj} is the (j, j)th element of the matrix $(\mathbf{X}'\mathbf{X})^{-1}$. If σ_u^2 was known, then we could use $\hat{\beta}_j - \bar{\beta}_j / \sigma_u \sqrt{\varsigma_{jj}}$ as our test statistic as this would follow the standard normal distribution. The problem is that σ_u^2 is typically not known. Therefore, we replace σ_u with the estimate $\hat{\sigma}_u = \sqrt{\sum\left(Y_i - \hat{Y}_i\right)^2 / (N-k)}$. This means that our test statistic becomes

$$t = \frac{\hat{\beta}_j - \bar{\beta}_j}{\hat{\sigma}_u \sqrt{\varsigma_{jj}}} \sim t_{N-k}, \tag{4.15}$$

which follows the t distribution with $N - k$ degrees of freedom under the null. We can then compare (4.15) with an appropriate critical value from the t_{N-k} distribution tables to make a decision as to whether to accept or reject the null hypothesis. Note that t tests of this form employ the Wald testing procedure because they involve estimation of the unrestricted model only.

Example: Consider our estimates of the demand curve for gasoline in the United States from Section 4.1. We already have estimates of the parameters and we can easily calculate the variance–covariance matrix of the coefficients in order to perform hypothesis tests. We have

$$\begin{aligned} V\begin{pmatrix} \hat{\beta}_2 \\ \hat{\beta}_3 \end{pmatrix} &= \hat{\sigma}_u^2 \begin{pmatrix} 0.03457 & 4.5055 \times 10^{-4} \\ 4.5055 \times 10^{-4} & 0.83355 \end{pmatrix}^{-1} \\ &= \hat{\sigma}_u^2 \begin{pmatrix} 29.0953 & -0.01538 \\ -0.01538 & 1.1997 \end{pmatrix}. \end{aligned} \tag{4.16}$$

The residual sum of squares is $RSS = 0.02753$, and therefore, $\hat{\sigma}_u = \sqrt{0.02753 / 64} = 0.02074$ is the standard error of the regression. Now suppose, for example, that we want to test the null hypothesis $H_0 : \beta_2 = 1$ against the alternative $H_1 : \beta_2 < 1$. The test statistic is $(0.7657 - 1) / \left(0.02074\sqrt{29.0953}\right) = -2.09$, and the 5% critical value for a one-tailed t statistic with 64 degrees of freedom is -1.669. Therefore, we reject the null hypothesis in favor of the alternative at the 5% level. Alternatively, we can calculate the p value of the test statistic as 0.02 which again indicates that we should reject the null at most reasonable levels of significance.

4.3.2 Testing a Hypothesis Involving Several Coefficients

In multivariate regression models, we often wish to test hypothesis which relates several of the model parameters. A typical example of this is where we wish to test the hypothesis that two coefficients have equal but opposite sign. Hypotheses of this type can be written as linear combinations of the model parameters. For example, in equation (4.1), we might want to test the null hypothesis $H_0 : \beta_2 = a\beta_3$ against the alternative $H_1 : \beta_2 \neq a\beta_3$ where a is a nonzero number. Following the discussion in the previous section, we could write a test statistic for this hypothesis as

$$\frac{\hat{\beta}_2 - a\hat{\beta}_3}{\sqrt{\operatorname{var}\left(\hat{\beta}_2 - a\hat{\beta}_3\right)}} \sim t_{N-k}. \tag{4.17}$$

It is straightforward to calculate $\operatorname{var}\left(\hat{\beta}_2 - a\hat{\beta}_3\right)$. From the definition of the variance, we have

$$E\left(\hat{\beta}_2 - a\hat{\beta}_3\right)^2 = \operatorname{var}\left(\hat{\beta}_2\right) + a^2 \operatorname{var}\left(\hat{\beta}_3\right) - 2a \operatorname{cov}\left(\hat{\beta}_2, \hat{\beta}_3\right). \tag{4.18}$$

Therefore, when we test hypotheses which relate different model parameters, the relevant variance depends on the off-diagonal elements of the variance–covariance matrix.

Example: Suppose we wish to test the null hypothesis $H_0 : \beta_2 = -\beta_3$ against the alternative $H_1 : \beta_2 \neq \beta_3$ in the gasoline demand model we estimated earlier. This is not necessarily an interesting economic hypothesis, but it will serve to illustrate the statistical procedure. The test statistic in this case can be written as

$$t = \frac{\hat{\beta}_2 + \hat{\beta}_3}{se\left(\hat{\beta}_2 + \hat{\beta}_3\right)}, \tag{4.19}$$

which, under the null hypothesis, follows a t distribution with 52 degrees of freedom. From (4.18), we have that

$$V\left(\hat{\beta}_2+\hat{\beta}_3\right)=(0.02074)^2(29.0953+1.1997+2\times-0.01538)$$
$$=0.0130.$$

Using the parameter values calculated earlier, we therefore have

$$t=\frac{0.7657-0.1006}{\sqrt{0.0130}}=5.83. \tag{4.20}$$

This is distributed as t_{64} under the null hypothesis and, since the test statistic is greater than the 5% critical value for this distribution, we reject the null in this case.

4.3.3 Testing Several Restrictions Simultaneously

With multivariate regression, we often wish to test several restrictions simultaneously. For example, if our model is (4.1), then we might wish to test $H_0:\beta_2=\bar{\beta}_2,\beta_3=\bar{\beta}_3$ against the alternative $H_1:\beta_2\neq\bar{\beta}_2$ and/or $\beta_3\neq\bar{\beta}_3$. Joint hypotheses of this type require the use of an F test. To calculate a test statistic, we run separate regressions, one imposing the restrictions given in the null hypothesis and one allowing the regression coefficients to be freely determined. This generates two values of the residual sum of squares. The *restricted residual sum of squares* is calculated with the restrictions imposed, that is, $RRSS=RSS\left(\beta_2=\bar{\beta}_2,\beta_3=\bar{\beta}_3\right)$, and the *unrestricted residual sum of squares* used the OLS estimated values for these coefficients $URSS=RSS\left(\beta_2=\hat{\beta}_2,\beta_3=\hat{\beta}_3\right)$. The F test is based on a comparison of these residual sums of squares. Under the null hypothesis, we have

$$F=\frac{(RRSS-URSS)/r}{URSS/(N-k)}\sim F_{r,N-k}, \tag{4.21}$$

where r is the number of restrictions we impose. Note that $RRSS\geq URSS$, therefore F must always be positive. We can interpret this test as an application of the likelihood ratio approach because it involves a comparison of unrestricted and restricted estimates.

Example: Suppose we wish to test the joint hypotheses $H_0 : \beta_2 = 1, \beta_3 = -1$ against $H_1 : \beta_2 \neq 1$ and/or $\beta_3 \neq -1$ for our gasoline demand model. We can estimate separate regressions in which first, neither restriction is imposed and second, in which both are imposed. The first regression is used to calculate the unrestricted residual sum of squares and the second, to calculate the restricted residual sum of squares. The values obtained for the residual sums of squares in this case are $URSS$ = 0.02753 and $RRSS$ = 0.7035. Therefore, the test statistic is

$$F = \frac{(0.7035 - 0.02753)/2}{0.02753/64} = 785.73. \tag{4.22}$$

Under the null hypothesis, this is distributed as $F_{2,64}$, and the 5% critical value for this distribution is 3.14. As the test statistic is greater than the critical value, we reject the null at the 5% level.

The joint test of linear restrictions described above has an important special case. This is the test of the joint significance of the regression coefficients, that is, a test of $H_0 : \beta_2 = \beta_3 = \cdots \beta_k = 0$ against the alternative that one or more regression coefficients is different from zero. Under the null hypothesis, we have

$$F = \frac{(TSS - RSS)/(k-1)}{RSS/(N-k)} \sim F_{k-1,N-k}, \tag{4.23}$$

where TSS is the sum of squared deviations of the y variable from its mean and RSS is the residual sum of squares from the regression. This is the F test which is frequently reported as part of the regression output in many econometrics packages.

4.4 GOODNESS OF FIT

So far, we have concentrated on the issue of hypothesis testing in the regression model. A related topic is the extent to which a regression model can be said to "explain" the variation in the data. This is the issue of *goodness of fit*. To measure goodness of fit, we first need to introduce the idea of *analysis of variation*. For any variable Y, we can divide up the variation into three parts: these are the total variation, the explained variation, and the residual variation. We define the following sums of squares:

$$
\begin{aligned}
TSS &= \sum \left(Y_i - \bar{Y}\right)^2 \\
ESS &= \sum \left(\hat{Y}_i - \bar{Y}\right)^2 \\
RSS &= \sum \left(Y_i - \hat{Y}_i\right)^2,
\end{aligned}
\tag{4.24}
$$

where TSS is the *total sum of squares* which is defined as the sum of the squared deviations of the observations of Y from their sample mean value, ESS is the *explained sum of squares* which equals the sum of the squared deviations of the fitted values from the regression equation from the sample mean of the data and, finally, RSS is the *residual sum of squares* which is the sum of the squared deviations of the observations of Y from the fitted values. Some simple algebra confirms that $TSS = ESS + RSS$, that is, the total sum of squares consists of the sum of the explained and residual sums of squares.

4.4.1 The Coefficient of Determination – *R*-squared

A natural way to measure the goodness of fit of an equation is to calculate the proportion of the total sum of squares which is accounted for by the regression. This gives the statistic known as the coefficient of determination or R-squared for a regression model. It can be written in two alternative ways as shown in equation (4.25),

$$
R^2 = \frac{ESS}{TSS} = 1 - \frac{RSS}{TSS}. \tag{4.25}
$$

The definition of R-squared implies a number of important properties. First, it is obvious that this statistic is bounded between zero and one. Since ESS and TSS are both positive numbers and $ESS \leq TSS$. Second, the closer R-squared is to one, then the more of the variation of y which is explained by the model and therefore the better is the fit of the model.

If R-squared measures goodness of fit, and increasing R-squared means an increase in fit, then should we always seek to choose a model which has the maximum value for this statistic? There are two reasons why this may be an unwise strategy. The first is that R-squared can always be increased by the addition of extra variables on the right-hand side of the regression equation – even if these are irrelevant to an explanation of the behavior of the variable in question. Thus, a strategy of maximizing R-squared will lead to models that are "overfitted," that is, include too many explanatory variables. The second reason is that the value of R-squared can depend on the way in which the regression equation is written (see the example which follows) and this can lead the inexperienced researcher to judge that one model fits "better" than another even when there is no difference between the two.

Example: For the gasoline demand model, we obtain $TSS = 0.0561$ and $RSS = 0.0275$, it follows that the R-squared for this regression can be calculated as $R^2 = 1 - 0.0196 / 0.0391 = 0.51$. Therefore, just over 50% of the variation in the left-hand side variable is being explained by the model. Now, consider what happens when we add an unrelated random variable as an extra regressor. The regression obtained is given in equation (4.26)

$$\begin{aligned} \Delta g_t &= \underset{(0.0045)}{-0.0055} + \underset{(0.1124)}{0.7827}\,\Delta y_t - \underset{(0.0226)}{0.1008}\,\Delta p_t - \underset{(0.0025)}{0.0030}\,z_t + \hat{u}_t \\ RSS &= 0.0269. \end{aligned} \tag{4.26}$$

Note that the unrelated variable z is statistically insignificant with a t-ratio of $-0.003 / 0.0025 = -1.2$. However, the residual sum of squares is lower than calculated previously and therefore, the R-squared increases. In this case, we have $R^2 = 1 - 0.0269 / 0.056 = 0.52$.

Next, consider what happens if we write the regression equation in a different form. Rather than regress the change in gasoline demand on the right-hand side variables, we regress the level of gasoline demand on the same right-hand side variables plus the lagged gasoline demand with a restricted coefficient value of one. The results are shown in equation (4.27) in which the coefficient estimates are identical to those we calculated earlier. However, the R-squared appears to increase substantially. The reason for this is that we now have a different left-hand side variable. The model "explains" substantially more of the variation of the level of this series than it explains the variation in its growth rate.

$$g_t = g_{t-1} - \underset{(0.0044)}{0.00397} + \underset{(0.1119)}{0.7657}\,\Delta y_t - \underset{(0.0227)}{0.1006}\,\Delta p_t + \hat{u}_t \tag{4.27}$$

$$TSS = 9.1357 \qquad RSS = 0.0275 \qquad R^2 = 0.997.$$

Historical Note: The first use of R-squared as a measure of goodness of fit is credited to an American geneticist and statistician, Sewell Wright, in a paper published in the *Journal of Agricultural Research* in 1921 [Wright1921].

4.4.2 Other Measures of Goodness of Fit

Although the R-squared statistic is the most frequently quoted measure of goodness of fit, most regression packages produce a range of other statistics. These are designed to deal with the pitfalls of the R-squared statistic in a variety of ways. If we wish to compare the fit of alternative regression models, the *standard error of the regression* provides a useful alternative. This is calculated as the square root of the residual sum of squares divided by the number of degrees of freedom available, that is,

$$\hat{\sigma}_u = \sqrt{\frac{\sum\left(Y_i - \hat{Y}_i\right)^2}{N-k}}. \tag{4.28}$$

This statistic cannot provide an absolute measure of goodness of fit since it is measured in the same units as the data series itself. However, what it can do is provide a basis for comparison of different models which is not sensitive to the way in which the models are written. For example, the standard errors of the regression for (4.26) remain unchanged when we transform the model to get equation (4.27).

Another measure of goodness of fit is the *adjusted R-squared statistic* or *R-bar squared.* This is designed to deal with the problem that the standard R-squared statistic always increases when we add variables to the model – even if these are irrelevant and statistically insignificant. The adjusted R-squared statistic is defined as

$$\bar{R}^2 = 1 - \left(1 - R^2\right)\frac{N-1}{N-k}. \tag{4.29}$$

Effectively the adjusted R-squared statistic penalizes the addition of irrelevant variables to the model. Unlike the simple R-squared statistic, it will fall in value if additional variables are not sufficiently significant. In extreme cases, the adjusted R-squared statistic can become negative.

Other goodness of fit statistics are based on transformations of the log-likelihood statistic. The log-likelihood is defined as

$$LL = -\frac{N}{2}\left(1 + \ln\left(2\pi\right) + \ln\left(\frac{RSS}{N}\right)\right). \tag{4.30}$$

Since the log-likelihood is a decreasing function of the residual sum of squares, and the residual sum of squares increases as we add extra variables, the value of the log-likelihood must always *increase* as we expand the number of explanatory variables. Thus, maximizing the likelihood function is likely to lead to overfitted models, in the same way as maximizing the R^2. However, a number of statistics based on the log-likelihood have been suggested as ways of selecting one model from a range of alternatives. These statistics penalize the addition of irrelevant variables by penalizing the loss of degrees of freedom.

Two of the most commonly used transformations of the log-likelihood, for the purposes of model selection, are the *Akaike Information Criterion* (AIC) and the *Schwartz Information Criterion* (SIC). The AIC is defined as

$$AIC = -\frac{2}{N}(LL - k). \tag{4.31}$$

whereas the SIC is defined as

$$SIC = -\frac{2}{N}\left(LL - \frac{k\ln(N)}{2}\right). \tag{4.32}$$

Note that in both these cases, the statistics are defined in such a way that a lower value implies a model that fits the data better. In each case, the inclusion of the k terms in the definition of the test statistic penalizes the addition of irrelevant or insignificant variables. As the number of variables on the right-hand side of the regression increases, the residual sum of squares falls meaning that the log-likelihood increases. However, this may not be enough to offset the increasing effect on the information criteria caused by the direct effect of k in equations (4.31) and (4.32). In both cases, the information criterion concerned penalizes the loss of degrees of freedom from the inclusion of extra variables and embodies a trade-off between this and the improved fit through the increase in the log-likelihood. The SIC penalizes the addition of irrelevant variables more than the AIC and will generally lead to the choice of a more *parsimonious* model, that is, one which contains fewer explanatory variables.

Example: Consider the following two equations for gasoline consumption in the United States. Each is estimated over the period 1951 to 2016. The only difference is that the second equation includes the lagged growth rate of US GDP as well as the current rate.

$$
\begin{aligned}
&\Delta g_t = \underset{(0.0044)}{-0.0032} + \underset{(0.1165)}{0.7305}\,\Delta y_t - \underset{(0.0227)}{0.1004}\,\Delta p_t + \hat{u}_t \\
&R^2 = 0.4821 \qquad \overline{R}^2 = 0.4657 \qquad \hat{\sigma}_u = 0.0207 \\
&LL = 163.75 \qquad AIC = -4.8714 \qquad SIC = -4.7718
\end{aligned} \tag{4.33}
$$

$$
\begin{aligned}
&\Delta g_t = \underset{(0.0053)}{-0.0051} + \underset{(0.1202)}{0.7127}\,\Delta y_t - \underset{(0.0229)}{0.1020}\,\Delta p_t + \underset{(0.1164)}{0.0759}\,\Delta y_{t-1} + \hat{u}_t \\
&R^2 = 0.4856 \qquad \overline{R}^2 = 0.4607 \qquad \hat{\sigma}_u = 0.02081 \\
&LL = 163.98 \qquad AIC = -4.8479 \qquad SIC = -4.7152
\end{aligned} \tag{4.34}
$$

If we compare these equations, we see that the additional variable has a t ratio of $0.0759/0.1164 = 0.65$ and is therefore insignificant at any standard level. However, its inclusion in the model produces an increase in both R-squared and the log-likelihood. Choosing a model using either of these statistics is therefore likely to lead to overfitting. The other goodness of fit statistics provide a more reliable basis for model selection. The standard error of the regression increases when we add the extra variable because the loss of one degree of freedom is more than enough to offset a small fall in the residual sum of squares. Similarly, the adjusted R-squared statistic falls in (4.34) reflecting the lower degrees of freedom. Finally, both AIC and SIC are lower for the simpler or more parsimonious model (4.33) indicating that we would choose this model rather than (4.34).

4.4.3 Goodness of Fit and Significance of the Regressors

The R-squared for a regression equation is closely related to the F statistic for the joint significance of the regressors. To see this, recall that the F statistic for a regression is defined as (4.23). We can think of this as a test statistic for a test of the $k - 1$ linear restrictions that the slope coefficients are all equal to zero $H_0 : \beta_2 = \beta_3 = \cdots = \beta_k = 0$. TSS is the restricted sum of squares when all the slope coefficients are set to zero, whereas RSS is the unrestricted sum of squares when all the slope coefficients are freely estimated. Using the relationship between the total, explained, and residuals sums of squares, we can also rewrite the F statistic in the form shown in equation (4.35).

$$
F = \frac{ESS}{TSS - ESS}\left(\frac{N-k}{k-1}\right) = \frac{R^2}{1-R^2}\left(\frac{N-k}{k-1}\right). \tag{4.35}
$$

Therefore, we can show that the F statistic and R-squared have an exact algebraic relationship. Moreover, it is straightforward to show that an increase in F will always produce an increase in R-squared.

4.5 MISSPECIFICATION

Misspecification refers to any situation in which one of the assumptions we make in setting up the regression model is incorrect. There are therefore many different ways in which a model can be misspecified, but the most basic is when the choice of explanatory variables is either incomplete or inappropriate. We now go on to consider both these possibilities and their implications for least squares regression. We will demonstrate that when we omit a relevant variable from our model, the estimates are typically biased. However, when we estimate a model with irrelevant variables, the estimates are unbiased, but inefficient.

Consider the case in which the "true" model takes the form $Y_i = \beta_1 X_{1i} + \beta_2 X_{2i} + u_i$. If we omit the X_2 variable from the model, then the effect is that the error term becomes $v_i = \beta_2 X_{2i} + u_i$. The OLS estimator of the β_1 coefficient takes the form

$$\hat{\beta}_1 = \frac{\sum X_{1i} Y_i}{\sum X_{1i}^2} = \beta_1 + \beta_2 \frac{\sum X_{1i} X_{2i}}{\sum X_{1i}^2} + \frac{\sum X_{1i} u_i}{\sum X_{1i}^2}. \tag{4.36}$$

Taking expectations demonstrates that $E\left(\hat{\beta}_1\right) \neq \beta_1$ unless $\beta_2 = 0$ (in which case X_2 should not have been in the model in the first place) or $\sum X_{1i} X_{2i} = 0$, that is, the correlation between the X variables is equal to zero. Therefore, except in very special cases, the omission of a relevant variable from the regression will lead to biased estimates of the remaining coefficients.

Given that the omission of relevant variables produces bias, it is tempting to adopt a strategy of including as many variables as possible to reduce the chance of accidentally introducing bias into our equation. However, this strategy also has its pitfalls. The problem we have is that, when the explanatory variables are *collinear*, the inclusion of extra right-hand side variables increases the variance of the OLS estimates, thus leading to inefficiency. Inefficiency of this kind is often described as *multicollinearity*. This can lead to problems even when we have a correctly specified model, in that collinearity between the right-hand side variables inevitably leads to some loss of efficiency.

Variables in a regression equation are said to be collinear if they are correlated with each other. A certain degree of collinearity is present in almost all econometric models since it is very rarely the case that the right-hand side variables of a model are completely uncorrelated (or orthogonal). However, collinearity becomes a serious problem if the extent of it is such that the **X** matrix has rank less than k, where k is the number of columns. Alternatively, if the **X** matrix has rank less than k, then at least one of the eigenvalues of the $(\mathbf{X}'\mathbf{X})$ matrix will be zero. Under these circumstances, $(\mathbf{X}'\mathbf{X})$ is not invertible and we cannot calculate the OLS estimator. Examples of situations in which this arises are when one variable is simply a scaled version of another $(X_m = \omega X_n;\ m \neq n)$ or when a linear combination of a subset of variables equals one of the other variables of the model. If either of these two situations is the case, then the OLS procedure breaks down.

The situations described in the previous paragraph can be defined as perfect collinearity. A less fatal, but still serious situation, could occur if two variables were very highly but not perfectly correlated. For example, we might have $X_m = \omega X_n + \varepsilon_t$, where σ_ε^2 was very small. OLS estimates could be calculated in this case but the correlation between the X variables would render these estimates extremely imprecise. This situation has been termed the multicollinearity problem.

Consider the following regression model with two variables on the right-hand side of the equation

$$Y_i = \beta_1 X_{1i} + \beta_2 X_{2i} + u_i. \tag{4.37}$$

The variance–covariance matrix of the slope coefficients of the OLS estimator can be written as

$$\begin{aligned} V\begin{pmatrix} \hat{\beta}_1 \\ \hat{\beta}_2 \end{pmatrix} &= \sigma_u^2 \begin{pmatrix} \sum (X_{1i} - \bar{X}_1)^2 & \sum (X_{1i} - \bar{X}_1)(X_{2i} - \bar{X}_2) \\ \sum (X_{1i} - \bar{X}_1)(X_{2i} - \bar{X}_2) & \sum (X_{2i} - \bar{X}_2)^2 \end{pmatrix}^{-1} \\ &= \frac{\sigma_u^2}{\Delta} \begin{pmatrix} \sum (X_{2i} - \bar{X}_2)^2 & -\sum (X_{1i} - \bar{X}_1)(X_{2i} - \bar{X}_2) \\ -\sum (X_{1i} - \bar{X}_1)(X_{2i} - \bar{X}_2) & \sum (X_{1i} - \bar{X}_1)^2 \end{pmatrix}, \end{aligned} \tag{4.38}$$

where $\Delta = \sum (X_{1i} - \bar{X}_1)^2 \sum (X_{2i} - \bar{X}_2)^2 - \left(\sum (X_{1i} - \bar{X}_1)(X_{2i} - \bar{X}_2)\right)^2$ is the determinant of the crossproduct matrix. Now it is easy to see that $\Delta = \sum (X_{1i} - \bar{X}_1)^2 \sum (X_{2i} - \bar{X}_2)^2 (1 - \hat{\rho}_{12}^2)$, where $\hat{\rho}_{12}$ is the sample correlation coefficient between the X_1 and X_2 variables. It follows that as the correlation coefficient approaches one, then Δ approaches zero. Since the variance–covariance matrix is a multiple of $1 / \Delta$, it also follows that the variances of the OLS estimator will increase in size.

To some extent multicollinearity is always present in econometric analysis. Economic data are rarely experimental and, as a result, different variables in the same regression model are usually correlated with each other. However, multicollinearity only becomes a serious problem when this correlation becomes very high. The main symptoms of a serious multicollinearity problem are individually insignificant variables (low t ratios) coupled with a high degree of joint significance (high values of R-squared and the F statistic). If such symptoms are detected, then examination of the covariance matrix of the right-hand side variables can prove useful in identifying which variables are most closely related.

Perfect multicollinearity – in which there is an exact linear relationship between a subset of right-hand side variables – should not be present in correctly specified models but can easily be introduced accidentally into a model. For example, if we have four separate quarterly dummy variables in a time series regression, then their sum will equal one. If we also include a constant, then there is a perfectly collinear relationship between this group of variables. Examples like this, in which the introduction of dummy variables produces perfect collinearity, are remarkably easy to generate accidentally in applied work. Most regression packages will generate an error message in cases like this – usually of the form "Near Singular Matrix" – and fail to generate any regression estimates.

4.6 INTERPRETING A MULTIVARIABLE REGRESSION EQUATION

We have now developed the statistical tools necessary to conduct a preliminary evaluation of a regression equation. To complete this chapter, we will look at the output from a regression package and discuss how we can interpret the results in a systematic and sensible manner. Figure 4.2 shows a typical set of regression results. To interpret these results, we need to address two separate but related issues – the first is the extent to which the equation has a sensible economic interpretation and the second is the extent to which the equation exhibits a reasonable statistical fit to the data.

Do the coefficient estimates make economic sense?

```
Ordinary Least Squares Regression Results
Sample period: 1950 to 2016
Dependent Variable DLOG(MOTOR)
Sample Size 67

Variable          Coefficient          Std Err          T-Ratio

C                   -0.003966         0.004351        -0.911479
DLOG(GDP)            0.765708         0.111874         6.844344
DLOG(RP)            -0.100632         0.022717        -4.429771

R-squared              0.5089    F-statistic           33.1615
SEE              0.207405 E-1    RSS              0.275308 E-1
Durbin-Watson          1.2352    LogL               166.135380
ARCH(1) Test           3.0439    AIC                 -4.869713
Jarque-Bera           10.1221    SIC                 -4.770995
```

How well does the model explain the data?

Are the coefficient estimates statistically significant?

FIGURE 4.2 Interpreting Regression Output – A First Look at the Results.

4.6.1 Economic Interpretation

The first question we often need to ask, when examining an equation, is whether the signs and magnitudes of the coefficient estimates are consistent with our expectations from economic theory. In this case, our equation is interpreted as a demand relationship, and therefore, we would expect to find a positive effect of income (GDP) and a negative effect of price (the ratio of gasoline price to the GDP deflator). Both estimated coefficients have their expected signs, so our equation is at least sensible from this perspective.

In addition to the signs of the regression coefficients we also need to consider their magnitudes. The equation is estimated in log-linear form and therefore the coefficient estimates are elasticities. The coefficient estimate for DLOG(GDP) of 0.7657 therefore indicates that a 1%-point increase in the growth rate of GDP will increase the growth rate of gasoline demand by 0.77% points. This is a reasonable magnitude in the sense that it does not lead to obviously absurd results. For example, a coefficient of 500 would be clearly ridiculous, in that it would imply a huge response of demand to a very small change in the scale variable. Similarly, a coefficient of 0.01 – although positive – would indicate a magnitude of response well below what a passing knowledge of this market would regard as reasonable. The coefficient estimate for DLOG(PGAS/PDEF) of –0.1006 is also reasonable in that it indicates a negative response of demand to relative price with a magnitude which is within plausible bounds.

If you have read this far, then you should have realized by now that assessment of an econometric equation embodies some of the qualities of an art rather than an exact science. It requires some prior knowledge of the context of the data, such that the investigator has some idea in advance of what constitutes a reasonable model. Models with wildly implausible parameter estimates can usually be rejected, either because the underlying theory has been proved inadequate or because there has been some statistical flaw in the calculation or estimation of the equation in question. However, it is rarely the case that economic theory provides hypotheses that are sufficiently tightly defined for the investigator to accept or reject a model without some degree of individual judgment.

4.6.2 Statistical Assessment of an Equation

When it comes to the statistical assessment of an equation, we are on somewhat safer scientific ground. First, we need to assess whether the coefficient estimates are statistically significant, either individually through the t ratios or jointly through the F test. Second, we need to assess the goodness of fit of the equation through the R-squared and other related statistics. Finally, anticipating the discussion of later chapters, we need to examine the residuals of the model for any obvious signs of misspecification. In particular, we need to assess if the model estimates indicate that it is consistent with the assumptions of the classical linear regression model. If this is not the case, then the interpretation of the coefficient estimates becomes problematic and we may need to look for an alternative specification. Examination of the equation in Figure 4.1 suggests a reasonable fit to the data. The coefficients appear to be significant and the overall fit is good (an R-squared close to 0.5 for data in difference form is quite reasonable).

4.7 PARTIAL CORRELATION

We have already introduced the correlation coefficient as a measure of the association between two variables. When we consider three or more variables, however, the correlation coefficient becomes harder to interpret. This is because the correlation between two variables may be the result of both being correlated with the third rather than a direct relationship. To deal with this issue, we introduce the idea of *partial correlation* and the *partial correlation coefficient*.

We will first consider the case of three variables X_1, X_2, and X_3 for which we have data $i = 1,\ldots,N$. This case generalizes very easily to more variables. We wish to assess the strength of the relationship between X_1 and X_2 while allowing for the possibility that both variables are related to X_3. To do this, we first "purge" X_1 and X_2 of the influence of X_3 by regressing each in turn on the X_3 variable and then calculating the correlation coefficient of the residuals from these regressions. This defines the *sample partial correlation coefficient* for variables X_1 and X_2. We will write this statistic as $\hat{\rho}_{X_1X_2|X_3}$. Like the standard correlation coefficient, $\hat{\rho}_{X_1X_2|X_3}$ must always lie in the range -1 to $+1$, with values closer to the extremes of the range indicating strong negative or positive correlation.

Let us consider an example. Suppose we wish to investigate the relationship between prices, output, and income in the market for oranges. We have annual data for the United States for the period 1980 to 2016 and, using this, we calculate the sample correlation coefficients shown in Table 4.2, where *LP* is the ratio of the price of oranges to the consumer price index, *LQ* is the quantity of oranges produced, and *LY* is real household income. In each case, we have taken the natural logarithm of the variable concerned.

TABLE 4.2 The Market for Oranges. Sample Correlations Based on Annual US Data for the Period 1980–2016.

	LP	*LQ*	*LY*
LP	1.0000		
LQ	-0.5987	1.0000	
LY	0.4539	-0.0543	1.0000

To calculate the sample partial correlations, we follow the procedure for the general case. For example, to calculate the partial correlation between price and quantity, we first regress each of these variables on income and calculate the residuals. We then calculate the sample correlation between these residuals. The result obtained is $\hat{\rho}_{LP,LQ|LY} = -0.64516$. Using this procedure, we can now calculate the partial correlations shown in Table 4.3.

TABLE 4.3 The Market for Oranges. Sample Partial Correlations Based on Annual US Data for the Period 1980–2016.

	LP	*LQ*	*LY*
LP	1.0000		
LQ	-0.6416	1.0000	
LY	0.5269	0.3047	1.0000

The partial correlations differ from the sample correlations in a number of ways. The partial correlations of price with quantity and income indicate a somewhat stronger relationship, with the absolute value of the sample partial correlations being higher than the sample correlations. The effect on the correlation between quantity and income is even more striking. The uncorrected sample correlation shows a weak negative relationship between these variables, with a value of –0.0543. The sample partial correlation, however, shows a moderately strong positive relationship, with a value of 0.3047. This illustrates the value of examination of the partial correlations when we wish to assess the relationships between groups of variables rather than simply making pairwise comparisons.

The process we have set out for the calculation of sample partial correlations is straightforward enough for small numbers of variables. However, when the number of variables increases, an easier method is available which uses the method of multiple regression to calculate the partial correlations. Suppose we wish to calculate the partial correlations of price with respect to output and income in our example. To do this, we first estimate a multiple regression equation linking the three variables with price as the dependent variable. The results are given in equation (4.39)

$$\begin{aligned} LP_t &= \underset{(-0.4617)}{-0.5918} - \underset{(-4.9236)}{0.3566}\, LQ_t + \underset{(3.6149)}{0.1857}\, LY_t + \hat{u}_t \\ R^2 &= 0.5365 \qquad T = 37, \end{aligned} \tag{4.39}$$

where t ratios are given in parentheses below coefficients. Let $t_{1,2}$ be the t ratio for LQ, the partial correlation between the dependent variable LP and LQ can be calculated using the formula given in equation (4.40)

$$\hat{\rho}_{12|3} = \frac{t_{12}}{\sqrt{\left(t_{12}^2 + dof\right)}} = \frac{-4.9236}{\sqrt{(-4.9236)^2 + 34}} = -0.6452, \tag{4.40}$$

where *dof* is the degrees of freedom for the regression, that is, the number of observations minus the number of estimated coefficients. Note that the result is identical to that obtained by our previous method. It can be shown that this method generalizes for all multivariable relationships. In practice, it offers an easier way of calculating the sample partial correlations which requires fewer regression estimates than the original method we set out.

> **Historical Note:** The first clear definition of partial correlation coefficients for many variables comes in a paper by George Udny Yule in 1907 [Yule1907]. However, the ideas behind the approach had been developed in earlier work by Yule and others.

EXERCISES

Excel files containing the data for these exercises are available as companion files for this book.

EXERCISE 4.1

An econometrician wishes to estimate the following model for US unemployment

$$\Delta UN_t = \beta_1 + \beta_2 \Delta GDP_t + \beta_3 t + u_t,$$

where UN is the percentage rate of unemployment, ΔGDP is the annual percentage change in Gross Domestic Product, t is a time trend, and u is a random error. He/she has annual data for the period 1950–2019 and has calculated the following set of sample moments.

	Mean	**Standard Deviation**
ΔUN	−0.044285	1.051518
ΔGDP	3.223957	2.286201
Trend	37.5	20.351085

Sample Correlation Matrix

	ΔUN	ΔGDP	Trend
ΔUN	1.00000		
ΔGDP	-0.70579	1.00000	
Trend	-0.03403	-0.34016	1.00000

Calculate the OLS estimates of the equation parameters.

HINT: While all the calculations in this exercise are straightforward, they are more than a little tedious. You might want to think about using a spreadsheet to avoid making avoidable arithmetic errors.

EXERCISE 4.2

An econometrician has estimated the following regression equation relating changes in GDP, government consumption, investment, and exports. The data are annual UK values for the period 1949–2011 and are measured in £m at 2005 prices. Standard errors are given in parentheses below coefficients.

$$\Delta Y_t = \underset{(1763.84)}{9071.12} + \underset{(0.3151)}{0.3788}\,\Delta G_t + \underset{(0.1797)}{1.7523}\,\Delta I_t + \underset{(0.1258)}{0.4530}\,\Delta X_t + \hat{u}_t$$

$$R^2 = 0.7562 \qquad T = 63.$$

Use these results to calculate the partial correlations of changes in GDP ΔY with changes in government spending ΔG, changes in investment ΔI, and changes in exports ΔX.

These exercises use the data in the Excel workfile FM.XLSX. This contains annual data for the US economy over the period 1959–2007. The data are taken from the Federal Reserve Board of St. Louis database (FRED). The expenditure variables are all in constant prices.

EXERCISE 4.3

The following regression results were obtained by regressing the change in consumption (DC) on the change in autonomous expenditures (DA) and the change in real money balances (DRM). Autonomous expenditures are defined as the sum of investment, government consumption, and exports. Real money balances consist of broad money (M2) deflated by the consumer price index (P). Data are measured in $bn at 2000 prices.

```
Ordinary least squares regression results
Sample period: 1960-2007
Dependent variable DC
Sample size 48
```

Variable	coefficient	Std Err	*t* ratio
C	50.841678	10.059046	5.054323
DA	0.571294	0.067305	8.488065
DRM1	0.392945	0.062496	6.287519
R-squared	0.7341		
SEE	41.676845		

a. Comment on the coefficient estimates. How can these be interpreted sensibly?

b. Using the information in the table calculate (i) the residual sum of squares and (ii) the F statistic for this regression.

c. Perform an F test for the joint significance of the two variables on the right-hand side of the equation using a 5% level of significance.

EXERCISE 4.4

The following sample moments are calculated using the data set given in the file FM.XLSX.

```
Sample period: 1960-2007 (Annual Data)

Sample moments

Variable                     DC           DA          DRM1
Mean                 139.545833    87.281250     98.845405
Standard deviation    79.088721    90.831278     97.821303
```

a. Using these sample moments, calculate estimates of the elasticity of consumption expenditure with respect to autonomous expenditures and with respect to real money balances.

b. Reestimate the original equation but this time use percentage changes in consumption, autonomous expenditures, and real money balances as the equation variables. Compare the coefficient estimates with the elasticities you estimated in part (a) and comment.

EXERCISE 4.5

Using the data in the workfile FM.XLSX, estimate an equation that allows for separate effects of the different components of autonomous expenditures, that is,

$$DC_t = \beta_1 + \beta_2 DI_t + \beta_3 DG_t + \beta_4 DX_t + \beta_5 DRM_t + u_t,$$

where I, G, and X are investment, government spending, and exports, respectively.

a. Examine the coefficients of your estimated model and assess which categories of autonomous expenditure have the most important effect on consumption expenditures.

b. Perform an F test for the hypothesis that the coefficients on the three categories of autonomous expenditure are equal.

REFERENCES

[Fisher1922] Fisher, R. A., "The Goodness of Fit of Regression Formulae and the Distribution of Regression Coefficients," *Journal of the Royal Statistical Society*, 1922, 85, pp. 597–612.

[Wright1921] Wright, S., "Correlation and Causation," *Journal of Agricultural Research*, 1921, 20, pp. 557–585.

[Yule1907] Yule, G. U., "On the Theory of Correlation for Any Number of Variables Treated by a New System of Notation," *Proceedings of the Royal Society of London Series A*, 1907, 79, pp. 182–193.

CHAPTER 5

SERIAL CORRELATION

The second Gauss–Markov assumption concerns the independence of the equation errors. The assumption that $E(u_i u_j) = 0$ for $i \neq j$ is necessary for the OLS estimator to be the best linear unbiased estimator. It is also helpful in the derivation of the statistical distribution of the parameter estimates. If this assumption does not hold, then the derivation of the distribution of the OLS estimator becomes much more difficult which, in turn, makes it harder to perform hypothesis tests and to derive confidence intervals for the coefficients of our model. Unfortunately, this assumption can fail in a variety of situations and is particularly problematic when we are working with time-series data. In discussing this assumption, we will therefore concentrate on time-series examples.

A data series is classified as a time series when it consists of observations of a random variable X made at different points in time. From a statistical point of view, however, time series cannot be treated as a random sample except under very special circumstances. If a sample of data is truly random, then it can be reordered or "shuffled" without loss of information. In the case of a time series, however, the ordering of the data contains important information. We can see this by simply plotting a data series like GDP. This will show a generally increasing trend, although with some random variation around the trend path. In other words, the level of GDP in one year is not independent of the level of GDP in the previous year.

Consider a regression model of the form $Y_t = \alpha + \beta X_t + u_t$, where Y and X are time-series variables. Given the nature of the data, it is highly likely that u, the error term in this relationship, will also behave like a time series variable in the sense that it will depend upon its own past values. In such circumstances, we say that there is *serial correlation* in the errors and therefore, because the errors are not independent of each other, we cannot assume that the distributional results we have derived under the assumption of independent errors are reliable. Serial correlation of the errors is a serious problem when dealing with time series data. In this chapter, we will discuss how econometricians deal with this problem. There are three basic stages to this as follows:

1. First, we discuss the causes and implications of serial correlation for least squares regression. We concentrate on the first two moments of the distribution of the least-squares regression estimates and show that serial correlation does not, in itself, mean that the least squares estimates will exhibit bias, although it will mean that the standard formulas for the variances of the least-squares coefficients will typically produce biased estimates.
2. Next, we discuss how we can test for the presence of serial correlation in a regression equation and the specific form it takes. This involves the construction of statistical tests for the presence of serially correlated errors and diagnostic tools that allow us to determine which of a range of possible types of serial correlation best describes the errors of our model.
3. Finally, we consider what to do if serial correlation is present. Although mechanical "corrections" are available to deal with the problem, we will argue that these are usually not the correct way forward. Serial correlation is often a symptom of a deeper problem with the estimated model and a better strategy is usually to consider how we can design models that avoid the problem in the first place.

5.1 CAUSES OF SERIAL CORRELATION

Consider a regression equation of the standard form $Y_t = \beta X_t + u_t$. We assume a model in mean deviation form to simplify the notation and the discussion. The errors of this model are said to be serially correlated if $E(u_t u_{t-k}) \neq 0$ for some $k \neq 0$. A natural question is why the errors might be correlated in this way? For the moment, we will simply assume that this is an intrinsic property of the data. That is, we assume that shocks to the equation are not random drawings from a distribution but instead depend upon their own past values. An alternative would be to assume that correlation in the errors arises because the model is misspecified in some way. However, this would complicate much of the discussion and we will avoid this assumption for the moment, on the understanding that it will be relaxed later.

There are many different forms that serial correlation might take. For example, the errors might follow a *first-order autoregressive (AR) process*. This would mean that the error process could be described by an equation of the form $u_t = \rho u_{t-1} + \varepsilon_t$, where ε_t is a truly random disturbance and $\rho \neq 0$. This is a very common and important case, but it is not the only form that serial correlation can take. An alternative is where the error term in the equation is an average over several time periods of the random disturbance ε_t. For example, we might have a *first-order moving average process* of the form $u_t = \varepsilon_t + \lambda\varepsilon_{t-1}$. Both error processes are said to be serially correlated but each produces different implications and problems for the modeler. However, in both cases, the problem of dealing with serial correlation is simplified because of the assumption that it is an intrinsic feature of the error themselves, that is, the problem is one of *error dynamics*. A more realistic conclusion might be that the errors are serially correlated because of some fundamental misspecification in the original equation.

The assumption of error dynamics is very convenient because it makes the serial problem entirely statistical. If this is assumption is true, then the basic equation is correctly specified and all we need to worry about is dealing with the serial correlation in the errors. The presence of serial correlation in the errors does mean that OLS will not be an efficient estimator and will have several other undesirable properties. However, these are essentially statistical problems that can be dealt with through mechanical procedures, such as the adjustment of the OLS estimator, or the use of alternative estimators. The problem with this approach is that if the serial correlation is the result of some other form of misspecification, then we may end up disguising the problem and therefore making unjustified claims for the adequacy of our original equation.

To illustrate how serial correlation can arise as the result of a misspecified model consider the case of *omitted variables*. Suppose the true regression model takes the form

$$Y_t = \beta_1 X_{1t} + \beta_2 X_{2t} + u_t, \tag{5.1}$$

but we estimate a model of the form

$$Y_t = \beta_1 X_{1t} + v_t. \tag{5.2}$$

It follows that the error from (5.2) is determined partly by the error from the true model and partly by the effects of the omitted variable. That is, we have $v_t = u_t + \beta_2 X_{2t}$. If the X_2 variable is itself serially correlated, which is very often the case with economic data, then the effect of omitting this variable is to introduce serial correlation into the errors of the misspecified model.

A more subtle form of the omitted variable problem is that of *dynamic misspecification*. This arises when the regression model contains the correct variables, but the true model allows for dynamic adjustment processes that are not present in the regression specification. Dynamic misspecification can arise for a variety of reasons. For example, consider the following model:

$$Y_t^* = \beta X_t + u_t \tag{5.3}$$

$$\Delta Y_t = \gamma\left(Y_t^* - Y_{t-1}\right). \tag{5.4}$$

The first equation (5.3) specifies the determination of an equilibrium or desired value of Y, which we label Y^*, whereas the second (5.4) describes how Y adjusts toward that desired value. Combining equations (5.3) and (5.4) gives a single equation of the form

$$Y_t = \beta\gamma X_t + (1-\gamma)Y_{t-1} + \gamma u_t. \tag{5.5}$$

This is the *partial adjustment model*, which has featured heavily in applied econometric research. From (5.5), it is clear that a simple regression of Y on X will be misspecified as it omits the lagged Y term. Moreover, because it is highly likely that Y will be serially correlated, it follows that a simple regression is likely to suffer from serial correlation in the errors.

5.2 CONSEQUENCES OF SERIAL CORRELATION

Now that we have established some of the reasons why serial correlation may arise in regression models, let us consider the implications for least squares regression analysis. Suppose we have a model in which the errors follow a first-order AR process as set out in (5.6)

$$\begin{aligned} Y_t &= \beta X_t + u_t \\ u_t &= \rho u_{t-1} + \varepsilon_t, \end{aligned} \tag{5.6}$$

where ε_t, $t = 1,\ldots,T$ are independent, identically distributed random disturbances with mean zero and constant variance. As we have seen, this is not the only possible type of serial correlation which may arise, but the results we derive for this model apply more generally to other forms of serial correlation.

The AR process defined in (5.6) can be written in moving average form. Using the method of backward substitution, we have

$$u_t = \varepsilon_t + \rho\varepsilon_{t-1} + \rho^2\varepsilon_{t-2}\ldots = \sum_{j=0}^{\infty}\rho^j\varepsilon_{t-j}. \quad (5.7)$$

This is an *infinite moving average* process. Providing $|\rho| < 1$, then the sequence defined in (5.7) will converge, in the sense that it will have a finite variance. To see this note that

$$E(u_t^2) = \sum_{j=0}^{\infty}\rho^{2j}E(\varepsilon_{t-j}^2) = \sum_{j=0}^{\infty}\rho^{2j}\sigma_\varepsilon^2 = \frac{\sigma_\varepsilon^2}{1-\rho^2} \quad (5.8)$$

Therefore, for the variance of the error term to be finite and positive, we need $|\rho| < 1$. If this condition holds, then the process is said to be *weakly stationary* and it can be shown that a general feature of stationary, finite AR processes is that they can be written as infinite moving average processes. Moreover, since $E(\varepsilon_{t-j}) = 0$ for all values of j, it follows that $E(u_t) = 0$. This is a useful property because we have already seen that the expected value of the OLS estimator can be written as follows: $E(\hat{\beta}) = \beta + \sum_{t=1}^{T} X_t E(u_t) / \sum_{t=1}^{T} X_t^2$. It therefore follows that $E(\hat{\beta}) = \beta$ and that the OLS estimator is unbiased even when the errors are serially correlated.

Our results so far indicate that the presence of serial correlation does not *in itself* indicate that the OLS estimator is unbiased. The phrase "in itself" needs to be emphasized, since the unbiasedness of the OLS estimator has only been demonstrated in the case where the serial correlation is due to pure error dynamics. More generally, if serial correlation is a symptom of some other misspecification, then we cannot rely on OLS remaining unbiased. For example, if we omit a serially correlated variable X_2 from the model, then it is straightforward to show that the OLS estimator will be biased, except in the special case, where X_2 is uncorrelated with the explanatory variables included.

The other important point to note is that even if the OLS estimator is unbiased, it will be inefficient. This follows from the fact that efficiency of the OLS estimator depends on all the Gauss–Markov assumptions holding. If the assumption of serially independent errors fails, then we can, in principle, design a more efficient estimator that takes into account this property. Therefore, in models with serially correlated errors, it is always possible, in principle, to design an estimator with a lower variance than the OLS estimator.

Perhaps, the most serious implication of serial correlation is that the OLS estimator of the standard errors of the coefficient estimates will be biased. From our earlier treatment of the OLS estimator, we have

$$V(\hat{\beta}) = E(\hat{\beta} - E(\hat{\beta}))^2 = E(\hat{\beta} - \beta)^2 = E\left(\frac{\sum_{t=1}^{T} X_t u_t}{\sum_{t=1}^{T} X_t^2}\right)^2. \tag{5.9}$$

In our earlier treatment of this problem, we made use of the Gauss–Markov assumption that $E(u_t u_{t-k}) = 0$ for all values of k not equal to zero. When (5.9) is expanded, then the resulting expression contains crossproduct terms including expressions of the form $u_t u_{t-k}$, if the errors are serially independent, then these have zero expectation and can be eliminated. This allows us to derive the standard OLS expression for the variance of the parameter estimate $V(\hat{\beta}) = \sigma_u^2 / \sum_{t=1}^{T} X_t^2$. If the errors are not serially independent, then this is no longer possible and the standard formula for the variance of the OLS estimator is biased. Both the sign and the magnitude of the bias will depend on the nature of the serial correlation process for the errors.

A very common scenario in applied econometric work is to find positive serial correlation in the errors of the equation in conjunction with an X variable which is itself positively correlated. Under these circumstances, the OLS variance is biased downward. To demonstrate this, consider the following three equation models:

$$\begin{aligned} Y_t &= \beta X_t + u_t \\ u_t &= \rho u_{t-1} + \varepsilon_t \\ X_t &= \varphi X_{t-1} + v_t, \end{aligned} \tag{5.10}$$

where ε and v are uncorrelated error processes, where each process consists of independent, identically distributed random variables with zero mean and constant variance. If we maintain Gauss–Markov Assumption 4, which the X variables are fixed in repeated samples, then the variance of the OLS estimator can be written as

$$V(\hat{\beta}) = \frac{1}{\left(\sum_{t=1}^{T} X_t^2\right)^2} E\left(\sum_{t=1}^{T} X_t u_t\right)^2. \tag{5.11}$$

In large samples, the denominator of this expression can be written as $\left(\sum_{t=1}^{T} X_t^2\right)^2 = T^2\sigma_X^4$. Expanding the numerator of (5.11) yields

$$E\left(\sum_{t=1}^{T} X_t u_t\right)^2 = E\left\{\sum_{t=1}^{T} X_t^2 u_t^2 + 2\sum_{t=2}^{T} X_t X_{t-1} u_t u_{t-1} + 2\sum_{t=3}^{T} X_t X_{t-2} u_t u_{t-2} + \ldots\right\}. \quad (5.12)$$

We now assume that the sample size becomes sufficiently large such that the change in the lower limit of the summations becomes unimportant. Taking expectations yields

$$E\left(\sum_{t=1}^{T} X_t u_t\right)^2 = T\sigma_X^2\sigma_u^2\left(1 + 2\varphi\rho + 2\varphi^2\rho^2 + 2\varphi^3\rho^3 + \ldots\right). \quad (5.13)$$

The expression in parentheses on the right-hand side of (5.13) is equal to 1 plus the sum of an infinite geometric progression, with initial term 2, and common ratio $\varphi\rho$. If both the error process and the X process are stationary, that is, $|\rho| < 1$ and $|\varphi| < 1$, then the common ratio is <1 in absolute value, and this converges to the expression shown in (5.14)

$$E\left(\sum_{t=1}^{T} X_t u_t\right)^2 = T\sigma_X^2\sigma_u^2\left[1 + 2\frac{\varphi\rho}{1-\varphi\rho}\right] = T\sigma_X^2\sigma_u^2\left(\frac{1+\varphi\rho}{1-\varphi\rho}\right). \quad (5.14)$$

Combining this with the expression, we have already derived for the denominator of (5.11) means that we can write

$$V\left(\hat{\beta}\right) = \frac{\sigma_u^2}{T\sigma_X^2}\left(\frac{1+\rho\varphi}{1-\rho\varphi}\right). \quad (5.15)$$

The standard formula for the OLS variance would give us an expression of the form $\sigma_u^2 / \left(T\sigma_X^2\right)$. Therefore, the presence of autocorrelation in both the errors and the X variable leads to bias in the standard OLS formula with the size of the bias being determined by the expression in parentheses. If ρ and φ are both positive and <1, then the true variance of the OLS estimator will be larger than the estimated variance. Moreover, the closer ρ and φ are to 1, then the larger will be the bias. In general, if ρ and φ have the same sign, then the OLS formula will underestimate the true variance of the estimator. If ρ and φ have opposite signs, then the OLS formula will overestimate the true variance.

> **Historical Note:** George Udny Yule [Yule1921 and Yule1926] provides an early warning to economists about the dangers of treating time-series data as it they were randomly sampled observations.

Example: To illustrate the implications of serial correlation, we have described; 1,000 regressions were run using artificially generated data for the model given below:

$$\begin{aligned} Y_t &= X_t + u_t \\ u_t &= 0.9u_{t-1} + \varepsilon_t \\ X_t &= 0.9X_{t-1} + \upsilon_t, \end{aligned} \tag{5.16}$$

where ε and υ are independent white noise errors. A typical regression taken from this simulation looks like this

$$\begin{aligned} &\hat{Y}_t = \underset{(0.27)}{-2.3872} + \underset{(0.12)}{0.7603}X_t \\ &R^2 = 0.31 \qquad \hat{\sigma} = 2.22 \qquad T = 100. \end{aligned} \tag{5.17}$$

The test statistic for the null hypothesis that the slope coefficient is equal to 1 is $t = (0.7603 - 1)/0.12 = -2.0$, and under the null hypothesis, this is distributed as t_{98}. Since the 5% critical value for a two-tailed test is ± 1.96, we therefore reject the null at the 5% level. This result is typical for this simulation. Out of 1,000 regressions, we reject the null that the slope coefficient is equal to 1 in 491 cases.

The reason why we reject the null so often is not because of any bias in the coefficient estimates. The distribution of the slope coefficient estimates from our simulation is illustrated in the histogram shown in Figure 5.1. The average slope coefficient estimate is 0.9955 which is very close to the true value of 1. Instead, the reason lies in the underestimate of the standard error of the slope coefficient which has resulted from the fact that both the errors and the X variable are serially correlated. To see this, compare the standard error of the slope coefficient from the regression equation (5.17) with the standard error of the slope coefficients from the simulation exercise shown in Figure 5.1. The latter provides an unbiased estimate of the standard error and, if we had used this to calculate our test statistic, we would not have rejected the null.

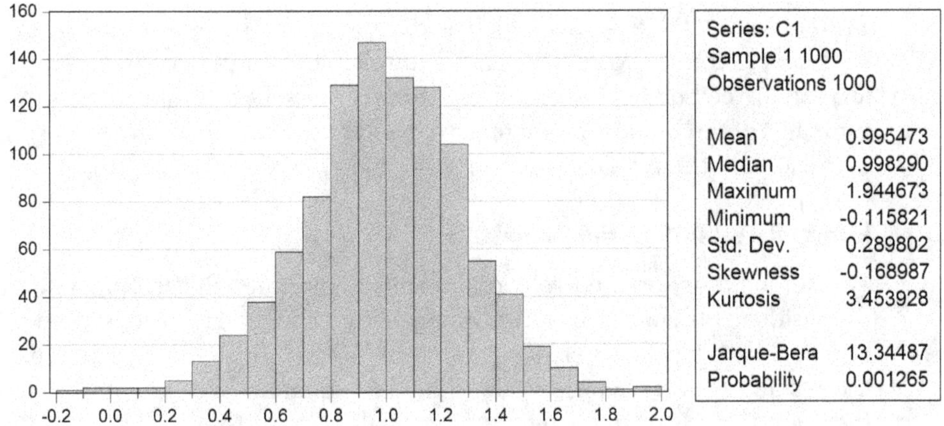

FIGURE 5.1 Distribution of Slope Coefficients from Monte Carlo Simulation with Positive Autocorrelation in Both the Errors and the *X* Variable

5.3 DETECTION OF SERIAL CORRELATION

Since serial correlation of the errors has been shown to have important implications for our interpretation of regression results, it becomes important to develop tests for whether it is present in the models we estimate. It will be helpful to have an example in mind as we develop such tests. Consider the following estimated consumption function for the US economy based on annual data from 1970 to 2019.

$$\begin{aligned} &\ln(C_t) = \underset{(0.0466)}{-0.2898} + \underset{(0.0052)}{1.0174}\ln(YD_t) + \hat{u}_t \\ &R^2 = 0.9987 \qquad \hat{\sigma}_u = 0.0154 \qquad T = 50 \qquad (5.18) \\ &DW = 0.5389, \end{aligned}$$

where C is total consumers' expenditure and YD is real personal disposable income. Both variables are measured in millions of dollars at 2012 prices.

On first inspection, this equation appears to have reasonable properties. The slope coefficient is statistically significantly different from zero and the coefficient of determination indicates a good fit. In this case, however, it is more interesting to test the null hypothesis that the slope coefficient is equal to 1. This is because the slope coefficient in this relationship measures the income elasticity of consumption expenditure. However, this null hypothesis is also rejected by our model. The t statistic for this test is

$(1.0174 - 1)/0.0052 = 3.35$ which leads us to reject the null at the 5% level. This conclusion is, however, may be unreliable if the errors in this model are serially correlated. Therefore, in order to assess the robustness of our estimates of the model parameters, we need to test for the presence of serial correlation in the residuals $\hat{u}$ from equation (5.18).

5.3.1 Informal Tests for Serial Correlation

We can look for serial correlation informally by simply inspecting a plot of the residuals. If runs of positive or negative residuals are obvious, then this is a sign that serial correlation is present. For example, Figure 5.2 shows the residuals for our consumption function equation and it is obvious that there are periods during which the residuals are consistently either positive or negative. This is a clear indication that the equation suffers from serial correlation. There are, however, forms of serial correlation, such as moving average errors, which are not so easily detected. Therefore, it is important to develop more formal tests as well as procedures for identifying the specific form of serial correlation that is relevant for this equation.

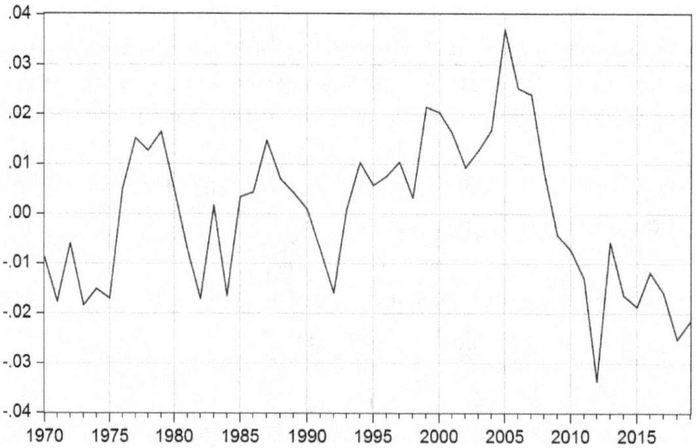

FIGURE 5.2 Residuals for Equation (5.18)

The *correlogram*[1] provides a more formal statistical method for the investigation of serial correlation. The correlogram is a table, or plot, of the

[1] We will return to the topic of the correlogram and discuss its construction in more detail in Chapter 9 on ARIMA modeling.

sample autocorrelations of the regression residuals. The sample autocorrelations are defined in equation (5.19), and Figure 5.3 gives both a table and a graph of the autocorrelations of the residuals from equation (5.18).

$$\hat{\rho}_k = \sum\nolimits_{t=k+1}^{T} \hat{u}_t \hat{u}_{t-k} \Big/ \sum\nolimits_{t=1}^{T} \hat{u}_t^2. \tag{5.19}$$

The correlogram output in Figure 5.3 also shows the *partial autocorrelations*. These allow for the presence of intermediate lags in the AR process in the same way that the partial correlations, which we discussed in Chapter 4, factor out the effects of other variables when calculating the correlation between two variables. The partial autocorrelations are helpful in identifying the nature of the serial correlation process in the residuals. Both the sample autocorrelations and the sample partial autocorrelations are constrained to lie on the interval $]-1,1[$ for $k \neq 0$. Under the null of no autocorrelation, both the autocorrelations and partial autocorrelations have expected value zero and variance $1/T$. This allows us to calculate the broken standard error bands shown in Figure 5.3, which define an approximate 95% confidence interval, and therefore allow us to judge the significance of the autocorrelations. Experienced modellers can identify the nature of the serial correlation process by inspection of the sample correlogram. In this case, the pattern of autocorrelations shows an initial positive value that declines quickly toward 0 and there is a single significant partial autocorrelation at lag 1. This is consistent with a first-order AR process.

Sample: 1970 2019
Included observations: 50

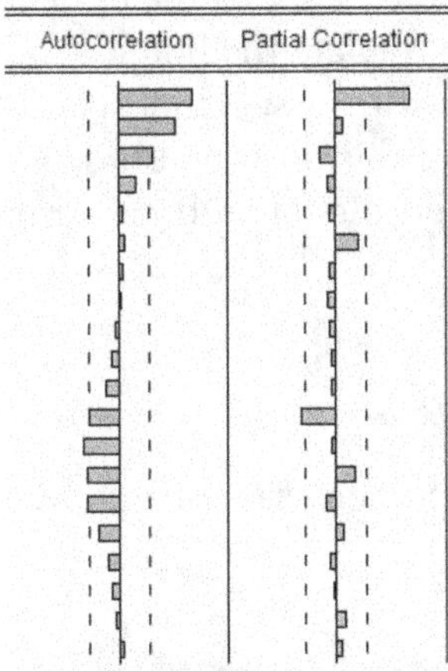

	AC	PAC	Q-Stat	Prob
1	0.707	0.707	26.508	0.000
2	0.537	0.075	42.146	0.000
3	0.328	-0.153	48.110	0.000
4	0.170	-0.071	49.746	0.000
5	0.037	-0.060	49.825	0.000
6	0.064	0.206	50.067	0.000
7	0.029	-0.057	50.119	0.000
8	0.011	-0.076	50.127	0.000
9	-0.028	-0.060	50.176	0.000
10	-0.067	-0.041	50.468	0.000
11	-0.132	-0.044	51.631	0.000
12	-0.289	-0.328	57.335	0.000
13	-0.348	-0.037	65.846	0.000
14	-0.307	0.183	72.637	0.000
15	-0.303	-0.091	79.454	0.000
16	-0.193	0.073	82.298	0.000
17	-0.105	-0.060	83.158	0.000
18	-0.073	-0.013	83.594	0.000
19	-0.044	0.085	83.759	0.000
20	0.025	0.053	83.814	0.000

FIGURE 5.3 Correlogram of Residuals from Equation.

> **Historical Note:** The term "correlogram" was first used by Herman Wold [Wold1938]. However, although he does not use the term, [Yule1926] provides a plot of the sample autocorrelations for an index of wheat prices based on annual data for the period 1545–1844.

5.3.2 Formal Tests for Serial Correlation

The *Durbin–Watson* (DW) test provides a formal test in which the null hypothesis is that the equation errors are serially uncorrelated and the alternative is that they follow a first-order autocorrelation process. This test was first introduced by Durbin and Watson in two papers published in Biometrika in 1950 and 1951 [Durbin1950] and [Durbin1951]. It is a standard part of the regression output for most econometrics packages. The DW test builds on a previous test developed by Von Neumann [VonNeumann1941] who developed a test for autocorrelation in a series of random variables with the null that the variables are independent random numbers. Unfortunately, this is not suitable when the series under examination comprises regression residuals, which are not independent by construction. Although Von Neumann's statistic has a relatively simple distribution, that is, the standard normal distribution, Durbin and Watson showed that the distribution of their test statistic was necessarily more complex. The nature of the test statistic means that it is not possible to derive unique critical values for a test of the null of no autocorrelation against the alternative of first-order autocorrelation. However, they did demonstrate that the critical values for their test were bounded and were able to tabulate the bounds for small sample sizes.

The DW test is concerned with a specific form of serial correlation, that is, first-order autocorrelation but is arguably sensitive to other forms. Consider the following regression model with an error that follows an AR process of order one:

$$\begin{aligned} Y_t &= \beta X_t + u_t \\ u_t &= \rho u_{t-1} + \varepsilon_t. \end{aligned} \tag{5.20}$$

Taking the residuals from an OLS regression of Y on X, we can construct the test statistic

$$DW = \sum_{t=2}^{T} \left(\hat{u}_t - \hat{u}_{t-1}\right)^2 / \sum_{t=1}^{T} \hat{u}_t^2. \tag{5.21}$$

DW can be seen as a test for the null hypothesis that $\rho = 0$ in (5.20). To see this, expand the numerator of (5.21) to obtain

$$DW = \frac{\sum_{t=2}^{T} \hat{u}_t^2 + \sum_{t=2}^{T} \hat{u}_{t-1}^2 - 2\sum_{t=2}^{T} \hat{u}_t \hat{u}_{t-1}}{\sum_{t=1}^{T} \hat{u}_t^2}. \tag{5.22}$$

Now for large T, we have $\sum_{t=2}^{T} \hat{u}_t^2 \approx \sum_{t=2}^{T} \hat{u}_{t-1}^2 \approx \sum_{t=1}^{T} \hat{u}_t^2$ and $\sum_{t=2}^{T} \hat{u}_t \hat{u}_{t-1} / \sum_{t=1}^{T} \hat{u}_t^2 \approx \hat{\rho}$, where $\hat{\rho}$ is the first autocorrelation of the least squares residuals. It follows that $DW \approx 2(1-\hat{\rho})$. Therefore, if there is no autocorrelation, then $E(DW) = 2$, if there is positive autocorrelation $(\rho > 0)$, then $E(DW) < 2$, and if there is negative autocorrelation $(\rho < 0)$, then $E(DW) > 2$. Note that the DW statistic is bounded between 0 and 4.

To use the DW statistic to conduct a test for autocorrelation, we need appropriate critical values. Unfortunately, there is a problem here in that the distribution of this statistic does not us to calculate unique critical values. Instead, the distribution gives us upper and lower bounds for the test statistic under the null. For example, consider the case in which we wish to test $H_0 : \rho = 0$ against the alternative $H_1 : \rho > 0$. That is, we wish to test the null that there is no autocorrelation against the alternative that there is positive first-order autocorrelation. If the test statistic is greater than d_U, then we can accept the null, if it less than d_L, then we reject the null but if $d_L < DW < d_U$, then we are in a region of indeterminacy that allows us to neither accept nor reject the null. If the DW statistic is >2, then the relevant upper and lower critical values become $4 - d_U$ and $4 - d_L$. Thus, the decision part of the testing procedure becomes rather more complicated than is usually the case. Figure 5.4 summarizes the possible situations that can arise when conducting the DW test.

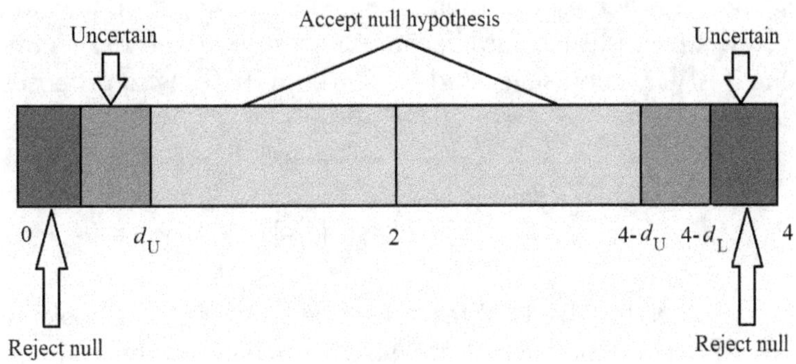

FIGURE 5.4 Possible Decisions with the *DW* Test.

In the case of our consumption function, we note that the *DW* statistic is 0.5389. From the *DW* tables, we find that the upper and lower bounds at the 5% level are $d_U = 1.59$ and $d_L = 1.50$, respectively, for $T = 50$. Since $DW < d_L$, we conclude that there is evidence of significant positive first-order autocorrelation at the 5% level. Tables of critical values, which give upper and lower critical bounds for different sample sizes and numbers of right-hand side variables, are included in most books of tables for econometric applications. The typical format of these tables is to give the critical bounds for a one-tailed test of $H_0 : \rho = 0$ against $H_1 : \rho > 0$ for different numbers of right-hand side variables. For example, the critical bounds for a 30 observations and one explanatory variable are $d_L = 1.35$ and $d_U = 1.49$. If we wish to test for negative autocorrelation, then the null hypothesis remains the same while the alternative becomes $H_1 : \rho < 0$. In these circumstances, we must calculate the critical bounds by subtracting the numbers reported in the tables from four. Thus, in our example, with one explanatory variable and 30 observations, the critical bounds become $d_U = 4 - 1.35 = 2.65$ and $d_L = 4 - 1.49 = 2.51$.

The DW test becomes problematic in regressions that include lagged endogenous variables. For example, consider a regression equation of the form

$$Y_t = \beta_1 X_t + \beta_2 Y_{t-1} + u_t. \tag{5.23}$$

Models of this kind are very common in time-series econometrics, where the lagged *Y* variable is included to capture dynamic adjustment of *Y* toward an equilibrium relationship with the *X* variable. The inclusion of the lagged *Y* variable will often reduce any residual autocorrelation. However, there is no guarantee that autocorrelation will be eliminated, and it is still important to test for its presence. This does, however, create a problem, in that the *DW* statistic can be shown to be biased toward acceptance of the null hypothesis when the equation estimated contains a lagged endogenous variable. An alternative test therefore becomes necessary under these circumstances. Such a test has been suggested by Durbin (1970), who proposes a test statistic of the form

$$h = \hat{\rho}_1 \sqrt{\frac{T}{1 - T \times \mathrm{V}(\hat{\beta}_2)}}, \tag{5.24}$$

Critical values for the DW test are widely available in books of statistical tables for economics. They are also available through numerous online sources.

where $\hat{\rho}_1 = \sum_{t=2}^{T} \hat{u}_t \hat{u}_{t-1} / \sum_{t=2}^{T} \hat{u}_t^2$ is the first-order autocorrelation coefficient, and $V(\hat{\beta}_2)$ is the estimated variance of the coefficient on the lagged endogenous variable obtained by estimation of (5.23) by least squares. Note that some presentations of this test use an approximation for the first-order autocorrelation coefficient in terms of the DW statistic which is given by the following expression: $\hat{\rho}_1 = (1 - DW/2)$. Durbin [Durbin1970] shows that this statistic is asymptotically distributed as normal with mean 0 and variance 1.

Example: To illustrate the use of Durbin's h test, let us consider a modified version of equation (5.18) in which we include a lagged endogenous variable. Estimation by least squares yields the following:

$$\begin{aligned} &\ln(C_t) = \underset{(0.0537)}{-0.1059} + \underset{(0.0937)}{0.5571}\ln(YD_t) + \underset{(0.0913)}{0.4479}\ln(CN_{t-1}) + \hat{u}_t \\ &R^2 = 0.9991 \quad \hat{\sigma}_u = 0.0127 \quad T = 49 \\ &DW = 0.7758 \quad \hat{\rho}_1 = 0.5985. \end{aligned} \tag{5.25}$$

To apply, Durbin's h test, we construct the test statistic

$$h = 0.5985\sqrt{\frac{49}{1 - 49 \times 0.091276^2}} = 5.45. \tag{5.26}$$

The 5% critical value for a test of the null hypothesis that the errors are uncorrelated against a one-sided alternative of positive autocorrelation is equal to 1.645. Thus, we conclude that there the inclusion of a lagged endogenous variable has not eliminated the autocorrelation for this model. Note that, in this case, the DW test would have sufficed. This is because the problem is that the DW test statistic is biased toward two due to the inclusion of the lagged endogenous variable. Even so, the test statistic is equal to 0.7758, which is less than the value of the lower critical bound (~1.46 in this case).

A more general test for serial correlation is provided by the Q *statistic*, also known as the *Box-Ljung* test statistic. The advantage of this test is that it allows for tests of higher-order serial correlation processes than the DW test. It is also designed to test for more general serial correlation processes such as moving average errors. This test statistic is defined as

$$Q = T(T+2)\sum_{j=1}^{J} \frac{\hat{\rho}_j^2}{T - J}. \tag{5.27}$$

Using this test statistic, we can test for general forms of serial correlation up to order J. We can show that the Q statistic is asymptotically distributed as χ_2 with J degrees of freedom under the null that $\rho_j = 0, j = 1,\ldots, J$. Note that this test can have low power when a large value of J is chosen.

Example: Suppose we wish to test for autocorrelation of either first or second order from the estimates of equation (5.18). The sample autocorrelations for this equation are equal to $\hat{\rho}_1 = 0.7067$ and $\hat{\rho}_2 = 0.5373$. The Q_2 statistic is therefore calculated as

$$Q_2 = 50 \times 52 \times \left(\frac{0.7067^2}{49} + \frac{0.5373^2}{48} \right) = 42.14. \tag{5.28}$$

Under the null hypothesis, this statistic is distributed as chi-squared with two degrees of freedom, and the 5% critical value is therefore equal to 5.991. It follows that we reject the null at the 5% level and conclude that autocorrelation of either first or second order is present.

Another test for serial correlation, which has become standard in the econometrics literature, is the Breusch-Godfrey test [Breusch1978] and [Godfrey1978]. The advantages of this test are (1) it is more general than the DW test, or Durbin's h test, in that it allows for higher-order serial correlation processes and is applicable to both autocorrelated and moving average error processes and (2) it can focus more easily on specific processes than the Box-Ljung test. The Breusch–Godfrey test is constructed by performing a regression of the least-squares residuals on their own lagged values plus the original regressors. We then test for serial correlation by testing the joint significance of the lags in this second-stage regression. This gives us a very flexible test for serial correlation.

To construct the Breusch–Godfrey test, let $\hat{u}_t, t = 1,\ldots,T$ be the residuals from a regression equation. If we now estimate an auxiliary regression of the form

$$\hat{u}_t = \gamma_0 + \sum_{i=1}^{p} \gamma_j \hat{u}_{t-j} + \sum_{j=1}^{q} \gamma_{p+j} X_{jt} + \varepsilon_t, \tag{5.29}$$

where $X_j, j = 1,\ldots,q$ are the original regressors. Let $Z = TR^2$, where R^2 is the coefficient of determination following estimation of (5.29), it can be shown that under the assumption of serially uncorrelated errors, Z is asymptotically distributed as chi-squared with degrees of freedom equal to p, the order of serial correlation for which we wish to test. The null hypothesis for this test is $H_0: \gamma_1 = \gamma_2 = \cdots = \gamma_p = 0$, and the X variables are included purely as controls with so that their coefficient values do not form part of the test. Although this is an asymptotic test, some authors have suggested using an F test on the grounds that it improves its performance in practice cf. [Kiviet1986]

Example: Suppose we wish to test for fourth-order autocorrelation in our consumption function equation (5.18). Having generated the residuals for this equation, we next estimate the auxiliary regression shown in equation (5.30)

$$\hat{u}_t = \underset{(0.0383)}{0.0369} - \underset{(0.0042)}{0.0041}\ln\left(YD_t\right) + \underset{(0.1543)}{0.6659}\,\hat{u}_{t-1} + \underset{(0.1847)}{0.2108}\,\hat{u}_{t-2} - \underset{(0.1817)}{0.1227}\,\hat{u}_{t-3} - \underset{(0.1567)}{0.0529}\,\hat{u}_{t-4} + \varepsilon_t \qquad (5.30)$$

$$R^2 = 0.5573.$$

To perform the chi-squared test, we calculate $TR^2 = 46 \times 0.5573 = 25.63$. Note that T is lower than the full sample size because we lose four observations by taking lags. The 5% critical value for a chi-squared distribution with four degrees of freedom is equal to 9.488, and therefore, we reject the null hypothesis at the 5% level. One advantage of this test is that, because we can inspect the individual t ratios for the coefficients on the lags in the auxiliary regression, we can more easily identify the of order of the serial correlation process. In this case, we see that it is only the first lagged residual that has a t ratio greater than the 5% critical value. This suggests that a first-order serial correlation process may be more appropriate than the fourth-order process we have assumed. Finally, we note that an F test for the joint significance of the lagged residuals in (5.30) also leads us to reject the null since it produces a test statistic of 12.16 which is greater than the 5% critical value of 2.606. In practice, the chi-squared and F versions of this test generally produce the same outcome.

5.4 DEALING WITH SERIAL CORRELATION

If serial correlation is present, then there are several ways to deal with the issue. Of course, the priority is to identify the nature, and hopefully the cause, of the serial correlation. If the root cause of the problem is the omission of a relevant variable from the model, then the natural solution is to include that variable. If it is determined that modeling of the serial correlation process is appropriate, then we have several different methods available for the estimation of such models by adjusting for the presence of serially correlation errors. It should be noted that mechanical adjustments, of the type we will describe in this section, are potentially dangerous. This process has been much criticized on the grounds that there is a risk that these methods disguise an underlying problem rather than dealing with it. McGuirk and Spanos [McGuirk2009] are particularly critical of mechanical adjustments to deal with autocorrelated arguments. In this paper, they show that unless we can assume that the regress and does not Granger-cause the regressors, adjusting for autocorrelation means that least squares yield biased and inconsistent estimates. However, these methods are still used and reported in applied work and it is therefore important that we consider how they work.

The first method we will consider is that of Cochrane–Orcutt estimation. This uses an iterative algorithm proposed by Cochrane and Orcutt [Cochrane1949] in which we use the structure of the problem to separate out the estimation of the behavioral parameters of the main equation from those of the AR process that describes the errors. Let us consider the case of an AR(1) error process as an example. Suppose we wish to estimate a model of the form (5.6). The two equations can be combined to give a single equation of the form

$$Y_t - \rho Y_{t-1} = \beta \left(X_t - \rho X_{t-1} \right) + \varepsilon_t, \tag{5.31}$$

that is, an equation in "quasi-differences" of the data. If ρ was known, then it would be straightforward to construct these quasi-differences and estimate the behavioral parameter β by least squares. In the absence of such knowledge, we make a guess at ρ and construct an estimate of β on this basis. We then generate the residuals $\hat{u}_t = Y_t - \beta X_t$ on this basis and calculate an estimate of ρ of the form $\hat{\rho} = \sum_{t=2}^{T} \hat{u}_t \hat{u}_{t-1} / \sum_{t=1}^{T} \hat{u}_t^2$. If, by some lucky chance, this estimate coincides with our assumption, then we stop. Otherwise, we use our estimate to recalculate the quasi-differences, reestimate β, and continue until our estimates of β and ρ converge. If a solution exists, then this provides a robust algorithm for estimation.

Example: Applying the Cochrane–Orcutt procedure to our consumption function data yields the following results:

$$\begin{aligned} &\ln(C_t) = -\underset{(0.1013)}{0.1778} + \underset{(0.0112)}{1.0050} \ln(YD_t) + \hat{u}_t \\ &\hat{u}_t = 0.7519\,\hat{u}_{t-1} + \hat{\varepsilon}_t \\ &R^2 = 0.9993 \qquad \hat{\sigma}_u = 0.0107 \qquad DW = 2.2277. \end{aligned} \tag{5.32}$$

The first thing to note is that convergence of the iterative process for this model is quite fast. Figure 5.5 shows the estimate of the AR parameter at each stage of the process, beginning with an initial guess of zero and stopping when the change in the parameter is $<|10^{-9}|$. From Figure 5.5, we see that the solution has very nearly converged after only three iterations, though it takes 13 in total for the convergence criterion to be achieved.

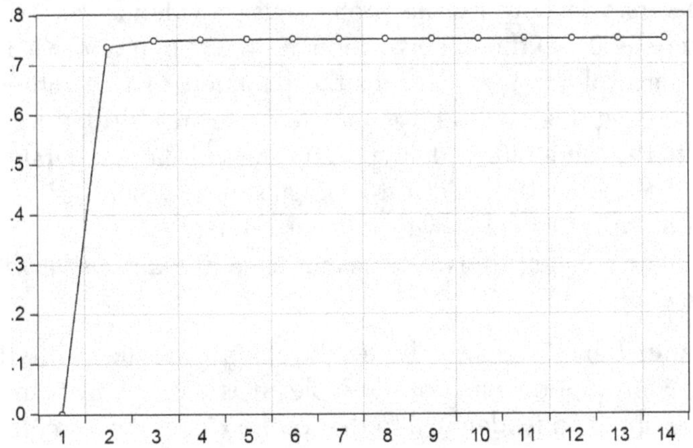

FIGURE 5.5 Convergence of the Autoregressive Parameter Using the Cochrane–Orcutt Procedure (Convergence Criterion = 10^{-9}).

Turning to the coefficient estimates we see that, in comparison with the simple OLS model, the point estimate of the slope coefficient has changed very little. Of course, we had established earlier that the presence of serial correlation did not lead to bias in coefficient estimates, and so this is not surprising. However, the standard error of the slope coefficient has increased, more than doubling from 0.0052 to 0.0112. This is because the simple OLS estimate of the standard error was biased by the combination of positive autocorrelation in the errors and in the right-hand side variable. By correcting for serial correlation, we have obtained a more realistic estimate of the standard error of the coefficient. The t test for the null hypothesis that the income elasticity of consumption is equal to one now gives a value of $(1.005-1)/0.0112=0.45$, and we no longer reject the null hypothesis. It is also worth noting that despite the increase in the standard errors of the coefficients, the standard error of the regression has fallen from 0.0154 to 0.0107, indicating an improvement in fit relative to the original model.

The Cochrane–Orcutt procedure looks for a minimum of the residual sum of squares by using the iterative method described earlier. An iterative method becomes necessary in this scenario because we must solve a pair of nonlinear equations in $\hat{\beta}$ and $\hat{\rho}$ in contrast with the linear equations for the standard least-squares model. A potential problem here is that iterative methods like this can lead to convergence to a local, rather than global, minimum point. An alternative method of solving for the regression parameters, which avoids this problem, is provided by the Hildreth and Lu [Hildreth1960] method. This adopts a grid search approach in which the algorithm explores the whole of the parameter space rather than a local search. Using this method, we start with a coarse grid search in which we estimate the model for widely spaced values of the AR parameter. We then locate the value with the lowest residual sum of squares and conduct a local search around this value with a lower interval value. This continues with smaller and smaller interval values until we find a solution with the chosen degree of accuracy.

Example: Table 5.1 shows the results of applying the Hildreth–Lu method to our consumption function example. It is evident that, in this case, the solution obtained is identical to that of the Cochrane–Orcutt method. This indicates that this solution is consistent with both a local and a global minimum of the residual sum of squares function.

TABLE 5.1: Hildreth–Lu Estimation of the Autoregressive Parameter.

Grid Search	Interval	Value of $\hat{\rho}$ that minimizes RSS	Minimum of RSS function ($\times 10^{-3}$)
1	10^{-1}	0.8	5.320922
2	10^{-2}	0.75	5.299408
3	10^{-3}	0.752	5.299374
4	10^{-4}	0.7519	5.299374
5	10^{-5}	0.75187	5.299374
6	10^{-6}	0.751868	5.299374
7	10^{-7}	0.751868	5.299374

Example: As a final example, we will consider the estimation of a demand curve for oranges for the United States. The form of the equation estimated is $\Delta P_t = \beta_1 + \beta_2 \Delta Q_t + u_t$, and we allow for an AR error of the form $u_t = \rho u_{t-1} + \varepsilon_t$. The data for this exercise consist of annual observations for the United States from 1980 to 2016. The data were expressed as percentage changes of the original data, and the price was adjusted for general inflation by dividing by the consumer price index prior to estimation. In Table 5.2, we give the results of estimating this equation by OLS, by autoregressive least squares (ARLS) using the Cochrane–Orcutt method and ARLS using the Hildreth–Lu method.

TABLE 5.2 Estimation of Demand for Oranges by Alternative Methods.

			$\hat{\rho}$	R^2	$\hat{\sigma}_u$
Least squares	Intercept slope	1.0169 (*1.9688*) -0.6014 (*0.1273*)	0	0.3961	11.7966
ARLS with Cochrane–Orcutt	Intercept slope	1.1196 (*1.3749*) -0.5843 (*0.1271*)	-0.3663	0.4739	11.1762
ARLS with Hildreth–Lu	Intercept slope	1.1196 (*1.3749*) -0.5843 (*0.1271*)	-0.3663	0.4739	11.1762

Standard errors are given in parentheses next to coefficients

As was the case with the consumption function, the choice of estimation method for ARLS makes no difference, and the results for the Cochrane–Orcutt and Hildreth–Lu methods are identical. In both cases, the estimated AR coefficient is negative. However, the right-hand side variable is also negatively autocorrelated, so the bias in the coefficient standard errors will be positive. The coefficient estimates change slightly relative to the OLS estimates.

Historical Note: The possibility of misleading relationships or "spurious regressions" arising because of autocorrelation in time series has been known for many years. In an early paper, "Student" [Gosset1914] proposed the "variate difference method" as a means of transforming the data so that correlations between variables reflect genuine relationships.

5.5 SERIAL CORRELATION AS A SIMPLIFYING ASSUMPTION

In Section 5.4, we saw that a regression equation with an AR error could be written as a single equation with lags of both the right-hand side variable and the endogenous variable. For example, suppose we have $Y_t = \beta X_t + u_t$ and $u_t = \rho u_{t-1} + \varepsilon_t$. From the error process, we have $u_t = \rho\left(Y_{t-1} - \beta X_{t-1}\right) + \varepsilon_t$, and therefore, the equations can be combined to write

$$Y_t = \beta X_t - \beta\rho X_{t-1} + \rho Y_{t-1} + \varepsilon_t. \tag{5.33}$$

Hendry and Mizon [Hendry1978] have pointed out that this is a restricted version of a more general autoregressive-distributed lag model (ARDL) of the form

$$Y_t = \beta_1 X_t + \beta_2 X_{t-1} + \beta_3 Y_{t-1} + \varepsilon_t, \tag{5.34}$$

where there is a nonlinear parameter restriction of the form $\beta_2 = -\beta_1\beta_3$. Thus, (5.33) can be viewed as a simplified version of (5.34) because it has fewer estimated parameters. Like any restriction, however, this should be tested before it is imposed.

Dynamic models such as equation (5.34) can conveniently be written using lag polynomial notation. We define L such that $L^k X_t = X_{t-k}$ and write (5.34) as

$$Y_t\left(1 - \beta_3 L\right) = \beta_1\left(1 + \frac{\beta_2}{\beta_1} L\right) X_t + \varepsilon_t. \tag{5.35}$$

If the restriction $\beta_2 = -\beta_1\beta_3$ is valid, then there is a *common factor* in the lag polynomials and tests of this restriction are therefore often referred to as common factor tests. The problem we have in testing this restriction is that the restriction imposed is nonlinear in the parameters. This means that the tests for linear restrictions we discussed in Chapter 4 are not appropriate here. However, Sargan [Hart1964] has demonstrated that this restriction can be tested using the following test statistic:

$$T \ln\left(\frac{RSS_R}{RSS_U}\right) \overset{a}{\sim} \chi_1^2, \tag{5.36}$$

where T is the sample size and RSS_R and RSS_U are the restricted and unrestricted residual sums of squares, respectively. Sargan shows that this statistic is asymptotically distributed as chi-squared with degrees of freedom equal to the number of restrictions.

Example: Consider the consumption function model we estimated earlier. Estimation of a general model including lags of both disposable income and consumption yields the following results:

$$\begin{aligned}\ln(C_t) &= \underset{(0.0473)}{-0.0314} + \underset{(0.0991)}{0.8434}\ln(YD_t) - \underset{(0.1313)}{0.6120}\ln(YD_{t-1}) \\ &\quad + \underset{(0.1024)}{0.7691}\ln(C_{t-1}) + \hat{u}_t \qquad (5.37)\\ RSS_U &= 0.004996,\end{aligned}$$

where standard errors are given in parentheses below coefficients and the model is estimated over the period 1971–2019. Estimation of the restricted model using the Cochrane–Orcutt method have already been reported in equation (5.32), and the residual sum of squares for this equation is $RSS_R = 0.005299$. To test if the restriction is valid, we construct the test statistic $T\ln(RSS_R / RSS_U) = 49 \times \ln(0.005299/0.004996) = 2.885$. Under the null, this statistic is distributed as chi-squared with one degree of freedom and the 5% critical value is equal to 3.841. In this case, therefore, we cannot reject the null hypothesis at the 5% level, and we conclude that the model with a first-order AR error is an acceptable simplification of the more general ARDL model.

The common factor approach generalizes straightforwardly to more complex ARDL processes. Consider the general ARDL model of the form

$$A(L)Y_t = B(L)X_t + \varepsilon_t, \qquad (5.38)$$

where $A(L)$ is a polynomial of order p, $B(L)$ is a polynomial of order q, and $\varepsilon_t, t = 1,\ldots,T$ are independent random errors. If these polynomials are factorized, then we can write

$$\begin{aligned}(1+\theta_1 L)(1+\theta_2 L)\ldots(1+\theta_p L)Y_t &= (1+\omega_1 L)(1+\omega_2 L) \\ &\quad \ldots(1+\omega_q L)X_t + \varepsilon_t.\end{aligned} \qquad (5.39)$$

Now, suppose we can identify a subset of the factors that are common to both polynomials, for example, $\theta_1 = \omega_1$ and $\theta_2 = \omega_2$. We can then use this property to write a restricted model of the form

$$D(L)Y_t = E(L)X_t + u_t, \qquad (5.40)$$

where $D(L)$ and $E(L)$ are lower-order polynomials than $A(L)$ and $B(L)$ and the error process now takes the form,

$$\begin{aligned} u_t(1+\theta_1 L)(1+\theta_2 L) &= \varepsilon_t \\ u_t &= -(\theta_1+\theta_2)u_{t-1} - \theta_1\theta_2 u_{t-2} + \varepsilon_t. \end{aligned} \tag{5.41}$$

That is, the model becomes one in which there are two fewer lags of both X and Y in the estimating equation, but the error now follows a second-order AR process. As with our example, the common factor restrictions necessary to move from (5.38) to (5.40) can be tested using Sargan's common factor test.

EXERCISES

Excel files containing the data for these exercises are available as companion files for this book.

EXERCISE 5.1

Consider the following model in with Y depends on its own lagged value and the error follows an AR(1) process

$$Y_t = \beta Y_{t-1} + u_t$$
$$u_t = \rho u_{t-1} + \varepsilon_t$$

where $|\beta|<1$, $|\rho|<1$, and $\varepsilon_t, t=1,\ldots,T$ are independent random disturbances. Show that $E(u_t Y_{t-1}) = \rho\sigma_u^2/(1-\rho\beta)$ and hence show that OLS estimation of the equation for Y will give a biased estimate of the β parameter.

EXERCISE 5.2

Consider a model in which the errors follow a fourth-order autocorrelation process of the form

$$u_t = \rho u_{t-4} + \varepsilon_t,$$

where $\varepsilon_t, t=1,\ldots,T$ are independent, identically distributed random errors with mean zero and variance σ_ε^2. Note that models like this might arise when estimating using quarterly data.

a. Derive an expression for the variance of u in terms of σ_ε^2 and ρ.

b. Show that the fourth-order autocorrelation is equal to ρ.

c. Show that the first, second, and third autocorrelations are all equal to zero.

EXERCISE 5.3

An econometrician has estimated the following model that relates UK investment expenditure I to GDP Y. The data are annual, measured in £m at constant prices and the estimation period is 1948 to 2005.

$$I_t = \underset{(2301.7)}{-24046.7} + \underset{(0.00339)}{0.1810}\,Y_t + \hat{u}_t$$

$$R^2 = 0.9807 \qquad DW = 0.2844 \qquad Q_1 = 43.3401 \qquad RSS = 2.293349 \times 10^9.$$

Standard errors are given in parentheses below coefficients and Q_1 is the Box–Ljung test statistic for first-order autocorrelation.

a. The econometrician then claims that this is an excellent model because it has a high R^2 and the t statistic for the slope coefficient is very large. Explain to him (firmly but politely!) why he is wrong.

b. Using the information given, perform two tests for the presence of first-order autocorrelation in the residuals.

EXERCISE 5.4

Following your explanation of his result, the econometrician is sufficiently worried about his model to perform a further test for autocorrelation. This involves running the following regression in which $\hat{u}_t$ are the regression residuals from the first regression.

$$\hat{u}_t = \underset{(1237.53)}{-398.6258} + \underset{(0.0711)}{0.8682}\,\hat{u}_{t-1} + \underset{(0.0018)}{0.0009}\,Y_t + \hat{\varepsilon}_t$$

$$R^2 = 0.7340 \qquad DW = 1.44823 \qquad RSS = 6.0760 \times 10^8.$$

a. Using these results, calculate the Breusch–Godfrey test statistic for the presence of first-order autocorrelation.

b. Set out the null and alternative hypotheses for the Breusch–Godfrey test, state the distribution for the test statistic under the null, and perform the test by comparing the test statistic with the 5% critical value under the null.

EXERCISE 5.5

Using the data in the *INVESTMENT.XLSX* spreadsheet, estimate a model of the following form

$$\Delta I_t = \beta_1 + \beta_2 \Delta Y_t + u_t.$$

a. Examine the residuals from this model. Is there any visual evidence of serial correlation?

b. Perform the *DW*, Box–Ljung, and Breusch–Godfrey tests for first-order autocorrelation. Do these tests produce the same result?

c. What does your analysis of the residuals of this model suggest about the properties of the OLS estimator? How does your answer differ from the conclusions you reached regarding the model estimated in question 3?

REFERENCES

[Breusch1978] Breusch, T. S., "Testing for Autocorrelation in Dynamic Linear Models," *Australian Economic Papers*, 1978, 17, pp. 334–355.

[Cochrane1949] Cochrane, D. and Orcutt, G. H., "Application of Least Squares Regression to Relationships Containing Auto-Correlated Error Terms," *Journal of the American Statistical Association*, 1949, 44, pp. 32–61.

[Durbin1950] Durbin, J. and Watson, G. S., "Testing for Serial Correlation in Least Squares Regression I," *Biometrika*, 1950, 37, pp. 409–428.

[Durbin1951] Durbin, J. and Watson, G. S., "Testing for Serial Correlation in Least Squares Regression II," *Biometrika*, 1951, 38, pp. 159–178.

[Durbin1970] Durbin, J., "Testing for Serial Correlation in Least-Squares Regression When Some of the Regressors are Lagged Dependent Variables," *Econometrica*, 1970, 38, pp. 410–421.

[Godfrey1978] Godfrey, L. G., "Testing Against General Autoregressive and Moving Average Error Models When the Regressors Include Lagged Dependent Variables," *Econometrica*, 1978, 46, pp. 1293–1301.

[Gosset1914] Gosset, W. S. ("Student") "The Elimination of Spurious Correlation Due to Positions in Time and Space," *Biometrika*, 1914, 10, pp. 179–180.

[Hart1964] Sargan, J. D., "Wages and Prices in the United Kingdom: A Study in Econometric Methodology (with Discussion)," in Hart, P. E., Mills, G., and Whitaker, J.K. (eds), *Econometric Analysis for National Economic Planning*, Vol. 16 of *Colston Papers*, London: Butterworth Co, pp. 25–63.

[Hendry1978] Hendry, D. and Mizon, G., "Serial Correlation as a Convenient Simplification, Not a Nuisance: A Comment on a Study of the Demand for Money by the Bank of England." *Economic Journal*, 1978, 88, pp. 549–563.

[Hildreth1960] Hildreth, C. and Lu, J., "Demand Relations with Autocorrelated Disturbances." Michigan State University AES Technical Bulletin 276.

[Kiviet1986] Kiviet, J., "On the Rigour of Some Misspecification Tests for Modelling Dynamic Relationships," *Review of Economic Studies*, 1986, 53, pp. 241–261.

[McGuirk2009] McGuirk, A. and Spanos, A., "Revisiting Error-Autocorrelation Correction: Common Factor Restrictions and Granger Non-Causality." *Oxford Bulletin of Economics and Statistics*, 2009, 71(2), pp. 273–294.

[VonNeumann1941] Von Neumann, J., "Distribution of the Ratio of the Mean Square Successive Difference to the Variance." *Annals of Mathematical Statistics*, 1941, 12(4), pp. 367–395.

[Wold1938] Wold, H., *A Study in the Analysis of Stationary Time Series*. 1938, Stockholm: Almqvist and Wiksell.

[Yule1921] Yule, G. U., "On the Time-Correlation Problem, With Especial Reference to the Variate-Difference Correlation Method." *Journal of the Royal Statistical Society*, 1921, 84, pp. 497–526.

[Yule1926] Yule, G. U., "Why Do We Sometimes Get Nonsense Correlations Between Time Series? A Study in Sampling and the Nature of Time Series." *Journal of the Royal Statistical Society*, 1926, 89, pp. 1–63.

CHAPTER 6

HETEROSCEDASTICITY, FUNCTIONAL FORM, AND STRUCTURAL BREAKS

Heteroscedasticity describes any situation in which the variance of the errors is not constant. In this chapter, we will discuss how heteroscedasticity arises in econometric models, how it can be detected, and how we can deal with it. We will also discuss the close relationship between heteroscedasticity and functional form. For example, we will show that if the functional form chosen for a model is incorrect, then this may result in heteroscedastic errors. Similarly, if heteroscedasticity is detected, then one of the ways in which we can deal with the problem is by modifying the functional form. It therefore makes sense to consider these topics together.

The presence of heteroscedasticity in the errors of a regression model is a failure of the third assumption of the Classical Linear Regression Model (CLRM). It does not, in itself, mean that least-squares estimates will be biased but it does mean that they will be inefficient and that the estimates of the standard errors of the OLS estimates will typically be biased. Although heteroscedasticity can be found in models estimated using time-series data, it is more commonly a problem that is associated with cross-section analysis. In many ways, the presence of heteroscedasticity in cross-section models creates problems similar to those found when serial correlation is present in time-series models.

In the following sections, we will investigate the causes and implications of heteroscedasticity in detail. We will begin by investigating situations in which the phenomenon can arise followed by a discussion of its consequences. This is followed by discussions of methods through which the presence of heteroscedasticity can be detected and how it can be dealt with. Although much of the discussion will be in the context of cross-section data, we will also discuss the *autoregressive conditional heteroscedasticity* (ARCH) model which is applicable to time-series data and which has featured prominently in the financial econometrics literature in recent years. Our discussion of heteroscedasticity is followed by a discussion of functional form specification. We consider how we can test for the correctness of the functional form chosen and show how an incorrect choice is linked to heteroscedasticity in the errors. This leads naturally to a detailed discussion of the choice between linear and log-linear functional forms.

> **Historical Note:** The terms *heteroscedasticity*, for situations in which the variance is not constant, and *homoscedasticity*, for situations in which in which it is constant, were first used by Karl Pearson [Pearson1905].

6.1 CAUSES OF HETEROSCEDASTICITY

Consider a regression model of the standard form $Y_i = \beta X_i + u_i$. Heteroscedasticity is said to occur in any situation in which $E(u_i^2)$ varies with the value of i. This is a very broad definition, and it is therefore useful to consider some more specific cases that may arise in practice. One case that frequently arises is where the variance of the error term is a function of the exogenous variable $E(u_i^2) = f(X_i)$. This naturally arises in many cross-section regression models. For example, suppose we wish to model of consumption expenditure for a cross section of households. Although we would reasonably expect the level of consumption to vary with the level of income, we might also expect the variability of consumption to depend on income levels. For example, we may observe a situation in which the variance of the error term is proportional to the square of the exogenous variable. This generates a model of the form

$$\begin{aligned} Y_i &= \beta X_i + u_i \\ E(u_i^2) &= \sigma_u^2 X_i^2. \end{aligned} \tag{6.1}$$

This model is a particularly easy case to deal with because it suggests a natural rescaling of the data which would eliminate heteroscedasticity. If divide both sides of (6.1) by X_i, then the model can be rewritten as $Y_i / X_i = \beta + v_i$, where $v_i = u_i / X_i$ and the heteroscedasticity is effectively eliminated. To see this, take expectations of the squared error term in the new form of the model to obtain $E(v_i^2) = E(u_i^2) / X_i^2 = \sigma_u^2$. This shows that the errors of the revised model have constant variance (i.e., are homoscedastic) and therefore that heteroscedasticity is no longer a problem.

> **Historical Note:** The spelling of the word *heteroscedasticity* is quite controversial. Some authors have argued that, because the word derives from the Greek word *skedatos* (capable of being scattered), it should be spelled *heteroskedasticity*. I have retained the established spelling with a "c" in this book, but you will frequently see the alternative in other texts.

Unfortunately, there is no guarantee that the heteroscedasticity we encounter will be of the convenient form found in equation (6.1). More generally, we might find a relationship between the error variance and the exogenous variable of the form $E\left(u_i^2\right)=\sigma_u^2 X_i^h$. If such a relationship exists, then we can still rescale the data by dividing through by $X_i^{h/2}$. This yields a relationship of the form

$$\frac{Y_i}{X_i^{h/2}}=\frac{\beta}{X_i^{h/2}}+v_i. \tag{6.2}$$

$v_i=u_i X_i^{-h/2}$ now has the desired properties for OLS estimation since $E\left(v_i^2\right)=\sigma_u^2$, but this now leads to a potential complication in that the parameter h is likely to be itself unknown and must therefore be estimated.

6.2 CONSEQUENCES OF HETEROSCEDASTICITY

Heteroscedasticity does not, in itself, imply that OLS estimation will lead to biased coefficient estimates. To demonstrate this, we note that the OLS estimator can be written as

$$\hat{\beta}=\frac{\sum_{i=1}^{N} X_i Y}{\sum_{i=1}^{N} X_i^2}=\frac{\sum_{i=1}^{N} X_i\left(\beta X_i+u_i\right)}{\sum_{i=1}^{N} X_i^2}=\beta+\frac{\sum_{i=1}^{N} X_i u_i}{\sum_{i=1}^{N} X_i^2}. \tag{6.3}$$

The fourth Gauss–Markov assumption allows us to write the expected value of this expression as

$$E\left(\hat{\beta}\right)=\beta+\frac{\sum_{i=1}^{N} X_i E\left(u_i\right)}{\sum_{i=1}^{N} X_i^2}, \tag{6.4}$$

and this, in turn, yields $E(\hat{\beta}) = \beta$, provided that the first Gauss–Markov assumption also holds. Therefore, providing that we can maintain Assumptions 1 and 4 of the CLRM, we can demonstrate unbiasedness. Since we have not had to invoke the assumption of homoscedastic errors, this assumption is not necessary for the proof. It therefore follows that heteroscedasticity is not *in itself* a reason to believe that OLS will produce biased estimates.

The derivation of the conditions under which heteroscedasticity does not lead to biased estimates parallels that in the previous chapter, where we showed that serial correlation does not, in itself, lead to bias in the coefficient estimates. However, a similar caveat applies in this case. As with serial correlation, heteroscedasticity may be the result of some other form of misspecification. For example, it may be the result of omitting a relevant variable from the model. If this is the case, then OLS estimates will be biased. This bias, however, is the result of omitting the variable *not* the heteroscedasticity which arises as a symptom of the underlying misspecification.

Although heteroscedasticity does not imply that OLS will be biased, it does imply that OLS will be inefficient. Consider the regression model expressed in matrix form

$$\begin{aligned} \boldsymbol{y} &= \mathbf{X}\boldsymbol{\beta} + \boldsymbol{u} \\ E(\boldsymbol{uu}') &= \sigma_u^2 \mathbf{V}. \end{aligned} \tag{6.5}$$

where $\mathbf{V}$ is a positive-definite matrix. This model is very general since it can allow for both heteroscedasticity and serial correlation in the errors. As a result, this framework is referred to as the *Generalized Least Squares (GLS) Model*. By virtue of the assumption that $\mathbf{V}$ is positive definite, it can be expressed as $\mathbf{V} = \boldsymbol{LL}'$, where $\boldsymbol{L}$ is a nonsingular matrix. If we premultiply (6.5) by the matrix $\boldsymbol{L}^{-1}$, then we obtain a model of the form

$$\boldsymbol{L}^{-1}\boldsymbol{y} = \boldsymbol{L}^{-1}\mathbf{X}\boldsymbol{\beta} + \boldsymbol{L}^{-1}\boldsymbol{u}. \tag{6.6}$$

Now considering the properties of the error term from this transformed model, we have

$$E\left(\boldsymbol{L}^{-1}\boldsymbol{uu}'\left(\boldsymbol{L}^{-1}\right)'\right) = \sigma_u^2 \boldsymbol{L}^{-1}\mathbf{V}\left(\boldsymbol{L}^{-1}\right)' = \sigma_u^2 \boldsymbol{I}. \tag{6.7}$$

What this shows is that an appropriate transformation of the data can produce a model which satisfies the CLRM assumptions of serial independence and homoscedasticity in the errors. It follows that application of OLS to the transformed data will produce the best linear unbiased estimates of the parameters of interest. In particular, OLS estimates of the parameters of interest based on the transformed model will have lower variance than OLS estimates based on the original data. Therefore, OLS estimates based on model (6.5) will be inefficient.

The GLS estimator appears an attractive methodology but it does suffer from the problem that the matrix $\boldsymbol{V}$ is typically unknown. However, it is often enough to know that such a matrix exists in principle to develop the theoretical properties of the estimator. There are also empirical methods by which the parameters of the $\boldsymbol{V}$ matrix can be estimated along with the $\boldsymbol{\beta}$ parameters.

In addition to the problem of inefficiency, we can also show that the use of OLS estimation in conjunction with heteroscedastic errors leads to problems for statistical inference, in that the estimates of the standard errors of the regression parameters are biased. If the variance of the error term is positively correlated that of the exogenous variable(s), then we can also show that these standard errors are biased downward, leading to incorrect size of tests based on t ratios. To show this, let us consider the regression model of the form $Y_i = \beta X_i + u_i$. The sampling variance of the OLS estimator is

$$E\left(\frac{\sum_{i=1}^{N} X_i u_i}{\sum_{i=1}^{N} X_i^2}\right)^2 = \frac{V(Xu)}{N\sigma_X^4}. \tag{6.8}$$

The numerator of this expression is the variance of the product of two zero-mean random variables X and u. By a standard result, we have

$$V(Xu) = \text{cov}\left(X^2, u^2\right) + V(X)V(u) - \left\{\text{cov}(X, u)\right\}^2. \tag{6.9}$$

We assume that $\text{cov}(X,u) = 0$, the X variable is uncorrelated with the equation error, but $\text{cov}\left(X^2, u^2\right) \neq 0$, the errors are heteroscedastic. Therefore, we can write the sampling variance of the OLS estimator as

$$\frac{\text{cov}\left(X^2, u^2\right) + \sigma_X^2 \sigma_u^2}{N\sigma_X^4} = \frac{\text{cov}\left(X^2, u^2\right)}{N\sigma_X^4} + \frac{\sigma_u^2}{N\sigma_X^2}. \tag{6.10}$$

The second part of the expression on the right-hand side is just the standard formula for the OLS variance. It therefore follows immediately that the OLS formula underestimates the true sample variance if $\text{cov}\left(X^2, u^2\right) > 0$, that is, if there is a positive correlation between the variance of the error and the variance of the X variable.

Example: The following example illustrates the problem of bias in the standard error of coefficient estimates when the equation errors are heteroscedastic. In this example, we have conducted 1,000 simulations of model (6.1) with β set equal to 1. The model estimated takes the form $Y_i = \alpha + \beta X_i + u_i$. The distribution of the slope coefficient estimates obtained is illustrated by the histogram shown in Figure 6.1.

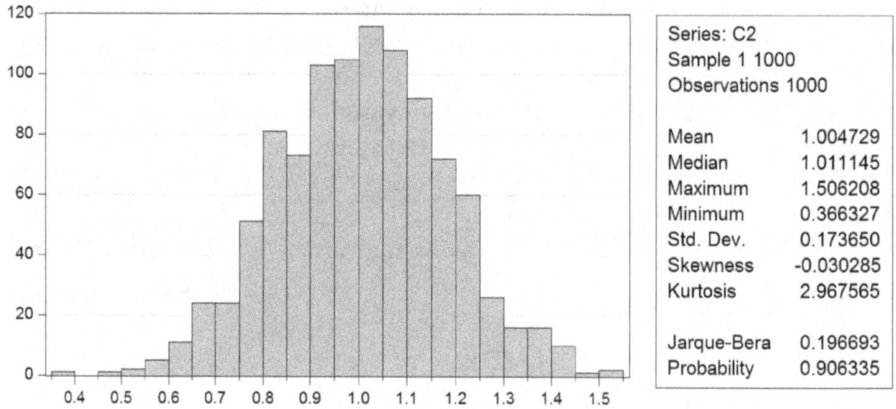

FIGURE 6.1 Distribution of Coefficient Estimates of Model with Heteroscedastic Errors.

From the information given in Figure 6.1, we can see that the OLS estimator is not biased. The average coefficient estimate is extremely close to the true value of 1. However, out of the 1,000 regressions we estimate, the null hypothesis is rejected in 265 cases in favor of the two-sided alternative $H_1 : \beta \neq 1$ when 5% critical values for the t test were used. If the test had the correct size, then we should observe rejection in about 50 cases for this simulation. The reason for this is that the standard error of the parameter estimate is typically underestimated. From Figure 6.1, we see that an unbiased estimate of this standard error is 0.17. However, the average standard error of the individual regression estimates was 0.098.

Now consider a transformed model of the form $Y_i / X_i = \alpha / X_i + \beta + u_i / X_i$ in which the parameter of interest is the intercept β. We again ran 1,000 simulations of this model and obtained the histogram of OLS parameter estimates shown in Figure 6.2. This shows that the OLS estimator is again unbiased with an average value which is very close to the true value of 1. The difference here is that estimates of the standard error of the parameter are now much more reliable. The average standard error from 1,000 regressions is 0.100 which is close to the standard error of the estimated coefficients of 0.101 shown in Figure 6.2. Moreover, we reject the null hypothesis $H_0 : \beta = 1$ in favor of the alternative $H_1 : \beta \neq 1$ only 59 times in this case. This is much closer to the 50 rejections predicted by the size of the test we have chosen.

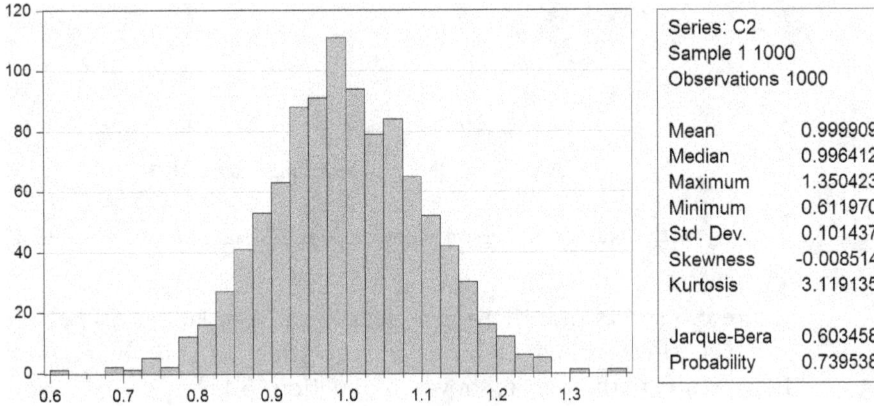

FIGURE 6.2 Coefficient Estimates Based on Transformed Model.

6.3 DETECTION OF HETEROSCEDASTICITY

The easiest case to deal with when testing for heteroscedasticity is the case in which the variance of the error term is known to be related to a one of the right-hand side variables of the model. In simple cases, this can be identified by plotting the data. For example, Figure 6.3 shows a scatter of points relating consumption per head and GDP per head for a sample of 182 countries. The data are taken from the Penn World Tables[1] and are measured in US dollars.

[1] *https://cid.econ.ucdavis.edu/pwt.html*

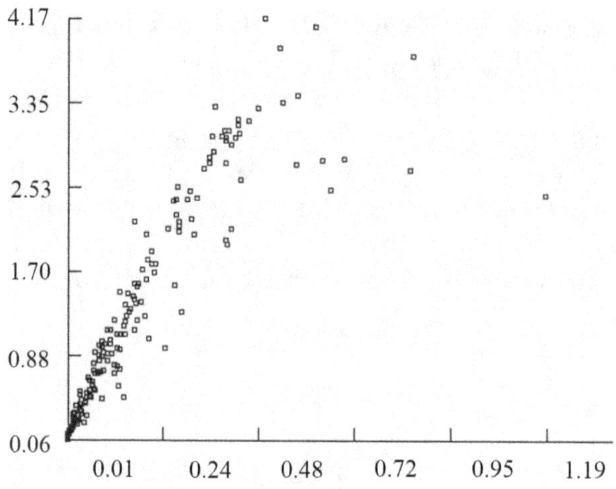

FIGURE 6.3 Scatter Diagram of Consumption per Head against GDP per Head.

Figure 6.3 shows that there is a clear, positive relationship between these variables. If we estimate a bivariate regression equation based on this data, then we obtain the results shown in equation (6.11)

$$CONS_i = \underset{(512.59)}{3805.88} + \underset{(0.0198)}{0.4842} GDP_i + \hat{u}_i \tag{6.11}$$

$$R^2 = 0.7692 \qquad RSS = 4.6308 \times 10^9.$$

It is also obvious from Figure 6.3 that the regression line we fit through this scatter of points will exhibit heteroscedasticity. This is evident because the scatter of points clearly becomes much wider as the size of the exogenous variable increases. This affects how we interpret our regression results because we know that the standard errors of the coefficient estimates will be biased under these circumstances. It therefore becomes important for us to test if heteroscedasticity is present in our regression models and deal with it appropriately if it is detected.

An early suggestion for a formal heteroscedasticity test is provided by Goldfeld and Quandt [Golfeld1965]. Their testing procedure works as follows:

1. Order the data according to the size of the exogenous variable.
2. Divide the sample into three sections of size $n, N-2n, n$, respectively, where n should be approximately equal to $(3/8)N$.

3. Estimate separate regressions for the first and last n observations and generate the residual sum of squares. Then use the following F statistic: $F = RSS_2 / RSS_1$ as the test statistic. Under the null hypothesis that the errors are homoscedastic, this statistic will be distributed as $F_{n-k,n-k}$.

Note that we have assumed that the variance increases with the size of the exogenous variable. It is also possible that the variance might decrease as x increases, in which case we would find $RSS_2 < RSS_1$. If this is the case, then the appropriate test statistic would be $F = RSS_1 / RSS_2$.

Example: If we apply the Goldfeld–Quandt test to our model, then we set the size of the subsamples at $n = 182 \times (3/8) \approx 68$. We then order the data according to the size of GDP per capita and estimate separate regressions for the lowest and highest values. This gives us residual sums of squares equal to $RSS_1 = 2.83167 \times 10^7$ and $RSS_2 = 2.37795 \times 10^9$. The test statistic is therefore equal to $F = 82.92$ and the 5% critical value is equal to $F^{5\%}_{66,66} = 1.504$. We therefore reject the null hypothesis that the variances in the subsamples are equal. The p value for this test statistic is effectively zero.

The Goldfeld–Quandt test is simple to apply in bivariate regression models because there is only one right-hand side variable and therefore, only one way in which we can order the data for the purposes of the test. In small samples, dividing the observations up in this way, and excluding a quarter of them, may produce a test with low degrees of freedom. A variety of tests have therefore been developed which avoid these problems. These all have the common feature that they are based on auxiliary regressions using the least squares regression residuals. These tests are listed in Table 6.1 with the form of the auxiliary regression. In each case, the test is based on the null of homoscedasticity which holds if the slope coefficients in the auxiliary regressions are equal to zero.

TABLE 6.1 Residual Based Tests for Heteroscedasticity.

Test	Form of Auxiliary Regression
Breusch–Pagan [Breusch1979]	$\hat{u}_i^2 = \gamma_0 + \gamma_1 X_i + \varepsilon_i$
Harvey [Harvey1976]	$\ln(\hat{u}_i^2) = \gamma_0 + \gamma_1 X_i + \varepsilon_i$
Glejser [Glejser1969]	$\lvert\hat{u}_i\rvert = \gamma_0 + \gamma_1 X_i + \varepsilon_i$
White [White1980]	$\hat{u}_i^2 = \gamma_0 + \gamma_1 X_i + \gamma_2 X_i^2 + \varepsilon_i$

The tests shown in Table 6.1 are all asymptotic in the sense that we can only derive large sample distributions for the test statistics. The test statistic in each case is NR^2, where N is the sample size and R^2 is the coefficient of determination for the auxiliary regression. The degrees of freedom is given by the number of slope coefficients in the auxiliary regression. Although it is not possible to derive small sample tests, it is also common practice to report F tests for the auxiliary regression. This is similar to the practice we noted for the Breusch–Godfrey test for serial correlation, which was justified by the Monte Carlo analysis of Kiviet [Kiviet1986] who showed that the small sample test had better properties than the asymptotic test. In practice, however, these tests usually produce the same results. To illustrate this testing procedure, Table 6.2 reports both chi-squared and F tests for the four procedures set out in Table 6.1. In each case, the distribution of the statistic under the null hypothesis is given below the statistic. Not surprisingly, given the results of the Goldfeld–Quandt test earlier, all the tests in this table give very strong evidence of the presence of heteroscedasticity in our estimating equation.

TABLE 6.2 Residual-Based Tests for Heteroscedasticity Based on Equation (6.11).

Test	Chi-Square Test Statistic	*F* Test Statistic
Breusch and Pagan (1979)	57.14 Distribution under $H_0 \sim \chi_1^2$ 5% critical value = 3.841	82.37 Distribution under $H_0 \sim F_{1,180}$ 5% critical value = 3.894
Harvey (1976)	42.99 Distribution under $H_0 \sim \chi_1^2$ 5% critical value = 3.841	55.67 Distribution under $H_0 \sim F_{1,180}$ 5% critical value = 3.894
Glejser (1969)	96.08 Distribution under $H_0 \sim \chi_1^2$ 5% critical value = 3.841	201.29 Distribution under $H_0 \sim F_{1,180}$ 5% critical value = 3.894
White (1980)	138.23 Distribution under $H_0 \sim \chi_2^2$ 5% critical value = 5.991	282.66 Distribution under $H_0 \sim F_{2,180}$ 5% critical value = 3.046

6.4 DEALING WITH HETEROSCEDASTICITY

We can deal with heteroscedasticity in several ways. We have already seen that it does not, in itself, imply that OLS will produce biased coefficient estimates, but that the estimates of the standard errors of the coefficients will typically be biased. Therefore, one method of allowing for the effects of heteroscedasticity is to adjust the coefficient standard errors to allow for this. Consider the general model (6.5), it is straightforward to show that $\hat{\boldsymbol{\beta}} - \boldsymbol{\beta} = (\mathbf{X'X})^{-1}\mathbf{X'}\boldsymbol{u}$ and therefore,

$$E(\hat{\boldsymbol{\beta}} - \boldsymbol{\beta})(\hat{\boldsymbol{\beta}} - \boldsymbol{\beta})' = (\mathbf{X'X})^{-1}\mathbf{X'}E(\boldsymbol{uu'})\mathbf{X}(\mathbf{X'X})^{-1}$$
$$= \sigma_u^2(\mathbf{X'X})^{-1}\mathbf{X'VX}(\mathbf{X'X})^{-1}. \tag{6.12}$$

If $\mathbf{V}$ is known, then we could calculate the variance–covariance matrix of the parameter estimates using this formula rather than the OLS formula. This is not typically the case. However, White (1980) shows that

$$\sum_{i=1}^{N} \hat{u}_i^2 \boldsymbol{x}_i \boldsymbol{x}_i' \tag{6.13}$$

is a consistent estimator of $\sigma_u^2\mathbf{X'VX}$, where $\boldsymbol{x}_i$ is a $k \times 1$ vector of observations for the right-hand side variables corresponding to observation i. This is the formula used to calculate the White standard errors reported by many regression packages. Adjusting the standard errors to allow for the presence of heteroscedasticity means that statistical inference will be more reliable than if we were to rely on the OLS standard errors. However, it does not deal with the problem of the inefficiency of the OLS estimator. This problem requires a reformulation of the model itself.

Example: Consider estimates of the model (6.11) but with White standard errors replacing the OLS standard errors. This yields the following results:

$$CONS_i = \underset{(924.79)}{3805.88} + \underset{(0.0635)}{0.4842}GDP_i + \hat{u}_i$$
$$R^2 = 0.7692 \qquad RSS = 4.6308 \times 10^9. \tag{6.14}$$

Note that the coefficient estimates and all the statistics other than the standard errors in (6.14) are unchanged relative to (6.11). The White standard errors for both the coefficients are higher than the OLS standard errors. This means that, in this case, confidence intervals based on the White standard errors will be wider and hypothesis tests will be less likely to reject any given null hypothesis. For example, using the OLS standard errors, the 95% confidence interval for the slope coefficient is (0.4454, 0.5230) but, using the White standard errors, the interval is (0.3597, 0.6087).

6.5 TESTING THE FUNCTIONAL FORM

So far, we have maintained the assumption of a linear functional form for our regression equation. However, there is no guarantee that linearity is appropriate in all cases and, like any other assumption, this should be tested and, if necessary, the equation should be adjusted to allow for nonlinear effects. As we will show, there is also a very close link between the presence of nonlinear effects in the model and heteroscedasticity when we estimate a linear specification.

One way to think about the assumption of linearity is in terms of a Taylor series approximation to a more general functional form. Consider a general model of the form $Y_i = f(X_i) + \varepsilon_i$, where $f(X_i)$ is a general nonlinear function and ε_i, $i = 1,\ldots,N$ are independent random errors. If we take a first-order Taylor series approximation to this relationship, then we can write

$$Y_i = \overline{Y} + f_x\left(X_i - \overline{X}\right) + u_i. \tag{6.15}$$

The errors u_i, $i = 1,\ldots,N$ include both the original random errors and the higher-order terms from the Taylor series expansion. For example, suppose the true relationship is quadratic, so that the true relationship is

$$Y_i = \beta_1 + \beta_2\left(X_i - \overline{X}\right) + \beta_3\left(X_i - \overline{X}\right)^2 + \varepsilon_i, \tag{6.16}$$

then the errors in the linear specification would take the form $u_i = \varepsilon_i + \beta_3\left(X_i - \overline{X}\right)^2$. This demonstrates the relationship with heteroscedasticity since the incorrect adoption of a linear functional form has generated an error process which is heteroscedastic.

This approach suggests a method for testing for functional form. We can think of the linear specification as a first-order Taylor series representation of a general functional form. Adding higher-order powers to the equation allows us to test if the first-order approximation is adequate by testing the significance of quadratic and possibly higher-order terms. This is reasonably straightforward for the case described here because it simply involves adding one extra regressor. However, it becomes progressively more difficult as we add extra variables to the model. Suppose we have two explanatory variables, so that the model takes the form $Y_i = f(X_{2i}, X_{3i}) + \varepsilon_i$. A second-order Taylor series for this function gives an equation of the form

$$Y_i = \beta_1 + \beta_2 X_{2i} + \beta_3 X_{3i} + \beta_4 X_{2i}^2 + \beta_5 X_{3i}^2 + \beta_6 X_{2i} X_{3i} + \varepsilon_i. \tag{6.17}$$

Testing for the null of linearity now involves testing three restrictions, that is, $H_0 : \beta_4 = \beta_5 = \beta_6 = 0$. If k is the number of regressors, then the number of extra parameters necessary to test for linearity increases rapidly as we add extra regressors to the model. For $k = 2$, the number of extra parameters equals three, for $k = 3$, it is equal to six, for $k = 4$, it is equal to ten and so on. Every time we expand the model by adding an extra variable, we lose degrees of freedom and we add extra variables to the model which are likely to be highly collinear with the existing variables. The problem becomes even worse if we consider higher order expansions such as cubic equations.

Ramsey [Ramsey1969] suggests a neat way of avoiding the problems of loss of degrees of freedom and collinearity when testing for linearity. This test has been named as Regression Equation Specification Error Test (*RESET*). Consider estimates of the linear specification of the two-variable model. We have

$$\hat{Y}_i = \hat{\beta}_2 X_{2i} + \hat{\beta}_3 X_{3i} \qquad \Rightarrow \qquad \hat{Y}_i^2 = \hat{\beta}_2^2 X_{2i}^2 + \hat{\beta}_3^2 X_{3i}^2 + 2\hat{\beta}_2 \hat{\beta}_3 X_{2i} X_{3i}, \tag{6.18}$$

that is, the square of the fitted values is a weighted average of the squared values of the explanatory variable and their cross product. Thus, we can test for the significance of nonlinear elements in our regression model by adding the squared fitted values from the linear specification to our regression equation and testing the restriction that the coefficient on this term is equal to zero. This has obvious economies in terms of limiting the loss of degrees of freedom from the inclusion of extra right-hand side variables as well as reducing the loss of efficiency in the estimates due to multicollinearity. The method also generalizes easily to test for cubic and higher-order nonlinear terms in the equation.

Example: Suppose we wish to test for nonlinearity in our cross-country consumption model. The linear specification is given in equation (6.11). Adding the squared fitted values from this equation to the model and reestimating yields

$$CONS_i = \underset{(347.98)}{262.338} + \underset{(0.0308)}{1.0428} GDP_i - \underset{(1.399\times 10^{-6})}{2.7 \times 10^{-5}} (CONS_i)^2 + \hat{u}_i \tag{6.19}$$

$$R^2 = 0.9258 \qquad RSS = 1.4876 \times 10^9.$$

The significance of the additional variable can be tested using the following F statistic

$$F = \frac{(RRSS - URSS)}{URSS}\frac{N-k}{1} = \frac{4.6308 - 1.4876}{1.4876} \times \frac{179}{1} = 378.2.$$

Under the null hypothesis, this statistic is distributed as F with (1, 179) degrees of freedom. The 5% critical value for this test is 3.894, and therefore, we reject the null hypothesis at the 5% level. Our conclusion is therefore that there is significant evidence that a linear specification is not appropriate for this model.

6.6 CHANGING THE FUNCTIONAL FORM

In the previous sections of this chapter, we noted the close relationship between the choice of functional form and the presence of heteroscedasticity. It should come as no surprise that adjustment of the functional form can potentially address both these issues simultaneously. For example, we have noted that our cross-country consumption equation fails tests for heteroscedasticity and the RESET test for functional form. The question therefore arises whether a change in the functional form can deal with either, or both, of these problems. One possibility is to scale the data by dividing through by GDP. This generates a regression equation of the form

$$\left(\frac{CONS}{GDP}\right)_i = \underset{(45.80)}{300.66}\left(\frac{1}{GDP}\right)_i + \underset{(0.0165)}{0.7611} + \hat{u}_i \tag{6.20}$$

$$R^2 = 0.1932 \qquad WHITE = 1.44 \qquad RESET = 18.32.$$

The slope coefficient in equation (6.11), now becomes the intercept in this regression. The question is whether this specification has dealt with the problems we identified in our previous estimates. The White test is distributed as chi-squared with two degrees of freedom under the null and therefore, comparing the test statistic with the 5% critical value of 5.991, we conclude that there is no longer evidence of heteroscedasticity. This means that we do not need to adjust the standard error of the coefficient estimate to account for heteroscedasticity, which means that we can generate a tighter 95% confidence interval than the original regression permitted. In this case, our 95% confidence interval becomes $(0.7288, 0.7934)$. The RESET test is distributed as F with (1, 179) degrees of freedom, under the null. Given the 5% critical value of 3.841, we therefore conclude that there remains significant evidence of nonlinearity. This equation is therefore a definite improvement on (6.11) in that it no longer fails the heteroscedasticity test but there remains a problem of nonlinearity.

The *log-linear* functional form provides another option when looking to deal with problems of heteroscedasticity and functional form misspecification. The general form for a bivariate regression is written $Y_i = BX_i^{\beta_2} \exp(u_i)$. Taking natural logarithms of this expression yields $\ln(Y_i) = \beta_1 + \beta_2 \ln(X_i) + u_i$, where $\beta_1 = \ln(B)$. This specification has the advantage that the slope coefficient gives us a direct estimate of the elasticity of *Y* with respect to *X*, which is constant when this functional form is used. This transformation can only be used when the data are always positive, but this is very common with economics data and, in some circumstances, is an advantage. For example, when modeling consumption expenditures, we would naturally wish to avoid models that could potentially generate negative predictions. In the context of the issues discussed in this chapter, we often find that models, which exhibit heteroscedasticity or fail the RESET test when estimated in linear form, are significantly improved by a transformation to a log-linear specification.

Before we present estimates of the log-linear version of our consumption model, it is interesting to examine a scatter plot of the log of consumption per year against log GDP per head. This is given in Figure 6.4. If we contrast this with Figure 6.3, which shows the levels data, we see immediately that there is no longer an obvious increase in the spread of the data as the size of the variable on the horizontal axis increases. The relationship also looks to be better approximated by a linear function.

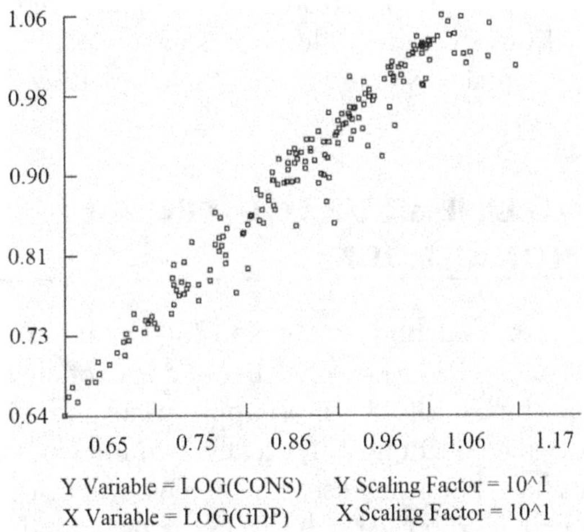

FIGURE 6.4 Scatter Diagram of Log Consumption per Head against Log GDP per Head.

Least-squares estimates of the log-linear model are given in equation (6.21). Both the White heteroscedasticity and the RESET tests indicate that the transformation has not completely removed these problems. The White test is asymptotically distributed as chi-squared with two degrees of freedom and the RESET test is distributed as F with 1 and 179 degrees of freedom. The critical values are therefore 5.991 and 3.894 and, therefore, we still reject the null in both cases. However, the absolute values of both test statistics are very much lower than was the case for the linear regression. This at least indicates that the test statistics are not as far out in the tails of the distributions than was the case for the linear model, which suggests that the performance of the model has at least improved relative to the earlier version.

$$\begin{aligned} &\ln(CONS_i) = \underset{(0.1317)}{1.0164} + \underset{(0.0142)}{0.8649}\ln(GDP_i) + \hat{u}_i \\ &R^2 = 0.9539 \qquad RSS = 9.6173 \\ &WHITE = 6.6715 \qquad RESET = 17.56. \end{aligned} \tag{6.21}$$

Finally, let us consider how we might test a hypothesis of interest using this model. Suppose we wish to test the null hypothesis that the elasticity of consumption with respect to GDP is equal to one. This can be tested using a t test of the form $\tau = (\hat{\beta}_2 - 1)/\mathrm{SE}(\hat{\beta}_2)$. We have $\hat{\beta}_2$ from equation (6.21) but the standard error reported will be biased because of the presence of heteroscedasticity. Using the White standard errors, we estimate $\mathrm{SE}(\hat{\beta}_2) = 0.0155$, which is slightly higher than the OLS standard error. The test statistic is therefore $\tau = -0.1351/0.0155 = -8.72$. This is higher than the 5% critical value of -1.645 for a one-tailed test, so we reject the null hypothesis that the elasticity is equal to one in favor of the alternative that it is <1.

6.7 TESTING LINEAR VS LOG-LINEAR FUNCTIONAL FORMS

We have seen that both linear and log-linear functional forms are widely used in econometric research. Theory does not always give us definite guidance as to which of these is most appropriate and it is therefore useful to test alternative specifications empirically. A number of alternative methods for doing this have been suggested, which we will discuss briefly in this section.

The first method we will discuss is described in a paper by Bera and McAleer [Bera1989]. They propose treating each specification in turn as the maintained hypothesis with the other as the alternative. We then test if information from the alternative model is sufficient to reject the null that the maintained hypothesis is true. This becomes clearer if we present the mechanics of the construction of the test. Suppose we wish to test the null of a log-linear specification against the alternative of a linear model. We can write the two functional forms as follows:

$$\ln(Y_i) = \beta_1 + \beta_2 \ln(X_i) + u_{0i}, \tag{6.22}$$

$$Y_i = \gamma_1 + \gamma_2 X_i + u_{1i}. \tag{6.23}$$

Let $\ln(Y_i)$ and $\tilde{Y}_i$ be the fitted values from these regressions, using these fitted values, we estimate two artificial regressions of the form

$$\exp(\ln(Y_i)) = \gamma_1 + \gamma_2 X_i + \eta_{1i}, \tag{6.24}$$

$$\ln(\tilde{Y}_i) = \beta_1 + \beta_2 \ln(X_i) + \eta_{0i}. \tag{6.25}$$

Finally, we estimate two further regressions of the form

$$\ln(Y_i) = \beta_1 + \beta_2 \ln(X_i) + \theta_0 \hat{\eta}_{1i} + u_{0i}, \tag{6.26}$$

$$Y_i = \gamma_1 + \gamma_2 X_i + \theta_1 \tilde{\eta}_{0i} + u_{1i}. \tag{6.27}$$

The residuals from the artificial regressions are included in these regressions to test if the alternative functional form can contribute any explanatory power to the dependent variable in each case. For example, if θ_0 is significant in (6.26), then it indicates that a linear combination of the variables contributes to an explanation of the log of the dependent variable. Similarly, if θ_1 is significant in (6.27), it indicates that a log-linear combination of the variables contributes to an explanation of the level of Y. The test statistic, in each case, is the t ratio for the coefficient of the residuals from the artificial regressions, which, as Bera and McAleer argue, follows a conventional t distribution.

Example: If we apply the Bera and McAleer test to our cross-country model of consumers' expenditure per head, then we obtain the following results. First, the t ratio for $H_0 : \theta_0 = 0$ in (6.26) is equal to 7.00, indicating that we reject the null of a log-linear specification. Second, the t ratio for $H_0 : \theta_1 = 0$

in (6.27) is equal to −3.32, indicating that we also reject the null of a linear specification. This highlights a problem with this approach. Because the hypotheses here are *nonnested*, that is, neither is a restricted version of the other, it is possible for the procedure to result in both hypotheses being rejected. Alternatively, there are circumstances in which neither hypothesis will be rejected. It is therefore possible that the testing procedure will leave us unable to make a choice between competing models.

An alternative approach, which goes back to the work of Box and Cox [Box1964], is to construct a model in which both the linear and the log-linear specifications are embedded as special cases. Given this, we can then test if either of these special cases is an acceptable simplification of the more general model. To construct such a model, consider the following function:

$$z_i(\lambda) = \begin{cases} \dfrac{z_i^{\lambda} - 1}{\lambda} \text{ if } \lambda \neq 0 \\ \ln(z_i) \text{ if } \lambda \neq 0. \end{cases} \tag{6.28}$$

This function is continuous since $\lim_{\lambda \to 0}(z_i^{\lambda} - 1)/\lambda = \ln(z_i)$. Using this function, we define a regression model of the form

$$Y_i(\lambda) = \beta_1 + \beta_2 X_i(\lambda) + u_i. \tag{6.29}$$

It is straightforward to estimate (6.29) for given values of λ but this is, of course, an unknown parameter. However, we can estimate (6.29) over a grid of values for λ and choose the value that matches some criterion, for example, maximizing the log-likelihood or minimizing the residual sum of squares.

Example: Application of the Box–Cox method to our model of consumer expenditure gives an estimated equation of the form

$$\frac{CONS_i^{0.77} - 1}{0.77} = \underset{(56.64)}{426.82} + \underset{(0.0192)}{0.5823} \frac{GDP_i^{0.77} - 1}{0.77} + \hat{u}_i \tag{6.30}$$
$$R^2 = 0.8353.$$

Note that the parameter estimate $\hat{\lambda} = 0.77$ lies some way between the extreme values of its possible range. This indicates that neither the linear nor the log-linear functional form provides an adequate approximation here which is consistent with the results of the Bera–McAleer test in which each specification rejected the alternative. The standard errors for the other coefficients are reported in parentheses below coefficients but these will be underestimated because the estimation method treats λ as a fixed parameter when, in fact, it is estimated through the grid search procedure.

It is interesting to compare the three functions we have estimated on the same graph. Figure 6.5 shows the three alternative functional forms with consumers' expenditure per capita on the vertical axis and GDP per capita on the horizontal axis. The curvature of the log-linear and Box–Cox functions is most noticeable for lower values of GDP. As expected, the Box–Cox function lies between the linear and the log-linear specifications.

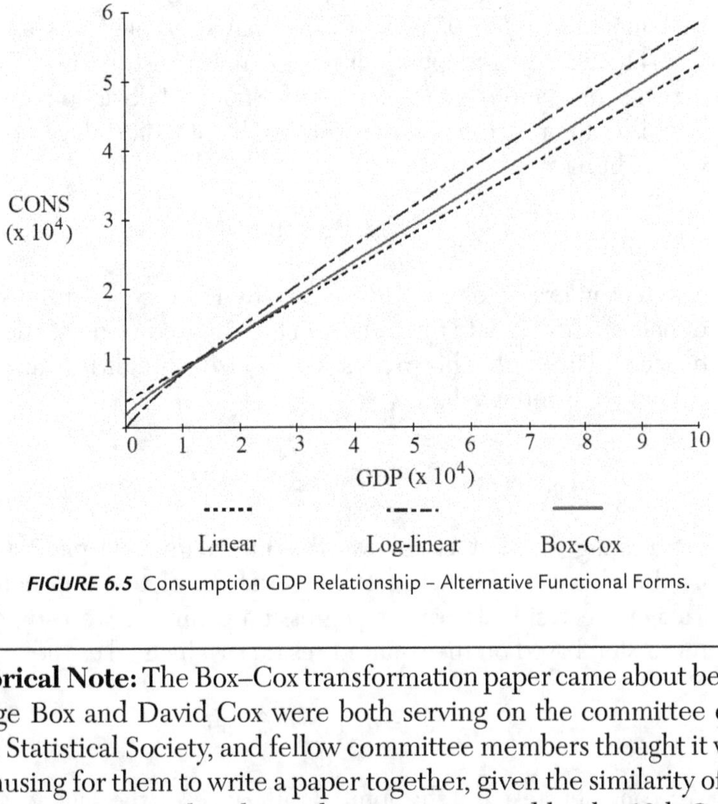

FIGURE 6.5 Consumption GDP Relationship - Alternative Functional Forms.

Historical Note: The Box–Cox transformation paper came about because George Box and David Cox were both serving on the committee of the Royal Statistical Society, and fellow committee members thought it would be amusing for them to write a paper together, given the similarity of their names. George Box also claims that it was inspired by the 19th Century comic opera *Cox and Box* or *The Long Lost Brothers*.

6.8 AUTOREGRESSIVE CONDITIONAL HETEROSCEDASTICITY

The discussion of heteroscedasticity in the previous sections has assumed that the model is estimated using cross-section data. This is because heteroscedasticity is often seen as a purely cross-section problem. Since the 1980s, however, there has been a developing literature concerning a type of heteroscedasticity which is relevant for time-series applications – this is the case of *Autoregressive Conditional Heteroscedasticity* or ARCH, which was first introduced by Engle [Engle1982]. ARCH modifies the error process of the regression model in an interesting way. Consider the standard[2] regression model $Y_t = \beta X_t + \varepsilon_t$. The error can be thought of as $\varepsilon_t = \sigma_t z_t$, where $z_t, t = 1,\ldots,T$ are independent, white noise disturbances with mean zero variance one, and σ_t is a time-varying standard deviation. Our main interest here is in the process that determines this standard deviation. For example, we might have

$$\sigma_t^2 = a_0 + a_1 \varepsilon_{t-1}^2, \tag{6.31}$$

which is an example of a first-order ARCH process. In this case, the conditional variance at date t depends on the squared value of the random disturbance at date t – 1. This process generalizes straightforwardly to the ARCH (q) model, where we have

$$\sigma_t^2 = a_0 + \sum_{i=1}^{q} a_i \varepsilon_{t-i}^2. \tag{6.32}$$

Testing for ARCH follows the Lagrange multiplier methodology we used earlier to construct tests for serial correlation and heteroscedasticity. Taking the residuals from a regression estimate, we estimate an auxiliary regression based on the squared residuals which takes the form

$$\hat{\varepsilon}_t^2 = a_0 + \sum_{i=1}^{q} a_i \hat{\varepsilon}_{t-i}^2 + v_t. \tag{6.33}$$

We can then test for the joint significance of the lagged squared regression residuals using either TR^2 from the auxiliary regression, which is

[2] Note that we have changed to use the subscript t for observation, since we this is a time-series model and that we have changed to the symbol ε for the equation error for consistency with the ARCH literature where this is the most commonly used notation.

asymptotically distributed as χ_q^2 under the null, or by using an F test for the joint restrictions $H_0: \alpha_1 = \alpha_2 = \cdots = \alpha_q = 0$. The F test is not strictly valid here, since it is a small sample test, and we do not know the small sample distribution. However, as with the Breusch–Godfrey test for serial correlation and the White test for heteroscedasticity, the F form of the test appears to perform well in practice and is commonly used in applied work.

> **Historical Note:** Robert Engle (1942-) won the Nobel prize for Economics in 2003 (awarded jointly to Engle and Clive Granger) for his work on ARCH processes.

Example: Consider the following OLS regression in which we have regressed the daily returns[3] from holding Cadbury Schweppes equity on a constant and the daily returns on the overall market index as measured by the FTSE100.

$$\Delta \ln(CS_t) = \underset{(4.6\times10^{-4})}{9.78\times10^{-5}} + \underset{(0.044)}{0.58}\Delta \ln(FT_t) + \hat{\varepsilon}_t \tag{6.34}$$

$$R^2 = 0.12 \qquad \hat{\sigma} = 0.016 \qquad DW = 1.98.$$

To test for a first-order ARCH process in the errors, we estimate the following regression based on the squared regression residuals from equation (6.34)

$$\hat{\varepsilon}_t^2 = \underset{(1.9\times10^{-5})}{2.22\times10^{-4}} + \underset{(0.027)}{0.19}\hat{\varepsilon}_{t-1}^2 + \hat{v}_t \tag{6.35}$$

$$TR^2 = 48.32 \qquad F = 50.11.$$

Under the null of homoscedastic errors, the test statistic TR^2 follows a chi-square distribution with one degree of freedom and, since the 5% critical value for the χ_1^2 distribution is 3.84, we reject the null using this test. Similarly, the F statistic follows an F distribution with 1 and 1,300 degrees of freedom under the null, and the 5% critical value for $F_{1,1300}$ is 3.84, meaning that this test also implies rejection of the null. In both cases, the test statistic is much larger than the 5% critical value, and therefore, it appears that there is strong evidence of ARCH effects in the residuals.

---

[3] The data are daily from 26/10/1995 to 23/10/2000 which gives a total of 1,303 observations when weekends and holidays during which no trading takes place are eliminated from the sample.

It is relatively straightforward to estimate models with a low-order ARCH process. As the number of lags increases, however, this becomes more difficult. For example, suppose we have a process of the form (6.32), where q is large. Unconstrained estimation of equations like this is problematic, because of the loss of degrees of freedom, and because the lagged squared residuals are likely to be highly correlated with each other, leading to imprecise estimates. As a response to these problems, Bollerslev [Bollserslev1986] introduced the Generalized Autoregressive Conditional Heteroscedasticity model (GARCH). This takes the form

$$\sigma_t^2 = a_0 + \sum_{i=1}^{q} a_i \varepsilon_{t-i}^2 + \sum_{i=1}^{p} \beta_i \sigma_{t-i}^2, \tag{6.36}$$

where σ_t is the conditional standard deviation at date t. This is the GARCH(p,q) model. In general, the order of the polynomial functions p and q will be quite low. For example, Hansen and Lunde [Hansen1965] have shown that the GARCH(1,1) fits very well in a wide variety of different models. The GARCH(1,1) model takes the form

$$\sigma_t^2 = a_0 + a_1 \varepsilon_{t-1}^2 + \beta_1 \sigma_{t-1}^2. \tag{6.37}$$

The key feature of the GARCH model is that the conditional variance evolves through time. Thus, the assumption of homoscedastic (constant variance) error terms is no longer valid. The conditional variance σ_t^2 in (6.37) depends on the squared lag of the unconditional variance ε_{t-1}^2 and its own lagged value σ_{t-1}^2. Backward substitution in the expression for the conditional variance allows us to write it as

$$\sigma_t^2 = \frac{a_0}{1-\beta_1} + a_1 \sum_{i=1}^{\infty} \beta_1^{i-1} \varepsilon_{t-i}^2. \tag{6.38}$$

Equation (6.38) shows that the conditional variance can be written as an infinite moving average of past values of the unconditional variance. Note that a necessary, but not sufficient, condition for this sum to converge is $|\beta_1| < 1$. The fact that this is not sufficient can be seen by examination of the behavior of the unconditional variance. Note that we can write $\varepsilon_t^2 = \upsilon_t + \sigma_t^2$, that is, the unconditional variance is equal to the conditional variance σ_t^2 plus a residual υ_t. Using this expression and substituting into (6.37), we can write the following equation for the unconditional variance:

$$\varepsilon_t^2 = a_0 + (a_1 + \beta_1)\varepsilon_{t-1}^2 + \upsilon_t - \beta_1 \upsilon_{t-1}. \tag{6.39}$$

Equation (6.39) indicates that the unconditional variance follows an ARMA(1,1) process. Moreover, it also demonstrates that, for this process to be stationary, we require $|\alpha_1 + \beta_1| < 1$.

We can deal with ARCH effects in two ways: we can either adjust the standard errors to allow for their presence or we estimate a model that explicitly allows for ARCH effects. We will consider each method in turn. First, we present a model with adjusted standard errors for the coefficients.

$$\Delta \ln(CS_t) = \underset{(4.25\times10^{-4})}{9.78\times10^{-5}} + \underset{(0.055)}{0.58}\Delta \ln(FT_t) + \hat{\varepsilon}_t \tag{6.40}$$

$$R^2 = 0.12 \qquad \hat{\sigma} = 0.016 \qquad DW = 1.98.$$

The results for equation (6.40) are very similar to those shown in equation (6.34). The parameter estimates are identical – the only difference is that the standard errors have been adjusted using the Newey–West [Newey1987] procedure. This has increased the standard error for the slope coefficient somewhat which, in turn, means that the confidence interval for the beta parameter will be somewhat wider.

An alternative approach to dealing with ARCH effects in the errors is to estimate a model that explicitly allows for their presence. To do this, we will need to assume a specific functional form for the ARCH effects. For example, the assumption of a GARCH(1,1) process allows us to estimate both the mean and the conditional variance equations given in

$$\begin{aligned} \Delta \ln(CS_t) &= \underset{(3.86\times10^{-4})}{3.98\times10^{-5}} + \underset{(0.035)}{0.62}\Delta \ln(FT_t) + \hat{\varepsilon}_t \\ \hat{\sigma}_t^2 &= \underset{(4.52\times10^{-7})}{2.09\times10^{-6}} + \underset{(0.0065)}{0.04484}\hat{\varepsilon}_{t-1}^2 + \underset{(0.00649)}{0.94906}\hat{\sigma}_{t-1}^2. \end{aligned} \tag{6.41}$$

When we estimate the model with a GARCH error, the parameter estimates for the mean equation do change, though the effects are small. If we examine the variance equation, then we see that the sum of the coefficients on the lagged squared residual and the lagged variance term is quite close to 1 (which would make the equation unstable). However, this is quite a common result in models with GARCH errors.

6.9 STRUCTURAL BREAKS

So far, we have assumed that the parameters of the regression model are constants. However, this is an assumption which we may wish to test under certain circumstances. For example, suppose there is a significant change in the economic environment, such as a major banking crisis, the outbreak of war, or a disease pandemic. It would be sensible to test whether such events have an effect on the parameters of the models which we estimate. A number of tests exist which allow us to do this, and we will consider some of the more commonly used tests here.

Tests for structural breaks, or parameter instability, differ according to whether or not we are aware of the nature of the division of the sample prior to estimation. The easier case to deal with is when we can identify the sample division in advance. When using cross-section data, we might wish to divide the sample into different subgroups and test if the parameters are constant across these groups. For example, if we have a sample of hours worked and wages paid, we could partition the sample according to some broad characteristics, say male and female workers, and test if the elasticity of hours worked with respect to the wage is the same for both groups. When working with time-series data, we might be aware of the date at which a major event occurred, which is the potential cause of parameter instability. Hence, we would divide the sample into observations prior to, and after, this date and test if the parameters are the same across subperiods.

The most commonly used tests for parameter instability, when the sample division is known in advance, are the *Chow Tests*. These are based on the work of Gregory Chow [Chow1960] who proposes two tests for the null hypothesis that parameters are constant across subsamples of the data. Suppose we wish to estimate a linear regression model of the form $Y_t = \alpha + \beta X_t + u_t$ based on a sample $t = 1,\ldots,T$ and that we believe a structural break may have occurred at date T_1. The first Chow test, known as Chow's Breakpoint test, is constructed by estimating separate regressions prior to the date of the break $t = 1,\ldots,T_1$ and after the break $t = T_1 + 1,\ldots,T$. From these, we calculate residual sums of squares RSS_1 and RSS_2. This allows both the intercept and the slope coefficients to differ in the two subperiods. We then estimate a single regression based on the whole sample and generate the residual sum of squares RSS. This imposes the restrictions that the intercept and the slope coefficients are equal across the whole sample. The test statistic for the null hypothesis that these restrictions are valid is given by

$$F = \frac{RSS - (RSS_1 + RSS_2)}{(RSS_1 + RSS_2)} \frac{T-4}{2}. \tag{6.42}$$

Under the null hypothesis, this test statistic is distributed as F with two and $T-4$ degrees of freedom. The degrees of freedom for the numerator is equal to two, that is, the number of restrictions, and the degrees of freedom for the denominator is equal to $T-4$, that is, the number of parameters in the unrestricted model. In more general models, with k parameters, the degrees of freedom would be equal to k and $T-2k$.

Example: Although we have set out the Chow Breakpoint test in terms of time-series data, it can just as easily be used in cross-section models where the sample is divided into subsets and we wish to test for parameter constancy across this division. To emphasize this, we will use a cross-section application to illustrate this test. The sample here is the cross-section consumption relationship we examined earlier in this chapter. Here, we wish to test if the parameters of the consumption relationship are different for low- and high-income economies. Therefore, we order the data by the size of income per capita and estimate separate regressions for low- and high-income countries. There are 182 data points in total, so we divide the sample in half and estimate separate regressions for the lowest 91, and highest 91, income economies. The results are given in Table 6.3 where c_i is consumption per head and y_i is GDP per head.

TABLE 6.3 Consumption Relationship Estimates 182 Economies.

Whole sample, $N = 182$	$\ln c_i = \underset{(0.1317)}{1.0164} + \underset{(0.0142)}{0.8649} \ln y_i + \hat{u}_i$ $R^2 = 0.9539 \quad \hat{\sigma}_u = 0.2311 \quad RSS = 9.6173$
Low-income economies, $N = 91$	$\ln c_i = \underset{(0.1962)}{0.4062} + \underset{(0.0238)}{0.9401} \ln y_i + \hat{u}_i$ $R^2 = 0.9462 \quad \hat{\sigma}_u = 0.1817 \quad RSS = 2.9378$
High-income economies, $N = 91$	$\ln c_i = \underset{(0.5248)}{2.1999} + \underset{(0.0513)}{0.7503} \ln y_i + \hat{u}_i$ $R^2 = 0.7062 \quad \hat{\sigma}_u = 0.2597 \quad RSS = 6.0003$

From Table 6.3, we see that the parameters certainly look quite different for the two subsamples. The intercept is much higher for the high-income economies, whereas the slope coefficient is much lower. If this difference significant? To test this, we compute the Chow Breakpoint test statistic as

$$F = \frac{9.6173 - (2.9378 + 6.0003)}{2.9378 + 6.0003} \times \frac{182 - 4}{2} = 6.763. \tag{6.43}$$

Since the 5% critical value for an F statistic with 2 and 178 degrees of freedom is equal to 3.047, we conclude that there is a significant difference in the parameters across this division of the sample. We note that this may be related to the functional form misspecification we detected earlier for this model.

The Chow Breakpoint test works well in situations where each subsample has sufficient observations to estimate the regression equation. In many situations, however, there may be very few observations in one of the subsamples, and we may be unable to estimate a regression equation in both cases. Even if we technically have enough observations for the regression equation to be computed, the degrees of freedom may be so low that the results are unreliable. In these cases, we can use Chow's second test which is described alternatively as *Chow's Forecast test* or the *Predictive Failure test*. This can be understood easily by examination of the test statistic which takes the form

$$F = \frac{RSS - RSS_1}{RSS_1}\frac{T_1 - k}{n}. \tag{6.44}$$

This test statistic is distributed as F_{n,T_1-k} under the null hypothesis that the parameters are constant. To implement this test, we estimate a regression for a subsample of data based on T_1 observations, where k is the number of regression parameters. We then compute the residual sum of squares for this sample RSS_1 and the residual sum of squares for the full sample of data RSS. The full sample differs from the subsample by the addition of n extra data points. The objective here is to test if the addition of the extra n data points increases the residual sum of squares by more than we would expect if the parameters are constant. This test works well in situations where the number of extra data points n is too small to permit estimation of a separate equation.

Example: Table 6.4 provides an example of Chow's Forecast test. Here, we estimate an Okun's law relationship between the change in percentage unemployment (DU) and the percentage change in GDP (DY), for the United States over the period 1950–2016. We then test, if the addition of three extra observations for the period 2017–2019, indicates a change in the parameters of this equation. The test statistic is calculated as

$$F = \frac{1.312}{36.9761} \times \frac{65}{3} = 0.7689. \tag{6.45}$$

The 5% critical value for an F test with 3 and 65 degrees of freedom is 2.746. Therefore, there is little evidence here of parameter instability from adding the extra three observations to the sample.

TABLE 6.4 Okun's Law Relationship Estimates for the United States.

Period 1950–2016	$DU_t = \underset{(0.1591)}{1.0434} - \underset{(0.0398)}{0.3291} DY_t + \hat{u}_t$ $R^2 = 0.5123 \quad \hat{\sigma}_u = 0.7542 \quad RSS = 36.9761$ $DW = 1.7776$
Period 1950–2019	$DU_t = \underset{(0.1558)}{1.0023} - \underset{(0.0395)}{0.3246} DY_t + \hat{u}_t$ $R^2 = 0.4981 \quad \hat{\sigma}_u = 0.7504 \quad RSS = 38.2881$ $DW = 1.7272$

The two Chow tests are most useful when we can identify the nature of the structural break prior to estimation. However, there may be situations in which this is not possible, and we would like to search for possible structural breaks without prespecification. A similar approach to the Chow Forecast test can be used here, by computing the one-step ahead prediction errors and using these to identify possible candidates for structural breaks. Consider the standard regression model in mean difference form, $Y_t = \beta X_t + u_t$, the one-step ahead prediction error is defined as

$$v_\tau = Y_\tau - \hat{\beta}_{\tau-1} X_\tau, \tag{6.46}$$

where $\hat{\beta}_{\tau-1}$ is the estimated parameter based on data up to period $\tau - 1$. The variance of the prediction error can be shown to be

$$\sigma_u^2 \left[1 + \frac{X_\tau^2}{\sum_{t=1}^{\tau-1} X_t^2} \right], \tag{6.47}$$

and the scaled prediction errors are defined as

$$\omega_\tau = \frac{v_\tau}{\left[X_\tau^2 / \sum_{t=1}^{\tau-1} X_t^2 \right]}. \tag{6.48}$$

Under the null hypothesis that the parameters are constant, the prediction errors follow a normal distribution with mean zero and variance σ_u^2. By estimating a succession of regression models, adding one observation to the sample in each case, we can compute the scaled prediction errors and assess if they indicate a significant deviation from their expected value. By doing, we identify possible periods in which there has been a change in the value of the regression parameters. Figure 6.6 shows the results of these calculation for our Okun's law relationship.

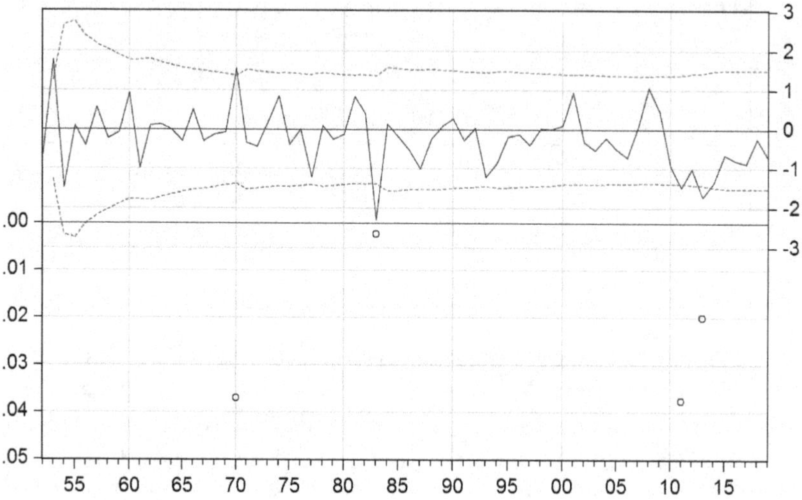

FIGURE 6.6 One-Step Ahead Prediction Errors for the Okun's Law Model

The broken lines in Figure 6.6 shown a 95% confidence interval for the prediction errors. Cases in which the prediction error falls outside these bounds are a potential indicator of parameter instability. The graph also gives the p values for such cases, indicating which are the most statistically significant. From the graph, we see that there are four cases in which the prediction error lies outside the confidence interval bands. However, with prediction errors in total, we would expect $65 \times 0.05 = 3.25$ to fall outside the confidence interval, simply due to our choice of interval. Therefore, there is very little evidence here of parameter instability.

EXERCISES

Excel files containing the data for these exercises are available as companion files for this book.

EXERCISE 6.1

Consider the regression model $Y_i = \alpha + \beta X_i + u_i$, where $E(u_i^2) = \sigma_u^2 \sqrt{X_i}$, and $X_i > 0$ for all values of i. Find a transformation of the model which has homoscedastic errors and therefore will permit efficient estimation by least squares.

EXERCISE 6.2

Consider a regression model of the form $Y_i = \alpha + \beta \ln X_i + u_i$, where the errors are homoscedastic. An econometrician estimates a linear model of the form $Y_i = \alpha + \beta X_i + u_i$. Using a second-order Taylor series approximation around $X = 1$, show that this will have heteroscedastic errors.

EXERCISE 6.3

An econometrician has estimated the following model that relates inflation to money growth for a sample of 83 economies. The data are average annual values for the period 1980–1993 and are taken from the 1995 *World Development Report*.

```
Ordinary Least Squares Regression Results
Sample period: 1 to 83
Dependent Variable INF
Sample Size 83
```

Variable	Coefficient	Std Err	*T* Ratio
C	-5.681642	0.704442	-8.065443
MG	1.046654	0.011553	90.589950

R-squared	0.9902	*F* statistic	8206.5392
SEE	5.657356	RSS	2592.460070
Durbin-Watson	1.7031	LogL	-260.595058
ARCH(1) test	3.3716	AIC	6.327591
Jarque-Bera	56.2514	SIC	6.385877

a. Comment on the regression results and say why the econometrician might argue that the model is both a good statistical fit and consistent with economic theory.

b. The data, which are stored in the Excel data file INF.XLSX, are ordered according to the rate of money growth. Given this you should easily be able to perform the Goldfeld–Quandt test for heteroscedasticity. Construct the test statistic and compare it with an appropriate critical value. What do the results show?

c. Perform the White test for heteroscedasticity. Are the results consistent with the Goldfeld–Quandt test.

d. Set out the implications of your test results for the interpretation of the OLS regression results given above.

EXERCISE 6.4

The regression output below gives the results of estimating a market model that relates the daily returns from holding shares in the Tesco company to the daily returns for the overall FTSE 100 index for the period January 2003 to May 2008.

```
Ordinary Least Squares Regression Results
Sample period: 2 to 1359
Dependent Variable RET_TESCO
Sample Size 1358
```

Variable	Coefficient	Std Err	*T* Ratio
C	0.045903	0.031952	1.436634
RET_MARKET	0.736342	0.032957	22.342221

R-squared	0.2690	*F* statistic	499.1748
SEE	1.176692	RSS	1877.524340
Durbin-Watson	2.0885	LogL	-2146.874490
ARCH(1) test	21.0253	AIC	3.164763
Jarque-Bera	532.8067	SIC	3.172442

a. Comment on the values taken by the slope coefficient and the R^2 for this regression.

b. Is there any evidence for an ARCH process in the residuals? Perform a formal test for the null hypothesis that the residuals do NOT exhibit ARCH.

c. Using the data in the Excel workfile SHARES.XLSX, estimate market models for the following companies: AstraZeneca, Lloyds Bank, and Vodafone. In each case, perform a test for the presence of ARCH in the residuals.

d. What are the implications of your results for the OLS estimates of the market model?

REFERENCES

[Bera1989] Bera, A. and McAleer, M., "Nested and Non-nested Procedures for Testing Linear and Log-Linear Regression Models." *Sankhya Series B*, 1989, 50, pp. 212–224.

[Bollerslev1986] Bollerslev, T., "Generalised Autoregressive Conditional Heteroskedasticity." *Journal of Econometrics*, 1986, 31, pp. 307–327.

[Box1964] Box, G. E. P. and Cox, D. R., "An Analysis of Transformations." *Journal of the Royal Statistical Society: Series B*, 1964, pp. 211–243.

[Breusch1979] Breusch, T. S. and Pagan, A. R., "A Simple Test for Heteroscedasticity and Random Coefficient Variation." *Econometrica*, 1979, 454, pp. 1287–1294.

[Chow1960] Chow, G. C., "Tests of Equality Between Subsets of Coefficients in Two Linear Regression Models." *Econometrica*, 1960, 28, pp. 591–605.

[Engle1982] Engle, R. F., "Autoregressive Conditional Heteroskedasticity with Estimates of the Variance of UK Inflation." *Econometrica*, 1982, 50, pp. 987–1008.

[Glejser1969] Glejser, H., "A New Test for Homoscedasticity." *Journal of the American Statistical Association*, 1969, 64 (235), pp. 315–323.

[Goldfeld1965] Goldfeld, S. M. and Quandt, R. E., "Some Tests for Homoscedasticity." *Journal of the American Statistical Association*, 1965, 60 (310), pp. 539–547.

[Hansen2005] Hansen, P. and Lunde, A., "A Forecast Comparison of Volatility Models: Does Anything Beat a GARCH(1,1)?" *Journal of Applied Econometrics*, 2005, 20, pp. 873–889.

[Harvey1976] Harvey, A. C., "Estimating Regression Models with Multiplicative Heteroscedasticity." *Econometrica*, 1976, 44 (3), pp. 461–465.

[Kiviet1986] Kiviet, J., "On the Rigour of Some Misspecification Tests for Modelling Dynamic Relationships." *Review of Economic Studies*, 1986, 53, pp. 241–261.

[Newey1987] Newey, W. K. and West, K. D., "A Simple Positive Semidefinite, Hetereoskedasticity Consistent Covariance Matrix." *Econometrica*, 1987, 55, pp. 703–708.

[Pearson1905] Pearson, K., "On the General Theory of Skew Correlation and Non-linear Regression." *Drapers' Company Res. Mem.* (Biometric Ser.) II, 1905.

[Ramsey1969] Ramsey, J. B., "Tests for Specification Errors in Classical Linear Least Squares Regression Analysis." *Journal of the Royal Statistical Society, Series B*, 1969, 31 (2), pp. 350–371.

[White1980] White, H., "A Heteroskedasticity-Consistent Covariance Matrix Estimator and a Direct Test for Heteroskedasticity." *Econometrica*, 1980, 50, pp. 1–25.

CHAPTER 7

BINARY DEPENDENT VARIABLES

Consider the standard regression model $Y_i = a + \beta X_i + u_i$. If we assume that the error follows a normal distribution, in which any real value is possible, then it follows that Y_i should also be able to take on any real value. However, the data we deal are often not consistent with this. In many cases, the dependent variable can only take on a limited number of values. One common example of this is when it is binary in nature. An example of this is survey data in which individuals are asked if they are employed or unemployed. Alternatively, we might observe a sample of companies some of which go into liquidation during a given period and some of which do not. In both these cases, the data can be coded so that the variable to be explained takes on only two possible values – 0 or 1.

There is nothing to prevent us from calculating a least squares regression equation even if the variable to be explained is coded as a 0–1 variable. However, the interpretation of such an equation becomes somewhat problematic. To illustrate this, let us consider a specific example. Suppose we have data for the share price of a company which is coded as 1 for days on which the share price rises, and 0 for days on which it remains constant or falls. We wish to examine whether there is a relationship between movements in the share price (coded in this way) and movements in the overall stock market index. As a first attempt, we estimate a least squares regression of the share price change variable (in our example, this is the price of British Airways (BA) shares) on a constant and the change in the Financial Times Stock Exchange (FTSE) market index. The results are given in equation (7.1),

$$BA_t = \underset{(0.0121)}{0.4959} + \underset{(1.2499)}{23.3086}\, FT_t + u_t \qquad (7.1)$$

$$R^2 = 0.2032 \qquad DW = 2.0407.$$

The results in equation (7.1) are consistent with what we might expect in that they show an apparently significant relationship between the change in the BA share price and the % change in the stock market index. This is indicated by the fact that the t ratio for the independent variable, which can be calculated as $23.3086/1.2499 = 18.65$, is above the 5% critical value of 1.96 by some margin. However, it is not immediately obvious how the coefficient estimate should be interpreted in this case. The FT variable in this case is the first difference of the logarithm of the stock market index. As such, a value of 0.01 for this variable is equivalent to a rise of 1% in the market index. Therefore, the coefficient of 23.3 indicates that a 1% rise in the overall stock market increases the expected value of the left-hand side variable by 0.233 – but what does this mean in economic terms when the dependent variable is binary?

We are used to thinking of regression coefficients as marginal responses and one way of interpreting the coefficients in models like this is as *marginal probabilities*. We can interpret the coefficient β as giving the increase in $P(BA_i = 1)$ associated with a unit increase in the value of X_i. In our example, the coefficient of 23.3 indicates that a rise of 1% in the overall stock market increases the probability of there being an increase in the value of BA shares by 0.233. This interpretation of the regression model is referred to as the *linear probability model* because it assumes a linear relationship between the probability of an event occurring and the set of explanatory variables on the right-hand side of the estimated equation.

The linear probability model is intuitively appealing, but it should quickly become apparent that this interpretation has a number of logical problems. The first concerns the nature of the probabilities estimated by the model. Since the left-hand side variable only takes on the values 0 or 1, the natural way to think of the data is as the outcomes of a series of Bernoulli trials or experiments which can either be successes ($BA_t = 1$) or failures ($BA_t = 0$). If we adopt the linear probability model interpretation of our regression, then the fitted values $\hat{\alpha} + \hat{\beta}FT_t$ are the estimated conditional probabilities of a success. Probability theory requires that these conditional probabilities should lie in the range 0–1 since negative probabilities, or probabilities >1, make no sense. However, there is nothing in the linear probability model that constrains the conditional probabilities to lie within this range. Thus, the linear probability model may easily produce results in which the predicted values are inconsistent with probabilities.

Examination of the actual and fitted values from our estimated model of BA share prices illustrates the inconsistency of the linear probability model. In Figure 7.1, we show the scatter of actual values against those of the right-hand side variable as well as the fitted values from the estimated regression equation. The actual values lie on two horizontal lines passing through 0 and 1, respectively. The fitted values lie on the line with a positive slope illustrated in the diagram. For a considerable number of values of X, the fitted values fall outside the range 0–1. Out of 1,366 fitted values, 30 are <0 and 29 are >1, meaning that 4.3% of the cases have fitted probabilities that lie outside the permissible range.

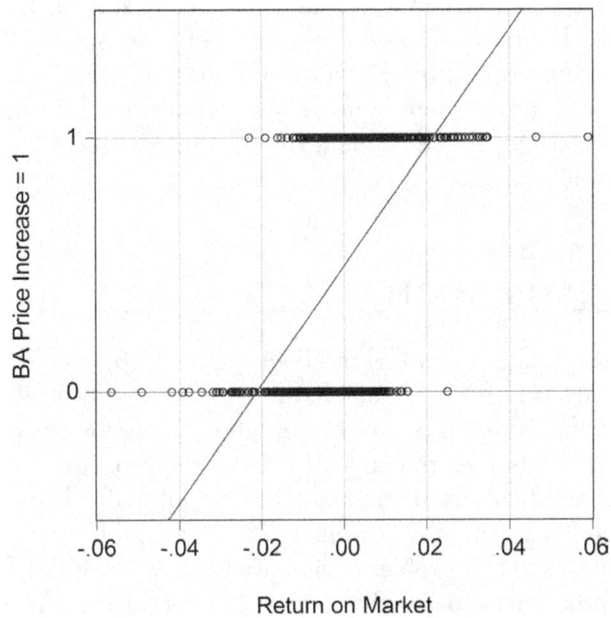

FIGURE 7.1 Actual and Fitted Values for the BA Share Price Model.

Another problem with the linear probability model is that the errors are heteroscedastic by construction. For simplicity, let us consider the model in mean deviation form $Y_i = \beta X_i + u_i$. This means that there is a single unknown parameter, allowing the algebra to be simplified. Given the binary nature of the Y variable, the error can take on only two possible values for given X.

$$u_i = \begin{cases} -\beta X_i & \text{if} \quad Y_i = 0 \\ 1-\beta X_i & \text{if} \quad Y_i = 1 \end{cases}. \tag{7.2}$$

The variance can be written

$$\begin{aligned} V(u_i) &= E(u_i^2) = \beta X_i(1-\beta X_i)^2 + (1-\beta X_i)(-\beta X_i)^2 \\ &= \beta X_i(1-\beta X_i). \end{aligned} \tag{7.3}$$

The error variance is a function of the X variable and therefore, heteroscedasticity is present. If this was the only problem with the linear probability model, then we could use the methods described in Chapter 6 to deal with this. For example, we could adjust the coefficient standard errors to allow for more reliable statistical inference, or we could try scaling the data to make the errors homoscedastic. However, the fact that we have both heteroscedasticity, and the problem of model predictions that lie outside the possibly range of values for the dependent variable, means that it becomes necessary to look for alternative models in this case.

7.1 LOGIT ESTIMATION

The linear probability model has been shown to be inconsistent in that the fitted values can often fall outside the range of theoretically feasible values. However, the principle of interpreting the regression results as giving estimates of probabilities remains a promising approach. What we need is to find an alternative formulation which does not suffer from the inconsistencies of the linear probability framework. Fortunately, there are several possible solutions to this problem which we will now consider.

Our problem is one in which we wish to model the effects of a variable X on the probability that a binary variable Y takes the value 1. Let us begin by assuming that this function takes the form $g(X_i) = P(Y_i = 1 | X_i)$. Note that it immediately follows that $P(Y_i = 0 | X_i) = 1 - g(X_i)$. This function should have the following properties:

1. The probability should always lie between the values 0 and 1 for any value of X.
2. The probability that $Y = 1$ should approach 0 for very small values of X.
3. The probability that $Y = 1$ should approach 1 for very large values of X.

Now, assuming that we can find a function that has these properties, we can write down the joint probability of observing any particular set of values of Y as

$$\prod_{i=1}^{N} g(X_i)^{Y_i}\left(1-g(X_i)\right)^{1-Y_i}. \tag{7.4}$$

The function $g(X_i)$ will typically depend on a number of unknown parameters. For example, if we take the linear probability function, then we would have $g(X_i)=\alpha+\beta X_i$. More generally, suppose we write the function in terms of its parameters as $g\left(X_i \middle| \alpha,\beta\right)$, then we can substitute this into the joint probability function to obtain the likelihood function

$$L(\alpha,\beta)=\prod_{i=1}^{N} g\left(\alpha,\beta \middle| X_i\right)^{Y_i}\left(1-g\left(\alpha,\beta \middle| X_i\right)\right)^{1-Y_i}, \tag{7.5}$$

or taking logarithms of (7.5), we have the log-likelihood function

$$LL(\alpha,\beta)=\sum_{i=1}^{N}\left\{Y_i \ln\left(g\left(\alpha,\beta \middle| X_i\right)\right)+(1-Y_i)\ln\left(1-g\left(\alpha,\beta \middle| X_i\right)\right)\right\}. \tag{7.6}$$

The method of maximum likelihood involves choosing estimated values of the parameters $\hat{\alpha}$ and $\hat{\beta}$, which maximize the function defined above in equation (7.6). In many cases, it will not be possible to find an analytical solution for the maximum-likelihood estimator. However, it will usually be possible to use numerical methods to find estimates of the unknown parameters.

Now, let us consider a particular functional form for the probability as shown in equation (7.7). This is known as the *logistic* function and is given in equation (7.7). This function has a characteristic sigmoid shape while is illustrated in Figure 7.2 for parameter values $\alpha = 0$ and $\beta = 1$,

$$g(X_i)=\frac{\exp(\alpha+\beta X_i)}{1+\exp(\alpha+\beta X_i)}. \tag{7.7}$$

Historical Note: The term "logit" was first used in 1944 by Joseph Berkson [Berkson1944] (1899–1982) to parallel that of "probit" previously used by Chester Bliss [Bliss1934]. Both authors were concerned with mapping variables that were limited to the range $[0,1]$ to $]-\infty,\infty[$ so they could be analyzed using the normal distribution.

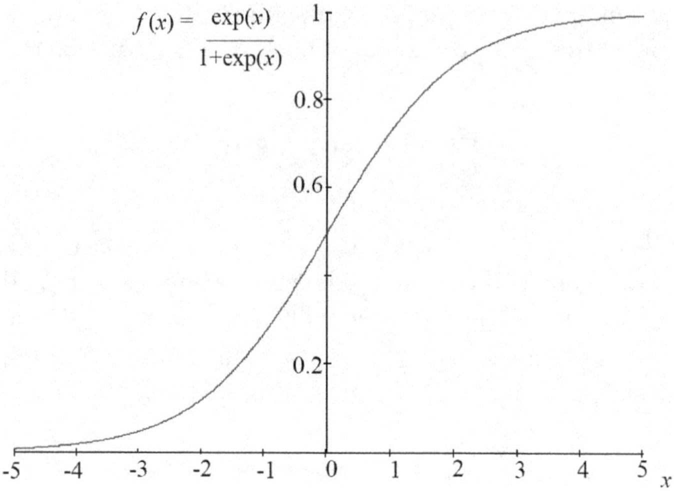

FIGURE 7.2 The Logistic Function.

Does (7.7) satisfy the properties we have set out for a probability function? The plot of equation (7.7) shown in Figure 7.2 indicates that this is the case. First, $g(X_i) > 0$ for any value of X, second, we have $\lim_{X\to-\infty} g(X) = 0$, and third, we have $\lim_{X\to\infty} g(X_i) = 1$. Thus (7.7) satisfies all the criteria for a probability function. Using this function, we derive the maximum likelihood as

$$\arg\max \sum_{i=1}^{N} \left\{ Y_i \ln\left(\frac{\exp\left(\hat{\alpha} + \hat{\beta} X_i\right)}{1 + \exp\left(\hat{\alpha} + \hat{\beta} X_i\right)} \right) + \left(1 - Y_i\right) \ln\left(\frac{1}{1 + \exp\left(\hat{\alpha} + \hat{\beta} X_i\right)} \right) \right\}. \quad (7.8)$$

Although we cannot find an analytical solution to this problem, it is relatively easy to find a numerical solution. For example, using our data set for the returns on British Airways shares, we obtain the results shown in Table 7.1.

TABLE 7.1 Logit Regression Results – British Airways Share Model.

	Coefficient	Standard error	*T* Ratio
$\hat{\alpha}$	−0.0426	0.0618	−0.6901
$\hat{\beta}$	148.9207	10.1124	14.7265
Mean of independent variable		0.000301	

The estimates in Table 7.1 indicate that the percentage change in the FTSE index has a significant effect on the probability of a rise in the value of British Airways shares. Moreover, this effect is positive, with a rise in the market as a whole increasing the probability of a rise in BA shares. It is difficult, however, to interpret the regression results in more detail without more work. This is because the estimated slope coefficient does not measure a marginal effect in the same way as that for the linear probability model. In the case of the linear probability model, the slope coefficient gives us a direct estimate of the marginal effect. Unfortunately, such a straightforward interpretation of the slope coefficient is not possible when we consider the logistic regression.

To interpret the slope coefficient of the logistic regression, let us consider once again the interpretation of the equation we have estimated. The parameters of the estimated equation determine the shape of the probability function. That is, the estimated probability of the variable Y being equal to 1 and is given by the formula

$$P(Y_i = 1) = \frac{\exp(\hat{\alpha} + \hat{\beta}X_i)}{1 + \exp(\hat{\alpha} + \hat{\beta}X_i)}. \tag{7.9}$$

How does the value of $\hat{\beta}$ affect this probability? If we differentiate (7.9) with respect to $\hat{\beta}$, then we have

$$\frac{\partial P(Y_i = 1)}{\partial \hat{\beta}} = \frac{\hat{\beta}}{\left(1 + \exp(\hat{\alpha} + \hat{\beta}X_i)\right)^2}. \tag{7.10}$$

This shows that the marginal effect on the probability is a decreasing function of the right-hand side variable X. In order to get some idea of the size of the marginal effect, we can evaluate (7.10) at the sample mean of X. In this case, we have

$$\frac{\hat{\beta}}{\left(1 + \exp(\hat{\alpha} + \hat{\beta}\bar{X})\right)^2} = \frac{148.92}{\left(1 + \exp(-0.0426 + 148.92 \times 0.000301)\right)^2} = 37.15\cdot \tag{7.11}$$

Again, we can express this in more intuitive terms by considering the effect of a 1% rise in the value of stock market. In this case, such a shock would increase the probability of an increase in the BA share price by an amount given by $37.15 \times 0.01 = 0.3715$.

It is interesting to see how the marginal probability varies with the independent variable by plotting the relationship between them over the range of values taken by the right-hand side variables. This is shown in Figure 7.3. It is clear that the size of the marginal effect is greatest close to the mean. As the value of the independent variable approaches either extreme of its range then the slope of the probability function, and hence the marginal effect on probability, approaches 0. The shape of the function shown in Figure 7.3 looks remarkably similar to that of the normal probability density function. This is not a coincidence since the logistic probability function is a cumulative probability function for a distribution which has a similar shape to that of the normal. Hence, its derivative will yield a function which resembles that of the normal probability density. The main difference is that the logistic function has somewhat heavier tails than those of the normal function. Thus, its shape more closely resembles that of a t distribution with a low number of degrees of freedom.

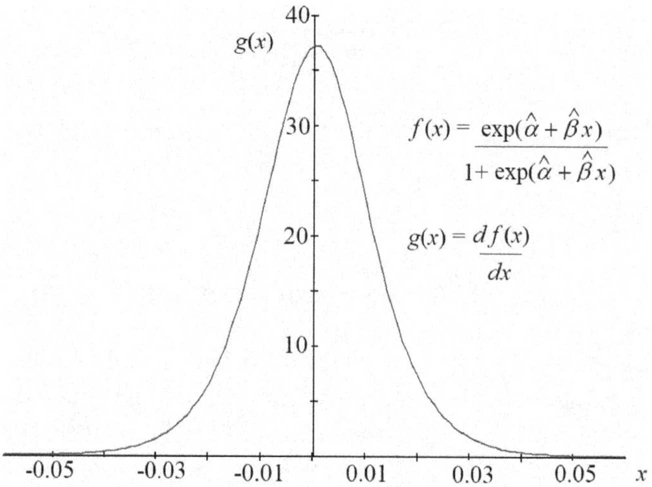

FIGURE 7.3 Marginal Probability as a Function of the Independent Variable.

7.2 GOODNESS OF FIT IN LIMITED DEPENDENT VARIABLE MODELS

The assessment of goodness of fit is difficult in models with binary dependent variables. For example, how do we compare a fitted value from such a regression with the actual values which can only take on the values 0 or 1? One method of calculating goodness of fit is to compare the value of the log-likelihood when the parameter β is fixed at 0 with that obtained when β is a free parameter. These values are then used to define a measure of goodness of fit known as McFadden's R^2. This is calculated as one minus the ratio of these two likelihoods.

$$\text{McFadden } R^2 = 1 - \frac{LL(\alpha,\beta)}{LL(\alpha|\beta=0)} \tag{7.12}$$

Example: For the BA share price model, we have $LL(\alpha|\beta=0) = -946.8156$ and $LL(\alpha,\beta) = -770.4887$.

$$\text{McFadden } R^2 = 1 - \frac{770.4487}{946.8156} = 0.1863 \tag{7.13}$$

McFadden's R^2 has similar properties to that of the coefficient of determination in more conventional models. In particular, it has an expected value of 0 when the X variable has no predictive value and is bounded above by 1, at which value the X variable would become a perfect predictor of the Y variable. The value of 0.1863 which we obtain in this case indicates that while the change in the FTSE index does provide some explanatory power, it is far from a perfect predictor and that there are likely to be other, company specific, factors that are responsible for movements in the price of BA shares.

Another possible way to assess goodness of fit is to calculate the percentage of correct predictions. For example, suppose we predict that BA share prices will rise if the fitted probability from our logit model is >0.5 and will fall if it <0.5. On this basis, we can assess the percentages of correct and incorrect predictions as shown in Table 7.2. From these results, we see that our model predicts the direction of the share price movement in 71% of cases. This is calculated by taking the sum of the diagonal elements in the table. The off-diagonal elements show the percentages of cases in which the model is incorrect. This is divided evenly between cases in which the model predicts a fall in share prices but prices rise (14%) and cases in which the model predicts a rise in share prices that are then observed to fall (15%). It is clear from these results that while our model is far from perfect, it does perform better than a naïve model in which we simply set the probabilities for share price movements equal to the marginal probabilities.

TABLE 7.2 Percentage Predictions and Outcomes for the Logit Model.

	Share price rises	Share price falls	Total (%)
$p > 0.5$	36	15	51
$p < 0.5$	14	35	49
Total (%)	50	50	100

7.3 AN ASIDE ON MAXIMUM LIKELIHOOD

We have already introduced the method of maximum likelihood in an earlier chapter. However, this method is of particular value when dealing with limited dependent variable models, and this is therefore, a good time to look at it in more detail. In doing so, we will introduce the idea of the *Fisher Information Matrix* and discuss how numerical methods can be used to calculate estimators when analytical solutions are not possible.

Maximum likelihood begins with assumption that we can write down the joint probability of observing a particular sample of data conditional on a set of parameters. For example, suppose we conduct a set of N independent Bernouilli trials. We observe k successes and $N - k$ failures. If the probability of a success is equal to p, then the joint probability of observing this outcome is given by the binomial distribution $p^k(1-p)^{N-k}$. Let us define the likelihood function for this case as

$$L(p) = p^k(1-p)^{N-k}. \tag{7.14}$$

This is a particularly easy likelihood function to work with because it depends on a single parameter p. It is usually easier to deal with a monotonic transformation of the likelihood function in the form of its logarithm which, in this case, we write as

$$LL(p) = k\ln(p) + (N-k)\ln(1-p). \tag{7.15}$$

The *score* is defined as the derivative of the log-likelihood with respect to its parameter(s). In this case, we have

$$\frac{dLL(p)}{dp} = \frac{k}{p} - \frac{N-k}{(1-p)}. \tag{7.16}$$

Assuming an interior solution, we can solve for the maximum-likelihood estimator by solving for the value of p which makes the score equal to 0. More generally, we will have a vector of parameters, meaning that we will have a set of first-order conditions which must be solved jointly for the solution. In this case, however, we have a single parameter solution of the form

$$\hat{p}_{ML} = \frac{k}{N}. \tag{7.17}$$

The *Fisher information* (or just the information) is defined as the expected value of the second derivative of the log-likelihood, that is,

$$I(p) = E\left[\left(\frac{dLL(p)}{dp}\right)^2\right]. \tag{7.18}$$

Again, this is simplified by considering a case in which we have a single parameter. In more general cases, the information will be defined as the outer product of the score vector with itself. This means that the information will be a square matrix with dimension equal to the number of parameters. If the log-likelihood function is twice differentiable, and if certain regularity conditions hold, then we can write the information as

$$I(p) = -E\left[\frac{d^2LL(p)}{dp^2}\right]. \tag{7.19}$$

If the number of parameters is >1, then this will be a matrix containing the second-order partial derivatives of the log-likelihood function on its diagonal and the cross-partial derivatives off the diagonal.

The information is useful because it allows us to calculate the variance of the maximum likelihood parameter estimates. In our example, we have

$$I(p) = -\frac{d^2LL(p)}{dp^2} = \frac{k}{p^2} + \frac{N-k}{(1-p)^2}. \tag{7.20}$$

The variance of the maximum-likelihood estimator is defined as the inverse of the information evaluated at the maximum-likelihood parameter, that is, $1/I(\hat{p}_{ML})$. In this case, we have

$$V(\hat{p}_{ML}) = \frac{k(N-k)}{N^3}. \tag{7.21}$$

More generally, for more than one parameter, the variance–covariance matrix of the maximum-likelihood estimator is defined as the inverse of the Fisher information matrix.

An advantage of this approach is that it lends itself naturally to numerical methods when analytical solutions are either not possible or prove intractable. Once we have determined the log-likelihood function, then it is straightforward to use numerical methods to solve for both the maximum-likelihood parameter estimates and their variances. This is particularly useful for problems involving limited dependent variables, where the log-likelihood function often has these characteristics, depending on the nature of the function relating the probabilities to the explanatory variables. In numerical analysis, the vector of first-order partial derivatives is referred to as the Jacobian vector, and the matrix of second-order partial and cross-partial derivatives is known as the Hessian matrix. These can often be calculated numerically and used to solve for a maximum of the likelihood function when analytical solutions are not available.

7.4 SOME ALTERNATIVE LIMITED DEPENDENT VARIABLE MODELS

So far, we have used the logit model to estimate conditional probabilities in a model with limited dependent variables. To do this, we have assumed that the probability that the right-hand side variable is equal to 1 can be written in terms of the formula $P(Y_i = 1) = \exp(a + \beta X_i) / 1 + \exp(a + \beta X_i)$. The problem is then one of using an appropriate estimation technique to estimate the unknown parameters a and β. The choice of the logit function was made simply on the basis that it has the necessary properties. In particular, it is always positive, always lies between 0 and 1, approaches 0 as $X \to -\infty$ and approaches 1 as $X \to \infty$. However, the logit function is by no means the only functional form which has these properties and there are other candidate functions we might consider. We will consider two alternatives. These are the *probit model* and the *extreme value model*.

The probit model is based on the cumulative distribution function for the normal distribution. Consider the function

$$\Phi(a + \beta X_i) = \int_{-\infty}^{a + \beta X_i} \varphi(s)\, ds, \tag{7.22}$$

where $\varphi(\ )$ is the probability density function of the normal distribution, then it is easy to see that (7.22) has all the necessary properties for a function that describes $P(Y_i = 1)$. Moreover, although (7.22) looks quite forbidding, the normal distribution is such a well-known distribution that calculation of the probabilities implied by it are quite straightforward (though again will require numerical methods.) Therefore, we can again use maximum-likelihood methods to obtain estimates of the unknown parameters α and β. If we apply the probit model to our data set for British Airways share prices, then we obtain the following results.

TABLE 7.3 Probit Regression Results - British Airways Share Model.

	Coefficient	**Standard error**	***T* Ratio**
$\hat{\alpha}$	-0.0250	0.0371	-0.673
$\hat{\beta}$	88.0622	5.5352	15.9095
Mean of independent variable		0.000301	
McFadden R-squared		0.1858	
	Share price rises	**Share price falls**	**Total (%)**
$p > 0.5$	36	15	51
$p < 0.5$	14	35	49
Total (%)	50	50	100

Although the individual coefficient estimates of the probit model look very different from those of the logit model, the models are, in fact, very similar. This is because these coefficients are parameters of the relevant likelihood functions and the functional forms for these are quite different. However, in terms of the accuracy with which the models fit the data, they are remarkably similar. This can be seen through the McFadden R^2 values which in both cases take a value of just over 0.18. This indicates that each model increases the value of the log-likelihood by a factor of about 18% relative to the naive model. Moreover, we can again evaluate the marginal effect of the right-hand side variable on $P(Y_i = 1)$ at its mean value as

$$\frac{\partial P(Y_i = 1)}{\partial X_i} = \varphi\left(\hat{\alpha}, \hat{\beta}, \bar{X}\right)\hat{\beta} = 35.136. \qquad (7.23)$$

This is very close to the value of 37.15 which we obtained for the logit model. Therefore, both the goodness of fit statistics and the marginal effects indicate that both models tell essentially the same story. This can be confirmed by examination of the extent to which these models accurately predict the direction of movements of BA shares as shown in Table 7.3. We see from this that the logit and probit models produce identical results. In each case, the proportion of correct predictions is 70% which compares with 50% from the naive model.[1]

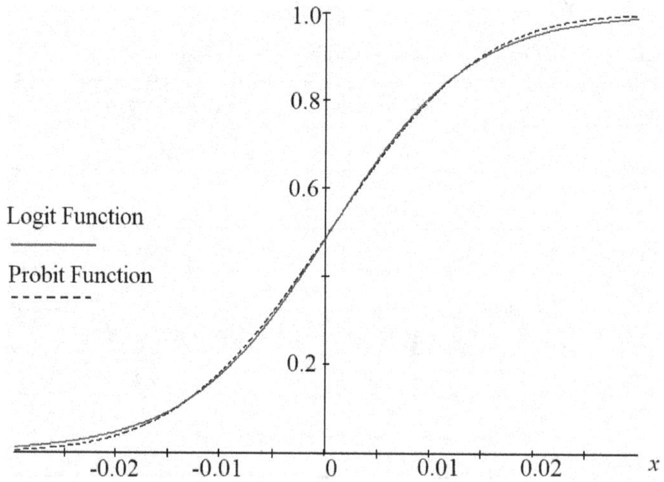

FIGURE 7.4 Logit and Probit Functions for Estimated Models.

The similarity of the logit and probit models can also be seen by plotting probability functions for the values of the estimated parameters. These functions are shown in Figure 7.4 that illustrates how similar are the results of the two models. For values of X in the middle of its range, there is virtually no difference between the two functions. The difference between the two functions tends to get a little larger for more extreme values of X.

[1] The 50% success rate of the naive model reflects the roughly even distribution of increases and decreases in the value of BA shares over the sample period. Prediction with the naive model is more or less equivalent to deciding whether the share value will rise or fall on the basis of the toss of a coin.

> **Historical Note:** The term "Probit" is first recorded as being used in a paper by Chester Bliss (1899–1979) in a paper in 1934 [Bliss1934]. The word is used as an abbreviation for "Probability Units." However, Bliss refers to the term as already being in use. If this is the case, then its previous use has not been found.

A third function that has been applied to the analysis of limited dependent variables is the *extreme value* which is also known as the *Gompit* or *Weibull* distribution. The functional form for the probability in this case can be written as

$$P(Y_i = 1) = \exp\left(-\exp\left(-\left(\alpha + \beta X_i\right)\right)\right). \tag{7.24}$$

This can also be shown to have the desirable properties for a probability distribution that $P(Y_i = 1) >$ for all values of X, that $P(Y_i = 1) \to 0$ as $X \to -\infty$, and that $P(Y_i = 1) \to 1$ as $X \to \infty$. However, this function differs from the logit and probit functions in being *asymmetric*. Consider the case in which $\alpha + \beta X_i = 0$, in the case of the logit and probit functions, we have $P(Y_i = 1 | \alpha + \beta X_i = 0) = P(Y_i = 0 | \alpha + \beta X_i = 0) = 0.5$. That is, the probabilities are evenly distributed around $\alpha + \beta X_i = 0$. However, this is not the case for the extreme value function where $P(Y_i = 1 | \alpha + \beta X_i = 0) = 0.3679$. This indicates that "successes" ($Y = 1$) are less likely than "failures" ($Y = 0$) with the extreme value model. When we estimate our model for BA share prices using the extreme value distribution, then we obtain the results shown in Table 7.4.

TABLE 7.4 Extreme Value Regression Results – British Airways Share Model.

	Coefficient	**Standard error**	***T* Ratio**
$\hat{\alpha}$	0.4024	0.0433	9.3029
$\hat{\beta}$	94.8987	5.8179	16.3115
Mean of independent variable		0.000301	
McFadden *R*-Squared		0.1819	
	Share price rises	**Share price falls**	**Total (%)**
$p > 0.5$	38	17	55
$p < 0.5$	12	33	45
Total (%)	50	50	100

The marginal effect for the extreme value regression can be calculated as

$$\hat{\beta}\exp\left\{-2\left(\hat{\alpha}+\hat{\beta}X_i\right)\right\}, \tag{7.25}$$

and evaluating this expression at the mean value of the right-hand side variable gives a value of 40.08. Therefore, an increase of 1% in the FTSE at the mean value will increase the probability of a rise in the BA share price by 0.4, close to the values obtained for the logit and probit models.

7.5 ANOTHER EXAMPLE: THE MARKET FOR ORANGES

Understanding the mechanics of the limited dependent variable model can be tricky. With this in mind, the following is another example of how this methodology might be applied in practice. We do not present any new results in this section. However, it does help bring together the important results we have derived already and, hopefully, it will help interested students understand the methodology better.

The example we consider concerns the market for oranges in the United States. Using the US Department of Agriculture Website (*www.usda.gov*), we have obtained data on total orange production in the United States as well as the price of oranges. This data have then been transformed into percentage changes with the price data being adjusted for general price movements by dividing by the consumer price index. The price data are then coded as 1 for a price increase and 0 for a price fall. The question we wish to address is whether an increase in orange production increases the probability that the price will fall. To do this, we estimate limited dependent variable models using the three methods we have discussion in the previous sections. The results are given in Table 7.5. Each method involves the estimate of an intercept α and a slope coefficient β for the probability function. Estimates of these coefficients are reported with their standard errors in parentheses below the estimate. ** indicates significance at the 1% level, that is, a p value <0.01, which is the case for the slope coefficient in all three cases. The McFadden R^2 is close to 0.22 in all three cases and, while the marginal effect evaluated at the mean is similar for the logit and probit regressions with a value close to –0.02, it is somewhat higher for the extreme value regression at –0.05.

TABLE 7.5 Comparative Limited Dependent Variable Regression Results US Market for Oranges 1981ñ2016.

	Logit	Probit	Extreme value
$\hat{\alpha}$	-0.3115	-0.1931	0.2338
	(0.3980)	(0.2376)	(0.2614)
$\hat{\beta}$	-0.0883**	-0.0554**	-0.0627**
	(0.0332)	(0.0196)	(0.0228)
McFadden R^2	0.2188	0.2252	0.2256
Marginal effect at mean	-0.0208	-0.0220	-0.0509

The comparison of results in Table 7.5 indicates that if our purpose is simply to assess if a given variable has a significant effect in the probability function, then it may not matter much which probability function we choose. All three functions show a significant effect of the change in quantity on the probability that prices will change. If we wish to calculate marginal effects on the probability, then the choice of functional form may be important. While the logit and probit models give very similar marginal effects when evaluated at the mean, the estimate from the extreme value model is noticeably different.

EXERCISES

Excel files containing the data for these exercises are available as companion files for this book.

EXERCISE 7.1

The probability distribution function for the Poisson distribution is given by the following expression:

$$f(\theta)=\theta^{x}\frac{\exp(-\theta)}{x!};\ x=1,2,\ldots$$

where θ is the parameter. An investigator has obtained a sample x_i, $i=1,\ldots,N$

a. Show that the maximum-likelihood estimator of the parameter is $\hat{\theta}_{ML}=\sum_{i=1}^{N}x_i/N$.

b. Show that the maximum-likelihood estimator of the variance of the parameter is $\hat{V}(\hat{\theta}_{ML})=\sum_{i=1}^{N}x_i/N^2$.

EXERCISE 7.2

An investigator tosses a coin ten times and observes eight heads and two tails. Using the method of maximum likelihood, test the hypothesis that this coin is fair (heads and tails equally likely) or biased (heads more likely than tails).

EXERCISE 7.3

An econometrician estimates two models for the US market for potatoes. The first is a linear regression of *DP* on *DQ*, where *DP* is coded as 1 when prices increase and 0 otherwise and *DQ* is the percentage change in production. The results are given in the table below:

```
OLS Regression Results
Sample period: 1976-2017
Dependent Variable DP
Sample Size 42
```

Variable	Coefficient	Std Err	*T* Ratio
C	0.457332	0.060100	7.609500
DQ	-0.052495	0.009826	-5.342234

R-squared	0.4164	*F* statistic	28.5395
SEE	0.384237	RSS	5.905530
Durbin-Watson	2.1209	LogL	-18.398031
ARCH(1) test	0.2323	AIC	0.971335
Jarque-Bera	1.1716	SIC	1.054081

He/she then estimates a logit model using the same data and obtains the following results:

```
Logit Estimates
Newton-Raphson Method
Dependent Variable DP
Sample Size 42
Iterations 4
```

Variable	Coefficient	Std Err	*T* Ratio
Constant	-0.239086	0.421524	-0.567195
DQ	-0.342064	0.102714	-3.330263

Mean of RHS variable	1.0014
SD of RHS variable	1.0136
Log-likelihood	-17.7023
Restricted LogL	-28.3456
McFadden *R*-squared	0.375483
Marginal effect	-0.085287

a. Plot the probability function $P\left(DP_t > 0 \middle| DQ_t\right)$ over the range $-14 \ldots 14$ (the approximate range for DQ).

b. Plot the marginal effect of changes in DQ for the logit model over the same range.

EXERCISE 7.4

Using the data in the Excel data file CADB.XLSX, estimate the linear probability model and the logit model relating the direction of movement of Cadbury–Schweppes shares to the change in the value of the stock market. Explain the meaning of the slope coefficient in each case and interpret your results.

EXERCISE 7.5

Using the data in the Excel data file CADB.XLSX, estimate logit, probit, and extreme value models for the direction of movement of Cadbury–Schweppes shares. Assess which of these provides the best predictor of the endogenous variable.

REFERENCES

[Berkson1944] Berkson, J., "Application of the Logistic Function to Bio-Assay." *Journal of the American Statistical Association*, 1944, 39(227), pp. 357–365.

[Bliss1934] Bliss, C. I., "The Method of Probits." *Science*, 1934, 79, pp. 38–39.

CHAPTER 8

Stochastic Regressors

The term "stochastic regressors" refers to any situation in which the right-hand side variables of the regression equation are themselves random variables. This is more consistent with the realities of econometric analysis than the classical assumption of regressors that are "fixed in repeated samples." Much of the discussion of stochastic regressors concerns the circumstances under which the right-hand side variable(s) in the regression can be treated as exogenous. Since the conditions for exogeneity tend to be more demanding for time-series data, we will, therefore, use time series notation in this section. So far, we have maintained the assumption that the only source of randomness in the regression model is the error term u. This is reasonable in the experimental sciences where the X variable consists of an input that is fixed by the person carrying out the experiment. However, it is unrealistic for most economic models in which the X variable is more likely to be a random variable which lies outside the control of the econometrician estimating the equation. The purpose of this chapter is to investigate the implications of working with stochastic regressors and to discuss some methods through which problems associated with this issue can be resolved.

8.1 EXOGENOUS REGRESSORS

Our first task is to define what we mean by the term exogeneity in the context of stochastic regressors. Consider the regression model $Y_t = \beta X_t + u_t;\ t = 1,\ldots,T$; Koopmans [Koopmans1950] defines exogeneity of the X variable as $E(X_t u_{t+i}) = 0$ for all values of i. That is, the current value of the X variable is uncorrelated with all values of the error term – past, present, and future. This condition has since been given the term *strict exogeneity* to distinguish it from other concepts of exogeneity that have been introduced into the econometrics literature. It is a very demanding condition, and it was quickly realized that it was not necessary for the purposes of estimation. A rather weaker condition is that the X variable is uncorrelated with the current and future values of the error, that is, $E(X_t u_{t+i}) = 0 \quad i \geq 0$. If X satisfies this condition, then it is said to be *predetermined*. Even this may be stronger than necessary for the purposes of estimation and some authors prefer to employ the assumption of *contemporaneous exogeneity*, which simply requires $E(X_t u_t) = 0$.

The clearest, and the most rigorous, definitions of exogeneity are provided by Engle, Hendry, and Richard [Engle1983]. In this paper, they argue that exogeneity cannot be defined independently of the purpose for which the assumption is being made. A variable may be exogenous for the purposes of estimation but may not be so when it comes to simulating a model for the purposes of forecasting. For this reason, they do not provide a unique definition of exogeneity, but rather a series of definitions that are applicable in different circumstances.

In the context of estimation, the Engle-Hendry-Richard (EHR) concept of exogeneity is termed *weak exogeneity*. It can be stated as follows. Consider the regression model $Y_t = \beta X_t + u_t$ where Y_t and X_t are jointly distributed random variables. The variable X_t can be considered weakly exogenous for the purpose of estimating the parameter β, if the parameters of the conditional distribution of Y_t given X_t do not depend on the marginal distribution of X_t. Note that weak exogeneity is defined in terms of a parameter of interest. In this situation described, we can always find a decomposition of the joint distribution into conditional and marginal distributions where the X_t variable is weakly exogenous for some parameter. The question is whether or not this is the parameter we wish to estimate, that is, the parameter of interest.

EHR introduces two other concepts of exogeneity that are relevant in other circumstances. The first is *strong exogeneity*. This is applicable when we wish to use the model we estimate for the purposes of forecasting. For this, we need the X variable to be weakly exogenous for the purpose of estimating the parameter(s) of interest and we also require that there be no feedback effects from lagged values of Y on the current value of X. This is often stated as the requirement that the X variable must not be *Granger caused* by Y. The concept of Granger causality will be discussed in more detail in a later chapter. Finally, there is the concept of *super exogeneity* which is relevant when the model is to be used for policy analysis. This requires that the parameters of the conditional distribution should be constant when the stochastic process determining the X variable changes. This is a demanding condition for which it is very difficult to test.

8.2 IMPLICATIONS FOR ORDINARY LEAST SQUARES ESTIMATION

Having discussed the concept of exogeneity, we now turn to the properties of the least-squares estimator when the right-hand side variable is stochastic. Let us consider the OLS estimator derived in an earlier chapter. Using the standard model in mean deviation form $Y_t = \beta X_t + u_t$ we have shown that the OLS estimator of the slope coefficient can be written $\hat{\beta} = \beta + \sum_{t=1}^{T} X_t u_t / \sum_{t=1}^{T} X_t^2$. To prove unbiasedness, we then applied the expectations operator to this expression using two of the standard Gauss-Markov assumptions – that $E(u_t) = 0$ and that X is non-stochastic. The combination of these two assumptions allows us to write $\sum_{t=1}^{T} E(X_t u_t) = \sum_{t=1}^{T} X_t E(u_t) = 0$, thus demonstrating unbiasedness under the Gauss-Markov conditions. When the X variable is itself stochastic, it is extremely difficult, and in most cases impossible, to prove unbiasedness in this way. This is because, when X is stochastic, it is not generally true that $E(X_t u_t) \neq E(X_t)E(u_t)$. Consider, for example, the case in which X and u both have expectation zero but are correlated with correlation coefficient ρ. In this case, we have $E(X_i u_i) = \rho\sigma_x\sigma_u$ which is clearly non-zero except for the special case $\rho = 0$.

Since it is no longer practical to prove unbiasedness when we relax the fourth Gauss-Markov condition, we instead make use of an alternative concept – that of *consistency*. Consistency is a large sample property and can be thought of as the requirement that the estimator $\hat{\beta}$ should converge toward the true value β as the sample size becomes large. Convergence is defined in terms of the *probability limit* of the estimator. The definition of a probability limit can be stated as follows. Let $\hat{\beta}_T$ be an estimator of an unknown parameter based on a sample of size T. $\hat{\beta}_T$ converges in probability to $\bar{\beta}$ if for any $\varepsilon > 0$ there exists a sample size T which is sufficiently large that the probability that the absolute difference between $\hat{\beta}_T$ and $\bar{\beta}$ exceeds ε is equal to zero. That is, we require

$$\lim_{T\to\infty} P\left(\left|\hat{\beta}_T - \bar{\beta}\right| > \varepsilon\right) = 0. \tag{8.1}$$

A more concise notation for this is to write plim $\hat{\beta} = \bar{\beta}$ and to note that consistency requires plim $\hat{\beta} = \beta$ where β is the true value of the coefficient.

It is easy to demonstrate that, if an estimator is asymptotically unbiased, and if its variance has a limiting value of zero, then the estimator is consistent. These are often useful conditions for demonstrating the consistency of any estimator but it should be noted that they are sufficient, but not necessary, for this to be true. The behavior of the probability density function for a consistent estimator is illustrated in Figure 8.1. This shows the PDF for an estimator of a coefficient whose true value is 0.5. Small sample estimates are biased as shown by the peak of the PDF which lies to the left of the true value. However, as the sample size increases, the peak shifts toward the true value and the variance falls. In the limit, the PDF collapses onto a vertical line passing through the true value of the parameter.

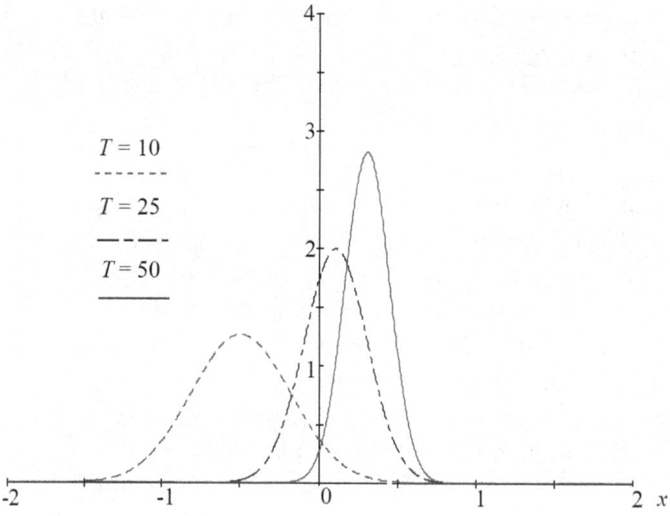

FIGURE 8.1 Behavior of the PDF for a Biased but Consistent Estimator.

The main advantage of working with probability limits is that they allow a number of mathematical operations that are not possible with the expectations operator. Assuming that a and b are random variables, we have the following

$$\begin{aligned}
&\text{plim}(a \pm b) = \text{plim}(a) + \text{plim}(b) \\
&\text{plim}(ab) = \text{plim}(a)\,\text{plim}(b) \\
&\text{plim}\left(\frac{a}{b}\right) = \frac{\text{plim}(a)}{\text{plim}(b)} \text{ for plim}(b) \neq 0 \\
&\text{plim}(g(a)) = g\,\text{plim}(a),
\end{aligned}$$

where g is a continuous function that does not involve the sample size. These properties allow us to demonstrate the consistency of the OLS estimator under a modified set of Gauss-Markov assumptions. We begin by assuming that the sample moments of the joint distribution of X and u converge to their true values, that is,

$$\operatorname{plim}\frac{1}{T}\sum_{t=1}^{T}X_t u_t = \operatorname{cov}(X_t, u_t) = \sigma_{Xu}$$

$$\operatorname{plim}\frac{1}{T}\sum_{t=1}^{T}X_t^2 = \operatorname{var}(X_t) = \sigma_X^2.$$

It is now possible to show that, if we replace Gauss-Markov assumption 4, that the X variable is non-stochastic, with the alternative assumption that it is stochastic but uncorrelated with the random disturbance, then the OLS estimator is consistent. Using the standard formula for the OLS estimator and taking probability limits gives

$$\operatorname{plim}\hat{\beta} = \beta + \frac{\operatorname{plim}\,(1/T)\sum_{t=1}^{T}X_t u_t}{\operatorname{plim}\,(1/T)\sum_{t=1}^{T}X_t^2} = \beta + \frac{\sigma_{Xu}}{\sigma_X^2}. \tag{8.2}$$

Next, using the modified Gauss-Markov assumption 4, we have $\sigma_{Xu} = 0, \sigma_X^2 > 0 \Rightarrow \operatorname{plim}\hat{\beta} = \beta$, which shows that OLS is a consistent estimator. This demonstration generalizes easily to the multivariate case.

When considering the properties of models with stochastic regressors we tend to rely on large sample properties, such as consistency, rather than exact small sample results. However, this creates a problem when it comes to comparing the distributions of alternative estimators since the variance of a consistent estimator goes to zero as the sample size gets large. The usual method for dealing with this problem is to consider the following a scaling of the distribution which gives a finite positive variance. Under the assumption that the sample variances and covariances converge in probability on the population values, we have

$$\sqrt{T}\left(\hat{\beta} - \beta\right)^{a} \sim N\left(0, \frac{\sigma_u^2}{\sigma_X^2}\right). \tag{8.3}$$

The asymptotic normality of this variable can be demonstrated using the central limit theorem (see the discussion in Greene [Greene1993] for more detail). The property that the variance is non-zero and finite means that this transformation can be used as the basis for a comparison of alternative estimators.

There are many estimators that are biased but nevertheless consistent. For example, consider the following estimator of the slope coefficient from a regression equation

$$\hat{\beta}_T = \frac{\sum_{t=1}^{T} X_t Y_t}{\sum_{t=1}^{T} X_t^2} + \frac{1}{T}. \tag{8.4}$$

This is clearly biased since $E\left(\sum_{t=1}^{T} X_t Y_t / \sum_{t=1}^{T} X_t^2 + 1/T\right) = \beta + (1/T)$. However, we can easily demonstrate that $\text{plim}\,\hat{\beta}_T = \beta$ meaning that the estimator is consistent. First, we have $E\left(\hat{\beta}_T\right) = \beta + 1/T$, and thus the limit of the expectation as T tends to infinity is clearly β. Second, we have $V\left(\hat{\beta}_T\right) = \sigma_u^2 / \sum_{t=1}^{T} X_t^2$, which tends to zero as T tends to infinity. Thus, this estimator meets sufficient conditions for consistency even though it is biased in small samples.

It is tempting to think that, because consistency is a large-sample property, that it is a weaker condition than unbiasedness, which holds in small samples. This is not true, however, since it is possible for an estimator to be unbiased but inconsistent. For example, consider the OLS estimator of the slope coefficient in the following regression model

$$Y_t = \beta\left(\frac{1}{t}\right) + u_t. \tag{8.5}$$

The OLS estimator is unbiased by virtue of the fact that $1/t$ is non-stochastic, and therefore, $E(u_t / t) = 0$. However, it is inconsistent since the variance does not tend to zero as T tends to infinity. This is because $\sum_{t=1}^{T} (1/t)^2$ converges to a finite limit as T tends to infinity. This case is admittedly very unusual, and in most circumstances, an unbiased estimator will also be consistent. However, the two concepts are logically separate and, as our examples show, there is no guarantee that either one implies the other.

8.3 ASYMPTOTIC DISTRIBUTION THEORY

Large-sample or asymptotic results are extremely important in econometrics because of the nature of economic data. This is because the stochastic nature of the regressors precludes the derivation of exact small sample results and we are forced to rely on large sample properties of our estimators. However, the use of large-sample properties creates a number of problems which we discuss below.

It is generally desirable that parameter estimates should "converge" on the true value of the parameter of interest as the sample size gets large.

However, there are several possible definitions of what constitutes such convergence. Let $\hat{\theta}_T$ be an estimator of an unknown parameter θ based on a sample size T. There are three possible definitions of convergence which we list below.

1. $\lim_{T\to\infty} E\left(\hat{\theta}_T\right) = \theta$
2. $E\left(\sqrt{T}\left(\hat{\theta}_T - \theta\right)\right) = 0$ as $T \to \infty$
3. $\text{plim}\left(\hat{\theta}_T\right) = \theta$

The first of these simply states that the limit of the expected value of the estimator should equal the true value as the sample size gets large, the second states that the mean of the limiting distribution of $\sqrt{T}\left(\hat{\theta}_T - \theta\right)$ should be zero, while the third states that the probability limit of the estimator should equal the true value. An estimator that satisfies either definition 1 or 2 is described as asymptotically unbiased, while an estimator that satisfies definition 3 is described as consistent. For many well-behaved estimators, all three of these properties will hold. However, it is not difficult to think of examples in which one or more of them fails. For example, there are cases in which the limit of the expected value simply does not exist. In practice, consistency is often the easiest concept to work with, but this does require very strong assumptions.

A second problem is that the asymptotic distribution of estimators is often degenerate. By this, we mean that the variance of the estimator goes to zero as the sample size becomes large. Indeed, this is a defining property of a consistent estimator. This creates a problem when it comes to comparing one consistent estimator with another since both have zero variance in large samples. The usual method of dealing with this problem is to work with a transformation of the distribution of the parameter in question which is not degenerate. For example, consider the sample mean as an estimator of the population mean. Assuming a normal distribution, we have $\bar{X}_T \sim N\left(\mu_X, \sigma_X^2 / T\right)$. This is a degenerate distribution since the variance of the sample mean goes to zero as $T \to \infty$. However, if we consider the transformed variable $\sqrt{T}\left(\bar{X}_T - \mu_X\right)$, then we can show that $\sqrt{T}\left(\bar{X}_T - \mu_X\right) \sim N\left(0, \sigma_X^2\right)$ has a non-degenerate distribution. We could, therefore, base efficiency comparisons of alternative estimators on the distribution of transformations of this kind rather than on the distribution of the estimator itself.

8.4 THE ERRORS IN VARIABLES MODEL

We have established that correlation between the right-hand side variables of the regression equation and the error term means that ordinary least squares estimates are inconsistent. In this section, we consider an important case in which such correlation naturally arises. This is the *errors in variables model* which arises when the right-hand side variables are measured with error. Consider the following model

$$Y_t = \beta X_t^* + u_t \tag{8.6}$$

$$X_t^* = X_t + \varepsilon_t, \tag{8.7}$$

where we assume $\sigma_{X^*u} = 0, \sigma_{u\varepsilon} = 0$. We would like to be able to estimate the parameter β from equation (8.6), but we do not observe X^* directly. Instead, we observe a proxy variable X which deviates from X^* by measurement error ε. The measurement error is uncorrelated with the error in the regression model and the error in the regression model is uncorrelated with X^*.

What happens if we estimate a regression equation in which the proxy variable is substituted for X^*? This situation arises frequently in applied econometrics when economic theory suggests the inclusion of variables for which no data is available, and we are forced, in such circumstances, to rely on proxy variables. Consider the regression equation (8.6), if we substitute for the right-hand side variable using (8.7), then we have the following equation

$$Y_t = \beta X_t + u_t + \beta\varepsilon_t. \tag{8.8}$$

The error term in (8.8) now comprises the original error term plus an additional component that depends on the measurement error from (8.7). Moreover, the X variable and the composite error are now correlated since $\sigma_{X\varepsilon} = -\sigma_\varepsilon^2$. This means that OLS will generate inconsistent estimates of the parameter of interest.

We can say more about the nature of the inconsistency of the OLS estimates by considering the probability limit of the OLS estimator. Substituting (8.8) into the standard formula for the OLS estimator yields the following expression.

$$\hat{\beta} = \beta + \frac{(1/T)\sum_{t=1}^{T} X_t u_t}{(1/T)\sum_{t=1}^{T} X_t^2} + \beta \frac{(1/T)\sum_{t=1}^{T} X_t \varepsilon_t}{(1/T)\sum_{t=1}^{T} X_t^2}. \tag{8.9}$$

Taking probability limits means that the second term in (8.9) can be discarded since $\operatorname{plim}(1/T)\sum_{t=1}^{T} X_t u_t = 0$ by assumption. However, the third term does not go to zero because of the correlation between X and ε. We have

$$\operatorname{plim}\hat{\beta} = \beta\left(1 - \frac{\sigma_{\varepsilon}^{2}}{\sigma_{X}^{2}}\right). \tag{8.10}$$

This illustrates several important points. The first is that the OLS estimator will underestimate the true regression parameter when the exogenous variable is measured with error. It also illustrates the point that the size of the inconsistency depends on the size of the variance of the measurement error relative to that of the variance of the explanatory variable X^* (since $\sigma_X^2 = \sigma_{X^*}^2 + \sigma_\varepsilon^2$). Thus, the more "noise" we introduce into the system in terms of measurement errors, the more inconsistent our estimate becomes. This gives us some insight into the circumstances in which the use of proxy variables may be acceptable and those in which they are likely to produce highly misleading results.

Example: The following model was used to generate artificial data sets using a random number generator.

$$X_t^* \sim N(0,1); \quad u_t \sim N(0,1); \quad \varepsilon_t \sim N(0,0.25)$$
$$\operatorname{cov}(X_t^*, u_t) = 0; \qquad \operatorname{cov}(u_t, \varepsilon_t) = 0$$
$$X_t^* = X_t + \varepsilon_t$$
$$Y_t = 1.0 + 0.5X_t^* + u_t$$

1,000 regressions of Y on X were then estimated using the data generated (in each case using a large sample of 1,000 observations). The distribution of the slope coefficient estimates was then examined. The result was an average slope estimate of 0.393. This is extremely close to the theoretically predicted value of the plim of the slope coefficient which can be calculated as $\operatorname{plim}\hat{\beta} = \beta\left(1 - \sigma_\varepsilon^2/\sigma_X^2\right) = 0.5(1 - 0.25/(1 + 0.25)) = 0.4$.

Historical Note: Adcock [Adcock1877] is often referenced as the first to discuss the errors in variables problem. However, Wald [Wald1940] provides the first treatment of the problem as we recognize it in modern econometrics.

8.5 THE INSTRUMENTAL VARIABLES ESTIMATOR

We have seen that the OLS estimator is inconsistent when a variable on the right-hand side of the equation is correlated with the error term. It, therefore, becomes important to find alternative estimators with superior properties. One possible alternative estimator is the *Instrumental Variables* (IV) estimator. To construct this estimator, we need to find a variable Z which has the properties that it is not correlated with the error but is correlated with the X variable, that is, $\text{cov}(Z_t, u_t) = 0$ but $\text{cov}(Z_t, X_t) \neq 0$. It may not be obvious where such a variable can be found but, if we proceed for the moment on the assumption that we have a suitable Z available, then we can demonstrate the properties of an estimator based around it. The issue of how to find, or construct, such a variable will be considered later.

The estimation problem we have is the standard one of finding an estimator of the unknown slope coefficient of an equation $Y_t = \beta X_t + u_t$. An estimator can be constructed using the variable Z which is defined as the *instrument* and the estimator itself is referred to as the *instrumental variable estimator*. It takes the form

$$\hat{\beta}_{IV} = \frac{\sum_{t=1}^{T} Z_t Y_t}{\sum_{t=1}^{T} Z_t X_t}. \tag{8.11}$$

Substituting for Y_i and expanding yields

$$\hat{\beta}_{IV} = \frac{\sum_{t=1}^{T} Z_t (\beta X_t + u_t)}{\sum_{t=1}^{T} Z_t X_t} = \beta + \frac{(1/T)\sum_{t=1}^{T} Z_t u_t}{(1/T)\sum_{t=1}^{T} Z_t X_t}. \tag{8.12}$$

Taking probability limits of the expression for the IV estimator in (8.12) yields $\text{plim}\,\hat{\beta}_{IV} = \beta + \sigma_{Zu} / \sigma_{ZX} = \beta$ since $\sigma_{Zu} = 0$ and $\sigma_{ZX} \neq 0$ by assumption. Therefore, the instrumental variable estimator can be shown to be consistent under the assumption that the instrument is uncorrelated with the equation error.

Next, consider the variance of the instrumental variable estimator. We have

$$\left(\hat{\beta}_{IV} - \beta\right)^2 = \left(\frac{\sum_{t=1}^{T} Z_t u_t}{\sum_{t=1}^{T} Z_t X_t}\right)^2. \tag{8.13}$$

Taking probability limits yields $\text{plim}\left(\hat{\beta}_{IV} - \beta\right)^2 = \sigma_u^2 \left(\sum_{t=1}^{T} Z_t^2 / T^2\right) \rho_{XZ}^2 \sigma_Z^2 \sigma_X^2 = \left(\sigma_u^2 / T\right) \rho_{XZ}^2 \sigma_X^2$, where ρ_{XZ} is the correlation coefficient between X and Z and can be written as $\rho_{XZ} = \sigma_{XZ} / \sigma_X \sigma_Z$. This shows that the distribution of the instrumental variable estimator is degenerate since its variance goes to zero as the sample size becomes large. However, multiplication by $\sqrt{T}$ produces a distribution that is not degenerate. Assuming that the errors are normally distributed, means that we can write an asymptotic distribution of the form

$$\sqrt{T}\left(\hat{\beta}_{IV} - \beta\right) \overset{a}{\sim} N\left(0, \frac{\sigma_u^2}{\rho_{XZ}^2 \sigma_X^2}\right). \tag{8.14}$$

It is clear from (8.14) that the instrumental variable estimator is less efficient than the OLS estimator since the asymptotic variance of $\sqrt{T}\left(\hat{\beta}_{OLS} - \beta\right)$ is σ_u^2 / σ_X^2. The presence of the correlation coefficient in the denominator of the variance term in (8.14) determines the degree of inefficiency. Note that $0 < \rho_{XZ}^2 < 1$, and therefore, the variance of the IV estimator exceeds that of the OLS estimator. The lower the correlation between X and Z (i.e., the closer ρ_{XZ}^2 is to zero), then the less efficient is the IV estimator.

The sample variance of the OLS estimator is calculated as

$$\hat{V}\left(\hat{\beta}_{IV}\right) = \hat{\sigma}_u^2 \frac{\sum_{t=1}^{T} Z_t^2}{\left(\sum_{t=1}^{T} Z_t X_t\right)^2} = \frac{\hat{\sigma}_u^2}{\hat{\rho}_{XZ}^2 \sum_{t=1}^{T} X_t^2}. \tag{8.15}$$

Therefore, the sample variance of the instrumental variable estimator is always higher than the sample variance of the OLS estimator. The difference between the two depends on the correlation between the instrument and the X variable. A high correlation between these variables will lessen the loss of efficiency associated with the instrumental variable method. Ideally, therefore, instruments should be uncorrelated with the regression error, to ensure consistency, but as closely correlated with the right-hand side variable as possible, to ensure efficiency. Instruments that are only weakly correlated with the right-hand side variable are referred to as "weak instruments."

Historical Note: The first use of the term "instrumental variable" comes in the dissertation by Olav Reiersøl [Reiersøl1945]. However, the first recorded use of the technique is in a book by Phillip Wright (father of Sewell Wright) in 1928.

Example: To illustrate the use of the instrumental variable estimator, we repeat the exercise for the errors in variables model. However, in this case, we assume the existence of a variable that has the desired properties for the instrumental variable estimator. In practice we define a new random variable $Z_t = X_t^* + v_t$ where $v_t \sim N(0,1)$. Given the existence of this variable, we then estimate 1,000 regressions using the OLS estimator and the IV estimator and compare the distribution of the parameter estimates obtained.

The results of this experiment are shown in Figure 8.2, which shows the distributions of the parameter estimates. Once again, the OLS estimator is clearly inconsistent. Even in a large sample of 1,000 observations the average slope coefficient is 0.4018. If we compare this with the distribution of the instrumental variable estimator, then we see that the mean value here is 0.4967 which is much closer to the true value of 0.5. However, this reduction in bias comes at a cost. The standard deviation of the OLS estimator is 0.029 which compares with 0.047 for the IV estimator. Thus the mean square error of the OLS estimator is $0.1^2 + 0.029^2 = 0.0108$ whereas that of the IV estimator is $0.047^2 = 0.0022$. Therefore, we would still prefer the IV estimator on the mean square error criterion. If the correlation between the X and the Z variables were lower, however, then it is possible that this could be reversed and we might choose the OLS estimator on the MSE criterion, even though it remains inconsistent.

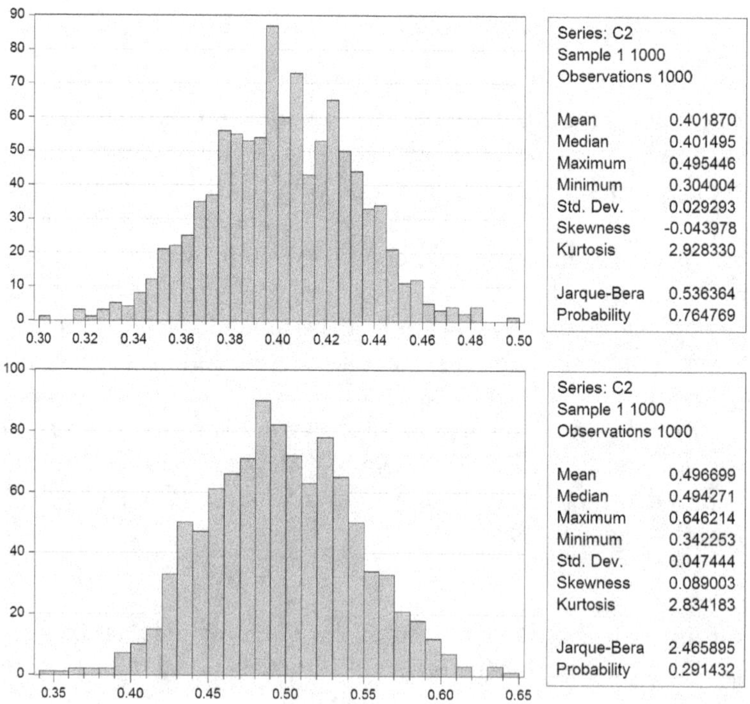

FIGURE 8.2 Comparison of OLS and Instrumental Variables Estimators in the Errors in Variables Model.

8.6 SIMULTANEOUS EQUATIONS

Another case in which stochastic regressors create problems for estimation is when the equation comprises one equation drawn from a system of simultaneously determined endogenous variables. For models of this kind, we must first establish whether it is even possible to estimate the parameters of interest. This is the issue of *identification*. It is useful to consider an example here before we attempt to look at the general case. Let us consider the system of equations defined by (8.16)

$$\begin{aligned} Y_{1t} &= \beta_{11} Y_{2t} + \gamma_{11} X_{1t} + \gamma_{12} X_{2t} + u_{1t} \\ Y_{2t} &= \beta_{21} Y_{1t} + \gamma_{21} X_{1t} + \gamma_{22} X_{2t} + u_{2t} \end{aligned} \tag{8.16}$$

X_1 and X_2 are weakly exogenous variables, u_1 and u_2 are independent errors, and Y_1 and Y_2 are jointly determined endogenous variables. This is typical of the type of structure derived from economic theory and is referred to as the *structural form* of the model. An example could be the demand and supply model in which Y_1 and Y_2 are price and quantity while X_1 and X_2 are independent variables such as incomes and weather conditions. The parameters of this system are referred to as the *structural parameters* of the model. These consist of the slope coefficients for other endogenous variables (the β coefficients), the slope coefficients for the independent variables (the γ coefficients), and the variances of the equation errors. It is usual to assume that the errors in the structural form are uncorrelated. However, this assumption can be relaxed if necessary. In general, the structural parameters of the model will be the parameters of interest for the purposes of estimation.

We can solve (8.16) to obtain equations each of which contains only one endogenous variable. This is referred to as the *reduced form* of the model. In this case, we have

$$\begin{aligned} Y_{1t} &= \pi_{11} X_{1t} + \pi_{12} X_{2t} + v_{1t} \\ Y_{2t} &= \pi_{21} X_{1t} + \pi_{22} X_{2t} + v_{2t} \\ \pi_{11} &= \frac{1}{\Delta}(\gamma_{11} + \beta_{11}\gamma_{21}) \quad \pi_{12} = \frac{1}{\Delta}(\gamma_{12} + \beta_{11}\gamma_{22}) \\ \pi_{21} &= \frac{1}{\Delta}(\beta_{21}\gamma_{11} + \gamma_{21}) \quad \pi_{22} = \frac{1}{\Delta}(\beta_{21}\gamma_{12} + \gamma_{22}) \\ v_{1t} &= \frac{1}{\Delta}(u_{1t} + \beta_{11} u_{2t}) \quad v_{2t} = \frac{1}{\Delta}(\beta_{21} u_{1t} + u_{1t}) \\ \Delta &= (1 - \beta_{11}\beta_{21}) \end{aligned} \tag{8.17}$$

The π parameters are referred to as the *reduced form parameters*. Since the right-hand side variables in (8.17) are weakly exogenous, it follows that we can obtain consistent estimates of the reduced form parameters by least-squares. However, these are not the parameters of interest. The question, therefore, becomes whether we can use the consistent estimates of the reduced form parameters to obtain consistent estimates of the structural parameters.

It is immediately obvious from (8.16) and (8.17) that it will not be possible to obtain estimates of all the structural parameter from the reduced form without some restrictions on the structural form. There are eight structural parameters in (8.16), in the form of the six slope coefficients, and the variances of the two errors. Allowing for a non-zero covariance between the two errors would increase the number of structural parameters to nine. In contrast, the reduced form contains only seven pieces of information in the form of the four slope coefficients plus the variances and covariance of the reduced form errors v_1 and v_2. It, therefore, becomes necessary to place restrictions on the structural form to get estimates of the parameters of interest. These restrictions can take the following forms:

1. Exclusion restrictions. For example, $\gamma_{11} = 0$ which excludes X_1 from the first structural equation.
2. Specific parameter values. For example, $\gamma_{11} = 1$ which ensures that X_1 enters the first structural equation with a unit coefficient.
3. Cross variable restrictions. For example, $\gamma_{11} + \gamma_{12} = 0$ which ensures that X_1 and X_2 have equal magnitude but opposite sign in the first structural equation.

It is the imposition of restrictions which *identifies* the structural parameters. That is, restrictions make it possible to find a mapping from the reduced form parameter estimates to the parameters of interest.

As an example, consider the case in which we impose the restriction $\gamma_{12} = 0$. This acts to exclude the X_2 variable from the first structural equation. The relationship between the reduced form and structural form coefficients now becomes

$$\begin{aligned} &\pi_{11} = \frac{1}{\Delta}(\gamma_{11} + \beta_{11}\gamma_{21}) \quad && \pi_{12} = \frac{1}{\Delta}\beta_{11}\gamma_{22} \\ &\pi_{21} = \frac{1}{\Delta}(\beta_{21}\gamma_{11} + \gamma_{21}) \quad && \pi_{22} = \frac{1}{\Delta}\gamma_{22} \\ &\Delta = (1 - \beta_{11}\beta_{21}). \end{aligned} \tag{8.18}$$

It is now possible to solve for the structural parameters of the first equation as

$$\beta_{11} = \pi_{11} / \pi_{22} \quad \text{and} \quad \gamma_{11} = \pi_{11} - \beta_{11}\pi_{21}. \tag{8.19}$$

Thus, the first structural equation is identified by the imposition of this restriction. Note, however, that the second structural equation remains unidentified without the imposition of further restrictions.

The process of solving for the structural parameters in terms of the reduced form parameters is referred to as the method of *indirect least-squares.* The estimates of the structural form parameters generated by this approach are consistent because the estimators of the reduced form are consistent. Even when this method is possible, however, it is difficult to apply and there are more straightforward methods to recover the parameters of interest. Our concern, at present, is simply to establish the conditions under which estimation of the structural parameters is possible rather than determining a method for actually computing the estimates.

Having established the nature of the problem, let us now consider the more general case. We can write the following expression for a general system of linear structural equations

$$\boldsymbol{B}\boldsymbol{y}_t + \boldsymbol{\Gamma}\boldsymbol{x}_t = \boldsymbol{u}_t, \tag{8.20}$$

where $\boldsymbol{y}$ is a $G \times 1$ vector of endogenous variables, $\boldsymbol{x}$ is a $K \times 1$ vector of weakly exogenous variables and $\boldsymbol{u}$ is a $G \times 1$ vector of random errors. The matrices $\boldsymbol{B}$ and $\boldsymbol{\Gamma}$ are $G \times G$ and $G \times K$ matrices of structural parameters. The reduced form of the system can be derived as

$$\boldsymbol{y}_t = -\boldsymbol{B}^{-1}\boldsymbol{\Gamma}\boldsymbol{x}_t + \boldsymbol{B}^{-1}\boldsymbol{u}_t = \boldsymbol{\Pi}\boldsymbol{x}_t + \boldsymbol{B}^{-1}\boldsymbol{u}_t. \tag{8.21}$$

Thus, the relationship between the structural and reduced form coefficients can be written as $\Pi = -\boldsymbol{B}^{-1}\Gamma$ or $\boldsymbol{B\Pi} + \boldsymbol{\Gamma} = \mathbf{0}_{G\times K}$, where $\boldsymbol{\Pi}$ is the $G \times K$ matrix containing the reduced form coefficients and $\mathbf{0}_{G\times K}$ is a $G \times K$ matrix of zeros. Alternatively, we can write

$$\begin{bmatrix} \boldsymbol{B} & \boldsymbol{\Gamma} \end{bmatrix} \begin{bmatrix} \boldsymbol{\Pi} \\ \boldsymbol{I}_k \end{bmatrix} = \mathbf{0}_{G,G+K}, \tag{8.22}$$

where $\boldsymbol{I}_K$ is the $K \times K$ identity matrix. Let $\boldsymbol{\alpha}_i$ be the ith row of $\begin{bmatrix} \boldsymbol{B} & \boldsymbol{\Gamma} \end{bmatrix}$. This contains the structural parameters of the ith equation of the model. We have

$$\boldsymbol{\alpha}_i \begin{bmatrix} \boldsymbol{\Pi} \\ \boldsymbol{I}_k \end{bmatrix} = \mathbf{0}_K, \tag{8.23}$$

which is a $K \times 1$ vector of linear equations. This defines K equations in a possible $G + K - 1$ unknown structural parameters. For a solution to be possible, we need the number of equations to be at least as many as the number of unknown parameters. Let g be the number of included endogenous variables and k be the number of included exogenous variables. For a solution to be possible we, therefore, need $g - 1 + k \leq K$ or $g - 1 \leq K - k$, that is, the number of included endogenous variables minus one in equation i must be less than or equal to the number of excluded exogenous variables. This is the *order condition for identification*. We have assumed here, that the only restrictions possible are exclusion restrictions. However, the framework is easily extended to deal with other forms of restriction.

The order condition is a convenient and quick way of assessing if a given equation is identified. In most situations, it will give us the correct answer. However, the order condition is necessary for identification but not sufficient. A more definitive answer to the question of whether an equation is identified is given by the *rank condition*. In practice, however, this is much more difficult to derive and is beyond the scope of this book. Using the order condition, and, assuming that it gives us the correct answer, we state the following

1. If the number of excluded exogenous variables is equal to the number of included endogenous variables is equal to one, $g - 1 = K - k$, then we say that the equation is *just identified*. In this case, there is a unique solution for the structural parameters in terms of the reduced form parameters.

2. If the number of excluded exogenous variables is greater than the number of included endogenous variables minus one, $g - 1 < K - k$, then we say that the equation is *over identified*. In this case, there are many solutions for the structural parameters in terms of the reduced form parameters.

3. If the number of excluded exogenous variables is less than the number of included endogenous variables minus one, $g - 1 > K - k$, then we say that the equation is *under identified*. In this case, there is no solution for the structural parameters in terms of the reduced form parameters.

8.7 ESTIMATION OF SIMULTANEOUS EQUATIONS MODELS

We have already seen two possible methods by which we can obtain consistent estimates of structural parameters in simultaneous equations models in the form of indirect least squares and instrumental variables. In this section, we will look at an example of these methods in practice and discuss how these relate to OLS estimation. The example we will consider is the Cobweb model, which is familiar from basic microeconomics modules. The structural form of this model can be written as follows

$$\begin{aligned} \Delta p_t &= \beta_{11} + \beta_{12}\Delta q_t + u_{1t} \\ \Delta q_t &= \beta_{21} + \beta_{22}\Delta p_{t-1} + u_{2t}, \end{aligned} \tag{8.24}$$

where Δp and Δq are percentage changes in price and quantity and u_1 and u_2 are random errors. The first equation here is the demand curve and the second is the supply curve. The model assumes a lagged response of quantity produced to the lagged price change which therefore acts as a predetermined, or exogenous, variable in the supply curve.

From the order condition, we see that both equations in this model are just identified. For the demand curve we have $g-1=K-k=1$ and, for the supply curve, we have $g-1=K-k=0$. This model is an example of a particular kind of simultaneous equations model known as a *recursive model*. This is because the supply curve features only one endogenous variable. More generally recursive models are structured so that we can order them so that each equation, in turn, contains one fewer endogenous variables, with the final equation containing a single such variable. We can show that it is possible to obtain consistent estimates of the structural parameters by OLS in models of this type. However, this does not stop us from using such a model to illustrate and discuss alternative estimators.

Let us first consider the relationship between the reduced form and structural form for this model. The reduced form of (8.24) can be written

$$\begin{aligned} \Delta p_t &= \pi_{11} + \pi_{12}\Delta p_{t-1} + v_{1t} \\ \Delta q_t &= \pi_{21} + \pi_{22}\Delta p_{t-1} + v_{2t} \\ \\ \pi_{11} &= \beta_{11} + \beta_{12}\beta_{21} \qquad & \pi_{12} &= \beta_{12}\beta_{22} \\ \pi_{21} &= \beta_{21} & \pi_{22} &= \beta_{22} \\ v_{1t} &= u_{1t} + \beta_{12}u_{2t} & v_{2t} &= u_{2t}. \end{aligned} \tag{8.25}$$

We, therefore, have a straightforward one-to-one mapping from the structural form to the reduced form parameters. The reduced form can be estimated consistently by least squares, and therefore, we can obtain consistent estimates of the structural parameters by the method of indirect least squares.

Example: Using annual data for the market for oranges in the United States for the years 1983–2016, we estimate the following pair of reduced form equations.

$$\begin{aligned}\Delta p_t &= 2.2363 - 0.3723\,\Delta p_{t-1} + \hat{v}_{1t} \\ \Delta q_t &= -0.3377 + 0.3389\,\Delta p_{t-1} + \hat{v}_{2t}.\end{aligned} \tag{8.26}$$

We have not reported standard errors or diagnostic tests for these equations because we are simply interested in these equations as a means of estimating the structural parameters. From (8.25) we can derive our structural form parameter estimates as follows

$$\begin{aligned}\hat{\beta}_{21} &= \hat{\pi}_{21} = -0.3377 \\ \hat{\beta}_{22} &= \hat{\pi}_{22} = 0.3389 \\ \hat{\beta}_{12} &= \frac{\hat{\pi}_{12}}{\hat{\pi}_{22}} = -\frac{0.3723}{0.3389} = -1.0986 \\ \hat{\beta}_{11} &= \hat{\pi}_{11} - \hat{\beta}_{12}\hat{\beta}_{21} = 2.2363 - (-1.0986 \times -0.3377) = 1.8653.\end{aligned} \tag{8.27}$$

Note that these parameter estimates are consistent but not efficient. This is because the recursive nature of the model means that OLS will also generate consistent estimates with lower variance in this case.

The example above makes clear that the method of indirect least squares (ILS) has a number of drawbacks. The algebra necessary to derive the relationship between the reduced form and the structural form is potentially difficult and this is something that will vary from one problem to another. Moreover, it is not easy to derive standard errors and other relevant statistics for the ILS estimates. This means that it is difficult to use the ILS method to conduct hypothesis tests. Finally, in the example given here, both equations are just identified. This means that the ILS method gives a unique solution. In cases where the equation is over-identified, there will be multiple solutions.

The method of instrumental variables provides an alternative method for estimation of the structural parameters with significant advantages relative to the indirect least-squares method. For our example, if we wish to estimate the demand curve, then we first need to find an appropriate instrument. This should have the properties that it is uncorrelated with the equation error but is correlated with the current right-hand side variable. An obvious candidate exists for just identified equations, in the form of the excluded exogenous variable. In the case of our demand curve Δp_{t-1} is a candidate instrument with the necessary properties. This is a general property of just identified equations because excluded exogenous variables are uncorrelated with the equation error by assumption but are correlated with the endogenous variables because the reduced form equations contain all the exogenous variables in the model.

Example: Using Δp_{t-1} as an instrument for Δq_t in the demand curve equation we obtain the following instrumental variables estimates for the demand curve.

$$\begin{aligned} \Delta p_t &= \underset{(2.3209)}{1.8652} - \underset{(0.4526)}{1.0988}\,\Delta q_t + \hat{u}_{1t} \\ \hat{\sigma}_u &= 13.5117 \qquad DW = 2.6843. \end{aligned} \tag{8.28}$$

An interesting property of this equation is that the parameter estimates are equal to those obtained using the ILS method (to the fourth decimal place). This is not a coincidence but rather, a standard property of IV estimates, when we use the excluded exogenous variables as the instruments in a just identified model. In this context, the IV method offers a much easier method of constructing consistent parameter estimates than the ILS method, with the added advantage that it allows us to calculate standard errors for the coefficient estimates, and therefore, to use the equation estimates for statistical inference.

We noted earlier that this system of equations is recursive. This means that it is not necessary to use ILS or IV to obtain consistent estimates in this case. Even though Δq_t is one of the endogenous variables of the model, it is uncorrelated with the error in the demand curve, and therefore, we can obtain consistent parameter estimates by the method of OLS. It is interesting to compare these with the IV estimates since this will illustrate the loss of efficiency from using IV in this context. Estimating the model by OLS yields the following results

$$\begin{aligned} \Delta p_t &= \underset{(2.0096)}{1.7408} - \underset{(0.1334)}{0.6653}\,\Delta q_t + \hat{u}_{1t} \\ \hat{\sigma}_u &= 11.7157 \qquad DW = 2.7814. \end{aligned} \tag{8.29}$$

Note that the coefficient standard errors are lower for the OLS estimates in (8.29), particularly that for the slope coefficient. This illustrates the value of using the most efficient estimator available. The overall fit of the OLS model is also better than that of the IV model, as evidenced by the lower value of $\hat{\sigma}_u$. This provides a better way of comparing the fit when we estimate the model by instrumental variables because the R-squared statistic is not a reliable measure of goodness of fit for the instrumental variable estimator and can fall outside the range zero to one.

> **Historical Note:** The problem of estimating structural parameters in simultaneous equations demand and supply models was first articulated clearly in papers by Holbrook Working [Working1925] and his brother Elmer Working [Working1927].

8.8 ESTIMATION OF OVER IDENTIFIED EQUATIONS

So far, we have assumed that the equation we wish to estimate is just identified. This simplifies things considerably because the number of instruments available to us is exactly equal to the number of right-hand side endogenous variables in the equation. There is, therefore, only one IV estimator available to us. Problems arise when we consider over identified equations because there are more excluded exogenous variables than right-hand side endogenous variables, and therefore, there is no longer a unique IV estimator.

Consider again, the simultaneous equations system defined in (8.16). Recall that the first equation of this system takes the form

$$Y_{1t} = \beta_{11} Y_{2t} + \gamma_{11} X_{1t} + \gamma_{12} X_{2t} + u_{1t}. \tag{8.30}$$

If we assume $\gamma_{11} = \gamma_{12} = 0$, then this excludes both X_1 and X_2 from the first equation. The equation is now over identified because there are two excluded exogenous variables and the number of included endogenous variables minus one is equal to one. Now both X_1 and X_2 are candidates to be instruments in the estimation of the remaining parameter β_{11}. Therefore, both $\hat{\beta}_{11}^{(1)} = \sum_{t=1}^{T} X_{1t} Y_{1t} / \sum_{t=1}^{T} X_{1t} Y_{2t}$ and $\hat{\beta}_{11}^{(2)} = \sum_{t=1}^{T} X_{2t} Y_{1t} / \sum_{t=1}^{T} X_{2t} Y_{2t}$ will yield consistent estimates of β_{11}. Moreover, these are not the only alternatives. X_1 and X_2 are suitable instruments because they have the property that $E(X_1 u_1) = E(X_2 u_1) = 0$ but, if this is the case, then any linear combination of these two variables $X = \theta_1 X_1 + \theta_2 X_2$ will also have the property that it will be uncorrelated with the error since $E(Xu) = \theta_1 E(X_1 u_1) + \theta_2 E(X_2 u_1) = 0$. It, therefore, follows that, for an over identified equation, there exists an infinite number of possible IV estimators corresponding to different linear combinations of the instruments.

The problem facing us is to choose between alternative consistent estimators. In these circumstances, we choose the most efficient, that is, we choose the estimator with the lowest asymptotic variance. Given the standard assumptions about convergence in probability of sample moments to population moments, the asymptotic distribution of the IV estimator can be derived as

$$\sqrt{T}\left(\hat{\beta}_{11}-\beta_{11}\right)^{a}\sim N\left(0,\frac{\sigma_u^2}{\rho_{ZY_2}^2\sigma_{Y_2}^2}\right), \tag{8.31}$$

where Z is the instrument and $\rho_{ZY_2}^2$ is the squared correlation between the instrument and the right-hand side endogenous variable. We, therefore, need to choose the instrument which maximizes this correlation in order to minimize the asymptotic variance. Assuming that X_1 and X_2 are candidate instruments, we regress Y_2 on the two instruments and calculate the fitted values, that is,

$$\hat{Y}_2=\hat{\theta}_1X_1+\hat{\theta}_2X_2. \tag{8.32}$$

This is a linear combination of the candidate instruments and is therefore itself a possible instrument. Moreover, it will have a higher correlation with the Y_2 variable than any other linear combination of X_1 and X_2 because the regression process acts to minimize the residual sum of squares from the regression of Y_2 on the two instruments. Y_2 is, therefore, the most efficient instrument we can use in this context.

The method described in the previous paragraph defines the *two-stage least-squares estimator* (TSLS). The terminology here is obvious. When we have an over-identified equation, we construct instruments through a first-stage regression in which we regress the right-hand side endogenous variable on all the possible instruments available to us. We then take the fitted values from this first stage regression, which constitute a linear combination of the available instruments, and use these as the instrument for a second stage IV regression. This will generate the most efficient possible IV estimator.

Example: As an example of the TSLS method, we will attempt to reproduce the results from the first simultaneous econometric model to be published. This is the classic paper by Girshick and Haavelmo [Girshick1947] in which they present a model of the demand and supply for food in the United States between 1922 and 1941. This paper contains the first statement of many of the principles of simultaneous equation estimation that we now take for granted, and provides an interesting example, which we now attempt to reproduce. Their model consists of five equations linking five endogenous variables to four exogenous or predetermined variables. The equation we will examine is their supply curve. This takes the form

$$y_1(t) = a_{22}\,y_2(t) + a_{24}\,y_4(t) + \gamma_{28}\,z_8(t) + a_{20} + u_2(t). \tag{8.33}$$

The notation is that of Girshick and Haavelmo. y variables are endogenous while z variables are exogenous. a coefficients are used for the intercept and endogenous variables while γ coefficients are used for exogenous variables. This is typical of the notation used for simultaneous equations models in much of the early literature on the topic. $y_1(t)$ is the consumption of food per capita; $y_2(t)$ is the retail price of food deflated by a general price index; $y_4(t)$ is production of agricultural food products per capita and $z_8(t)$ is a time trend. The exogenous variables not included in this equation are $z_6(t)$ lagged price received by farmers for food products; $z_7(t)$ net investment per capita and $z_9(t)$ lagged disposable income per capita. The equation is over identified because $g-1=2$ and $K-k=3$.

TABLE 8.1 Estimates of Supply Curve for Food US Data 1921–1941.

Ordinary Least Squares Estimates
$y_1(t) = \underset{(0.0508)}{0.1388}\,y_2(t) + \underset{(0.0599)}{0.5589}\,y_4(t) + \underset{(0.0492)}{0.3076}\,z_8(t) + \underset{(7.9447)}{24.9841} + \hat{u}_2(t)$
$\hat{\sigma}_u = 1.1106 \qquad DW = 1.7759$
Two-Stage Least Squares Estimates
$y_1(t) = \underset{(0.0613)}{0.1311}\,y_2(t) + \underset{(0.0877)}{0.6689}\,y_4(t) + \underset{(0.0556)}{0.3332}\,z_8(t) + \underset{(10.1809)}{13.9982} + \hat{u}_2(t)$
$\hat{\sigma}_u = 1.2218 \qquad DW = 1.9564$
Girshick and Haavelmo's Estimates (Limited Information Maximum Likelihood)
$y_1(t) = 0.157y_2(t) + 0.653y_4(t) + 0.339z_8(t) + 13.319 + \hat{u}_2(t)$

Table 8.1 presents three sets of regression results based on Girshick and Haavelmo's data, the first are the OLS estimates of (8.33), the second gives estimates obtained using two-stage least squares estimates using all available instruments and the third gives the estimates reported in the original paper, obtained using the method of Limited Information Maximum Likelihood (LIML). There is a clear difference between the OLS and TSLS estimates. In particular, the coefficient on $y_4(t)$ is noticeably larger. There is also a loss of efficiency with the standard errors for all the slope coefficients being larger for all three slope coefficients. It is also interesting to compare the

TSLS results with those reported in the original paper. The LIML estimator is closely related to the TSLS estimator and the values of the coefficients reported are very close to those we obtain. In the original paper, Girshick and Haavelmo state that "*A theory of confidence intervals for the parameters has not yet been worked out. Such a theory is essential in order to judge the reliability of the estimates.*" Thankfully, such methods are now available and, therefore, we are able to report the standard errors given for our TSLS estimates.

Historical Note: Girshick and Haavelmo [Girshick1947] is the first published empirical simultaneous equations model. However, it builds on theoretical work published earlier by Haavelmo [Haavelmo1944]. This latter paper is arguably the key paper that established econometrics as a distinct branch of statistical theory.

EXERCISES

Excel files containing the data for these exercises are available as companion files for this book.

EXERCISE 8.1

An econometrician has estimated the following equation which relates the growth rate of consumption expenditures to the growth rate of GDP. The data are annual values for the UK economy for the period 1949–2005

$$\Delta c_t = \underset{(0.3202)}{0.5103} + \underset{(0.1043)}{0.8478}\,\Delta y_t + \hat{u}_t$$

$$R^2 = 0.548 \qquad \hat{\sigma}_u = 1.3933.$$

a. Explain to your econometrician why the slope coefficient estimate may suffer from the problem of simultaneous equations bias.

b. Using your knowledge of the simple Keynesian income-expenditure model, suggest possible instruments that might be used to construct an instrument variable estimator for the same equation.

EXERCISE 8.2

Using the data set contained in the Excel workfile NAC.XLSX obtain the instrumental variable estimator of the relationship given in the equation for Exercise 8.1, using the change in investment expenditure as an instrument for the change in GDP. What effect does this have on the parameter estimates you obtain?

EXERCISE 8.3

Consider the following pair of simultaneous equations

$$Y_{1t} = \beta_{11} Y_{2t} + \gamma_{11} X_{1t} + u_{1t}$$
$$Y_{2t} = \beta_{21} Y_{1t} + \gamma_{22} X_{2t} + u_{2t},$$

where Y_1 and Y_2 are endogenous variables, X_1 and X_2 are exogenous variables and u_1 and u_2 are independent random errors.

a. Using the order condition, show that both equations are just identified.

b. Write down the reduced form of the system and solve for the structural parameters as functions of the reduced form parameters.

EXERCISE 8.4

Consider the following system of three equations

$$Y_{1t} = \beta_{11} Y_{2t} + \beta_{12} Y_{3t} + \gamma_{11} X_{1t} + u_{1t}$$
$$Y_{2t} = \beta_{22} Y_{3t} + \gamma_{21} X_{2t} + u_{2t}$$
$$Y_{3t} = \gamma_{31} X_{3t} + u_{3t}$$

where Y_1, Y_2 and Y_3 are endogenous variables, X_1, X_2 and X_3 are exogenous variables and u_1, u_2 and u_3 are independent random errors. Show that the parameters of all three equations can be estimated consistently by ordinary least-squares.

REFERENCES

[Adcock1877] Adcock, R. J., "Note on the Method of Least Squares." *The Analyst*, 1877, 4(6), pp. 183–184.

[Engle1983] Engle, R., Hendry, D., & Richard, J., "Exogeneity." *Econometrica*, 1983, 51(2), pp. 277–304.

[Girshick1947] Girshick, M. A. and Haavelmo, T., "Statistical Analysis of the Demand for Food: Examples of Simultaneous Estimation of Structural Equations." *Econometrica*, 1947, 15(2), pp. 79–110.

[Greene1993] Greene, W. H., *Econometric Analysis*. 1993, 2nd edition. New York: Macmillan.

[Haavelmo1944] Haavelmo, T., "The Probability Approach in Econometrics." *Supplement to Econometrica*, 1944, 12.

[Koopmans1950] Koopmans, T. *Statistical Inference in Dynamic Economic Models, Cowles Commission Monograph 10*. 1950.

[Reiersøl1945] Reiersøl, O., "Confluence Analysis by Means of Instrumental Sets of Variables." *Arkiv for Mathematic, Astronomi, och Fysik. 32A*, 1945. Almquist & Wiksells.

[Wald1940] Wald, A., "The Fitting of Straight Lines if Both Variables Are Subject to Error." *Annals of Mathematical Statistics*, 1940, 11, pp. 285–300.

[Working1927] Working, E.J., "What Do Statistical 'Demand Curves' Show?" *Quarterly Journal of Economics*, 1927, 41, pp. 212–235.

[Working1925] Working, H., "The Statistical Determination of Demand Curves." *Quarterly Journal of Economics*, 1925, 39, pp. 503–543.

[Wright1928] Wright, P., *The Tariff on Animal and Vegetable Oils*. 1928, New York: MacMillan.

CHAPTER 9

Dynamic Models

In this chapter, we consider the estimation of dynamic econometric models. These are concerned with modeling relationships in which there is adjustment over time and fall into two main categories. *Distributed lag models* mean that the response of a variable Y to another variable X is spread out over a number of time periods while *autoregressive models* allow Y to be affected by its own past values. In practice, we may wish to allow both elements to be present in the same equation.

There are two types of distributed lag model we need to consider. *Finite distributed lag models* include a limited number of past values of the explanatory variable on the right-hand side of the equation. For example, we might have a relationship of the form

$$Y_t = \beta_1 X_t + \beta_2 X_{t-1} + \beta_2 X_{t-2} + u_t. \tag{9.1}$$

Equations like this do not, in principle, create any new problems for us since they involve the estimation of a limited number of parameters. In practice, however, the current and lagged values of the X variable are likely to be highly correlated, meaning that multicollinearity may become an issue. An early attempt to deal with this was provided by Almon [Almon1965], who set out a method that involves constraining the coefficients in finite distributed lag models to lie on a specific polynomial function. This method was frequently used during the 1960s and 1970s but has fallen out of use since then. The reasons for this are that econometricians have tended to concentrate more on infinite distributed lag models and have found more flexible ways of modeling the distributed lag relationships that these imply.

An infinite distributed lag relationship takes the form

$$Y_t = \sum_{i=0}^{\infty} \beta_i X_{t-i} + u_t. \tag{9.2}$$

Clearly, it is impossible to estimate such a model with a finite data set. Even if we had infinite data, the model would be problematic because of the likely high degree of collinearity of X with its own lags. It follows that some restrictions on the lag coefficients become necessary before we can even attempt to estimate a relationship of the form (9.2). One popular method is to assume that the β_i coefficients in (9.2) can be modeled as a geometric progression with a constant ratio less than one. Thus, we have $\beta_i = \beta\lambda^i$ where $0 < \lambda < 1$. We now only need to estimate two parameters, an initial coefficient on the current X value β, and the rate at which the parameters decline λ. This leads naturally to models with the *Koyck lag* structure – named for Koyck [Koyck1954] who first used this structure to model the aggregate investment function. Consider the infinite distributed lag model with exponentially declining weights. We have

$$Y_t = \sum_{i=0}^{\infty} \beta\lambda^i X_{t-i} + u_t. \tag{9.3}$$

Lagging everything in (9.3) by one period and multiplying by λ gives $\lambda Y_{t-1} = \sum_{i=1}^{\infty} \beta\lambda^i X_{t-i} + \lambda u_{t-1}$. We can use this to replace all the lagged X terms in (9.3) and write the model as

$$Y_t = \beta X_t + \lambda Y_{t-1} + v_t, \tag{9.4}$$

where $v_t = u_t + \lambda u_{t-1}$. This is the familiar Koyck lag specification. We now have an equation with only two unknown parameters. However, the error now follows a first-order moving average process. Moreover, since $E(Y_{t-1}u_{t-1}) \neq 0$, it follows that OLS estimates of the parameters of (9.4) will be both biased and inconsistent.

In practice, the Koyck lag structure may sometimes be overly restrictive. The distributed lag relationship it implies is one in which the coefficients decline exponentially to zero. However, this does not allow for the possibility that the weights on past values of X may plausibly increase for a number of time periods before starting to decline. A method for dealing with general models like this was suggested by Jorgenson [Jorgenson1966] in the form of the *Rational Distributed Lag Model*. Let us assume that the coefficients in the general model (9.2) lie on a possibly infinite polynomial function written in terms of the lag operator, that is,

$$Y_t = A(L)X_t + u_t. \tag{9.5}$$

We assume that the roots of this polynomial function lie within the unit circle. The rational distributed lag model uses the fact that we can approximate any polynomial function of this type as the ratio of two lower-order polynomial functions. We can, therefore, write $A(L)=B(L)/C(L)$, and this allows us to write (9.5) as

$$Y_t = \frac{B(L)}{C(L)} X_t + u_t, \tag{9.6}$$

or, multiplying through by $C(L)$ yields

$$C(L)Y_t = D(L)X_t + v_t. \tag{9.7}$$

As with the Koyck lag, the errors in (9.7) will no longer be independent. Instead, we have $v_t = C(L)u_t$, and therefore, the errors in the transformed equation will follow a moving average process. The order of the process depends on the order of the polynomial function $C(L)$. This creates a problem because the OLS estimator will be biased and inconsistent due to the correlation of the error with the lagged Y terms in (9.7).

For both equation (9.4) and equation (9.7) we obtain an autoregressive equation in Y as the result of a transformation used to simplify the distributed lag relationship between Y and X. However, it is possible that autoregressive elements enter the relationship independently of the distributed lag relationship. This would be the case, for example, if the adjustment of the Y variable is costly and is therefore spread over a number of time periods. In such cases, it makes sense to begin with a specification that allows for both autoregressive and distributed lag components. This is the *Autoregressive Distributed Lag Model* or ARDL model. For example, we might begin with a specification of the form

$$A(L)Y_t = B(L)X_t + u_t. \tag{9.8}$$

Another advantage of this approach is that, by choosing sufficiently general lag polynomials $A(L)$ and $B(L)$, it is normally possible to ensure that the error u is serially uncorrelated. It may be possible to reduce the order of these polynomials in cases where they contain common factors. For example, if both $A(L)$ and $B(L)$ contain a common factor $(1-\varphi L)$, then we can write (9.8) as $C(L)Y_t = D(L)X_t + v_t$ where $v_t = \varphi v_{t-1} + u_t$ and $C(L)$ and $D(L)$ are polynomials which are one degree lower than $A(L)$ and $B(L)$, respectively. This is the COMFAC restriction which Hendry and Mizon [Hendry1978] put forward as a possible rationale for estimation with autocorrelated errors. While this approach offers a potential efficiency gain because it involves fewer parameters to be estimated, there is no guarantee that the restrictions involved are valid, and the added complication of estimation means that it is rarely used in practice.

9.1 MODELS WITH EXPECTATIONS

Distributed lag relationships arise naturally in models with adaptive expectations. This is because the adaptive expectations hypothesis is that agents form expectations about variables of interest by taking a weighted average of past values of the variable in question. This approach has been criticized because it implies (1) that agents ignore relevant information when forming expectations and (2) that agents make predictable, and easily correctable, errors in expectations. However, this is not the time or place to enter this debate, and we will simply consider the implications of the hypothesis for econometric modeling. First, consider a simple model in which expectations play a role. Suppose we wish to estimate an equation of the form

$$Y_t = \beta X_{t+1}^e + u_t, \tag{9.9}$$

where the e superscript indicates an expectation. Equations like this can be found in many areas of economics, particularly in macroeconomics. For example, consumption expenditure is often argued to depend on expectations of income rather than actual income. Another example is the case of price adjustment where the expectations of future inflation enter as one of the determinants of the current rate of inflation in the Phillips curve.

Historical Note: Adaptive expectation is a key assumption in generating the price and quantity dynamics of the cobweb model of agricultural markets and in the wage-price dynamics of Milton Friedman's analysis of the Phillips curve. However, the approach has fallen out of favor since the introduction of the rational expectations hypothesis in an important paper by Muth [Muth1961]. This paper was neglected for over a decade but was later rediscovered and formed the basis of the rational expectations revolution of the 1970s.

The problem facing the econometrician is that the expectations term in (9.9) is not usually observable directly. Instead, we must make use of an auxiliary model for the determination of expectations. One such model is provided by the adaptive expectations hypothesis. This states that agents revise their expectations based on past errors made in forecasting the variable in question. Expectations, therefore, adjust according to the following relationship

$$X_{t+1}^{e} - X_{t}^{e} = \gamma\left(X_{t} - X_{t}^{e}\right). \tag{9.10}$$

Using the lag operator, we can write (9.10) as

$$X_{t+1}^{e} = \frac{\gamma X_{t}}{1-(1-\gamma)L} = \gamma X_{t}\left(1+(1-\gamma)L+(1-\gamma)^{2}L^{2}+\ldots\right), \tag{9.11}$$

that is, the expected future value of X is an infinite distributed lag of current and past values of X with geometrically declining weights (assuming $0<\gamma<1$). Substituting for X_{t+1}^{e} and solving means that we can derive the following representation of the model

$$Y_{t} = \beta\gamma X_{t} + (1-\gamma)Y_{t-1} + u_{t} - (1-\gamma)u_{t-1}. \tag{9.12}$$

The two variables on the right-hand side of (9.12) (X_t and Y_{t-1}) are both observable, and therefore, (9.12) can, in principle, be estimated. Thus, the adaptive expectations model leads naturally to a Koyck lag specification in which the estimating equation contains a lagged endogenous variable and the equation error follows a first-order moving average process.

Example: Turner and Benavides [Turner2001] estimate a version of this model for the demand for money in Mexico using quarterly data for the period 1980 to 1999. The model consists of two equations

$$\begin{aligned} (m-p)_{t} &= \beta_{1}y_{t} - \beta_{2}\pi_{t}^{e} + u_{1t} \\ \pi_{t}^{e} &= (1-\gamma)\pi_{t-1}^{e} + \gamma\pi_{t-1}, \end{aligned} \tag{9.13}$$

where m, p, and y are narrow money, the price level and output all in the form of natural logarithms, $\pi = \Delta p$ and the superscript e denotes an expectation. This pair of equations is solved to eliminate the expectation from the first equation which gives a single equation of the form

$$(m-p)_{t} = \beta_{1}y_{t} - (1-\gamma)\beta_{1}y_{t-1} - \beta_{2}\gamma\pi_{t-1} + (1-\gamma)(m-p)_{t-1} + u_{1t} + \mu_{1}u_{t-1}. \tag{9.14}$$

Equation (9.14) allows for an unrestricted moving average error but the restriction implied by the model, $\mu_{1} = 1-\gamma$, can be tested using an F-test comparison of the residual sums of squares from restricted and unrestricted versions of the model. The model was estimated using an iterative search routine for the moving average coefficient (details can be found in the paper) and the results shown in Table 9.1 were obtained. The income and interest rate elasticities were both significant with the correct sign, but the moving average coefficient was insignificantly different from zero. Moreover, the F-test for the restriction $H_0 : \mu_1 = 1 - \gamma$ strongly rejected the null. Thus, the inclusion of a moving average error in this model appears to introduce an unnecessary complication in a model that otherwise fits the data reasonably well[1].

TABLE 9.1 Demand for Narrow Money in Mexico 1982q1 to 1999q1.

	β_1	β_2	γ	μ_1
Coefficient Estimate	0.5094	–7.6289	0.0688	0.1073
Absolute Value of t-ratio	1.67	2.77	2.52	0.70

9.2 COSTS OF ADJUSTMENT

Consider a model of the form

$$Y_t^* = \beta X_t + u_t. \tag{9.15}$$

The variable Y_t^* is the equilibrium or desired value of Y for a given value of X. Let us suppose we are interested in estimating the parameter β which determines the equilibrium response of Y to changes in X. The problem is that equation (9.15) cannot be estimated directly because it contains an unobserved variable Y_t^*.

To enable us to estimate we need to develop a theory of the adjustment process. There are many reasons why agents might not immediately adjust the actual value of Y to its equilibrium value. If adjustment is costly (which it almost always will be) then it pays agents to make the adjustment gradually

[1] In the same paper, we estimated and tested a similar Koyck lag model of the Phillips curve relationship between inflation and the output gap, with very similar results. Our conclusion is that the moving average term adds nothing to the empirical fit of the model.

rather than all in one go. As an example, think of the case of a firm adjusting its capital stock in response to an increase in demand for its product. It takes time and resources to install new machinery and it will pay the firm to spread this process out over a period of time rather than attempt to do this immediately. Let us suppose that the agent responds to a gap between the actual and equilibrium values of Y by changing this variable by some fraction of the difference, that is,

$$Y_t - Y_{t-1} = \gamma\left(Y_t^* - Y_{t-1}\right), \tag{9.16}$$

where $0 < \gamma < 1$ measures the fraction of any disequilibrium which is eliminated within one time period. Again, we can think of γ as measuring the speed of adjustment. Values of γ close to zero indicate slow adjustment while values of γ close to 1 indicate fast adjustment. This specification can be motivated by the assumption of a quadratic cost function in which agents trade off the costs of being away from the equilibrium against the cost of changing the decision variable.

If we substitute our equation for the equilibrium value of Y into (9.16) then we obtain

$$Y_t = \beta\gamma X_t + (1-\gamma)Y_{t-1} + u_t, \tag{9.17}$$

which only contains observable variables, and which can, therefore, be estimated. The specification in (9.17), and the Koyck lag model we derived earlier in equation (9.12) for the expectations model, are very similar in that both include the current value of X and the lagged value of Y on the right-hand side. However, they have been developed from different theoretical models of the relationship between the two variables. In the first case, the distributed lag relationship arose because Y depended on the expectation of X rather than its actual value, while in the second case, it arose because the adjustment of Y towards its equilibrium value was not instantaneous. In principle we might be able to distinguish between the two models by looking at the autocorrelation properties of the residuals – equation (9.12) has a moving average error while equation (9.17) does not. This is not always easy, however, particularly in cases where the parameter γ is close to one, and therefore, the coefficient on the moving average term in (9.12) (i.e., $1-\gamma$) is close to zero. This illustrates an important feature of dynamic economic modeling in that it shows that it is often hard to pin down the exact causes of distributed lag relationships between variables. In real-world applications, we may observe a distributed lag response, but this may be due to a mixture of causes rather than one particular explanation.

Example: Suppose we wish to model the demand for US exports. Two plausible explanatory variables are the overall level of trade in the world economy and the real exchange rate for the US relative to other currencies. Therefore, we can write an equilibrium relationship of the form

$$\ln X_t^* = \beta_0 + \beta_1 \ln W_t + \beta_2 \ln E_t + u_t, \tag{9.18}$$

where X^* is equilibrium US exports, W is the level of world trade and E is the real effective exchange rate. u is a random error that we assume has the normal classical properties. The equation is written in the log-linear form so that the coefficients can be interpreted as elasticities. Estimation of this model by OLS using quarterly data for the United States for the period 1975q1–2008q2 yields the results shown in equation (9.19)

$$\begin{aligned} &\ln(X_t) = \underset{(0.8407)}{-9.0624} + \underset{(0.0375)}{1.1343}\ln(W_t) + \underset{(0.0856)}{0.2086}\ln(E_t) + \hat{u}_t \\ &R^2 = 0.9532 \qquad \hat{\sigma}_u = 0.0908 \qquad DW = 0.5880 \qquad \hat{\rho}_1 = 0.70. \end{aligned} \tag{9.19}$$

Equation (9.19) clearly suffers from serial correlation. This is confirmed by the value of the Durbin-Watson statistic which, at 0.59, is well below the 5% lower bound of 1.63. There is, therefore, strong evidence of first-order serial correlation. The implications of this are that, while the coefficient estimates may not be biased, they are certainly inefficient. Moreover, the estimates of the standard errors are most likely biased downwards, meaning that we cannot rely on t-tests for the significance of the individual coefficients or the F-test for their joint significance. We should also note that the estimates reported in equation (9.19) are problematic from the point of view of economic theory as well as their statistical properties. The coefficient for world trade is plausible – it indicates that a 1% rise in world trade is associated with a 1.13% rise in US exports. However, the sign of the coefficient for the real exchange rate runs counter to our expectations. We would expect that an appreciation of the real exchange rate for the dollar should lead to a fall in US exports. Our coefficient estimate is positive indicating that the direction of the effect is counter to the predictions of theory. We should not, however, read anything into the apparent significance of this coefficient since this is most likely a product of the underestimation of the standard error due to the presence of serial correlation.

We can attempt to deal with both the economic and statistical problems of our equation by allowing for a dynamic relationship between the variables. In other words, we acknowledge that the effects on US exports of changes in world trade and the real exchange rate are not instantaneous and we seek to model these explicitly. Let us modify our estimating equation to include the lagged endogenous variable. This means that the coefficient estimates for world trade and the real exchange rate will no longer provide direct estimates of the equilibrium elasticities. Estimates of the model with a lagged endogenous variable are given in equation (9.20)

$$\ln(X_t) = -\underset{(0.2564)}{0.8044} + \underset{(0.0220)}{0.1467}\ln(W_t) - \underset{(0.0200)}{0.0388}\ln(E_t) + \underset{(0.0175)}{0.8620}\ln(X_{t-1}) + \hat{u}_t \qquad (9.20)$$

$$R^2 = 0.9976 \qquad \hat{\sigma}_u = 0.0204 \qquad DW = 1.4562 \qquad \hat{\rho}_1 = 0.26.$$

This equation can be interpreted as a partial adjustment model. It shows a noticeable improvement in statistical terms. Note, for example, that the Durbin-Watson statistic has risen from 0.59 for the simple regression to 1.46 in this case. This indicates that the extent of the serial correlation has fallen noticeably. This is confirmed by the estimate of the first-order autocorrelation coefficient which has fallen from 0.70 to 0.26. Although there is still significant first-order autocorrelation, the reduction in the magnitude of the autocorrelation coefficient means that the bias in the standard errors of the coefficients will have been reduced. Moreover, this equation has better properties in terms of economic theory, in that the coefficient on the real exchange rate now has the correct (negative) sign. We will see later that it is possible to improve on this equation further. However, for the moment it does reflect a significant improvement on the simple regression model and emphasizes the importance of allowing for a distributed lag relationship between the variables.

9.3 ASSESSING THE DYNAMICS

The coefficient estimates in equation (9.20) give the *impact effects* of changes in the right-hand side variables on US exports. For example, the coefficient for LW indicates that a 1% rise in world trade immediately increases US exports by just under 0.15%. Similarly, the coefficient on LE tells us that a 1% appreciation of the real exchange rate immediately reduces US exports by about 0.04%. However, we are often more interested in the dynamic path of US exports in response to changes in world trade and the exchange rate and in the long-run or equilibrium effects of changes in these variables rather than the impact effects. From our estimated equation, we can calculate the effects of sustained increases in the explanatory variables as $\eta_i(t) = \beta_i \sum_{t=0}^{\infty} \beta_3^t$ where $i = 1,2$ are the impact coefficients for the two variables and β_3 is the coefficient on the lagged endogenous variable. These, in turn, allow us to calculate the long-run elasticities as $\bar{\eta}_i = \beta_i / (1 - \beta_3)$. Thus, an equilibrium solution is only possible if $\beta_3 \neq 1$.

The dynamic paths of US exports in response to sustained changes in world exports and the real exchange rate are illustrated in Figures 9.1 and 9.2. From Figure 9.1, we see that there is a positive impact effect of the increase in world trade which grows over time, until eventually reaching a new equilibrium value determined by the long-run multiplier which is calculated as $0.1467 / (1 - 0.862) = 1.063$. From Figure 9.2, we see that a 1% real exchange rate appreciation has a long-run effect equal to $-0.0388 / (1 - 0.862) = -0.2812$. In both cases, we have assumed a sustained increase in the value of the explanatory variable. If the increase had been temporary, then the effects of any change would eventually die out and US exports would return to the original equilibrium. It should also be noted that we have assumed that there is no feedback from US exports to the variable in question. To borrow a piece of terminology that we will discuss later, we have assumed that US exports do not "Granger cause" world exports or the US real exchange rate. Relaxation of this assumption requires a multi-equation approach such as that of Vector Autoregression.

One way of assessing the speed of adjustment in dynamic models is to calculate the *half-life* of a shock. This is the length of time it takes for half the adjustment process to be completed. From standard results for geometric progressions, we can find the half-life of a shock by finding the value of t such that $\beta_3^t = 0.5$ or $t = \ln 0.5 / \ln b_3$ which, in this case, yields $t = \ln(0.5) / \ln(0.862) = 4.667$. Since the data here are quarterly, this indicates that 50% of the adjustment process is completed in just over one year.

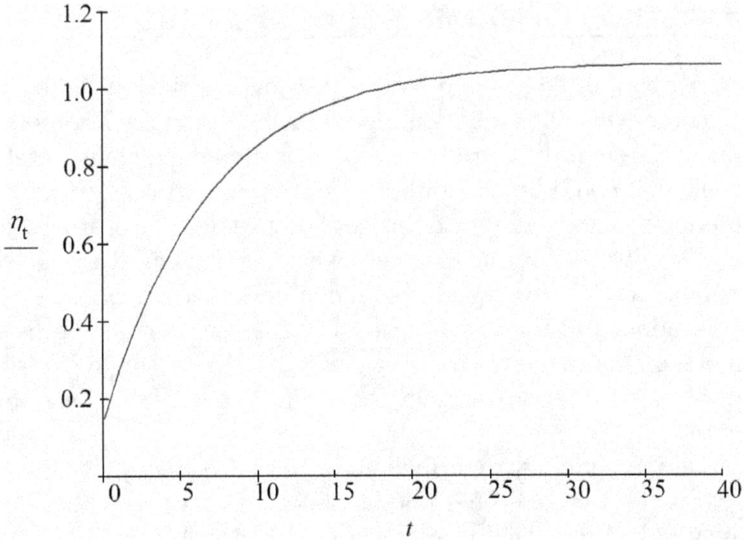

FIGURE 9.1 Dynamic Path of US Exports in Response to a 1% Increase in the Exports of Other Industrialized Economies.

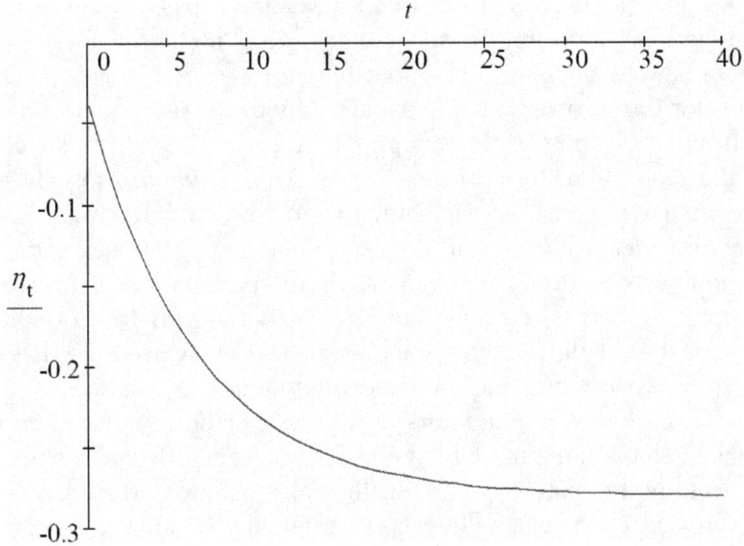

FIGURE 9.2 Dynamic Path of US Exports in Response to a 1% Appreciation in the Real Exchange Rate.

9.4 MODELING DYNAMIC RELATIONSHIPS

In the example considered in the previous section, we showed that the dynamic misspecification identified in a simple regression model could be reduced through the introduction of a lagged endogenous variable into the model. This could be interpreted in economic terms as arising from either the effects of expectations or costs of adjustment. The procedure adopted was, therefore, to estimate a regression model based on an equilibrium economic model, investigate that model for any evidence of statistical misspecification, and then revise the model if necessary. There is an obvious temptation to proceed in this way generally when fitting models to the data, but this is a dangerous strategy for several reasons which we now go on to discuss.

The methodology described above is that of *specific to general* modeling. It is attractive to the economist because it begins with a model that is closely linked to equilibrium economic theory. However, it runs into the problem that most such models will be statistically misspecified. This is because economic theory rarely offers a complete description of all the factors which can lead to distributed lag relationships between the variables of the model. Therefore, the specific to general methodology implies that we almost always begin with a misspecified model which creates statistical problems for the investigator. The reason why this is problematic is that all statistical tests begin with the assumption that we have a well-specified model as the basis of our tests. If this is not the case, then any tests based on misspecified models are unreliable. A second problem is that misspecification of one form can produce multiple types of failure in misspecification tests. For example, a structural break (change in the parameters of the model during the sample period) can produce results that appear to show the presence of serial correlation in the model residuals. Finally, we note that there is no unique way of modifying a misspecified model to produce a well-specified model. Different investigators using the specific to general approach can begin with the same model but end up with very different looking models as they attempt to patch up misspecifications that are detected.

For all the reasons described earlier, the specific to general methodology is not regarded as a good practice among modern econometricians. Instead, the *general to specific* methodology is widely regarded as providing a sounder basis for empirical work. This approach was pioneered by David Hendry in the 1970s and can be summarized as follows:

1. Use economic theory to determine the nature of the equilibrium relationship between a set of variables of interest.
2. Estimate the most general model possible, including many lags of the regression variables to maximize the chances of obtaining a statistically well-specified model.
3. Test if it is possible to simplify the model by eliminating irrelevant variables. If the restrictions involved in eliminating variables from the general model are not rejected, then proceed to the simpler model or *parsimonious specification*, as it is often referred to in this literature.
4. At all stages of the analysis test for evidence of misspecification in the model by examination of the residuals for evidence of serial correlation, heteroscedasticity, and other signs that the model does not provide an adequate statistical description of the data.
5. When a parsimonious specification has been identified, test restrictions on the equilibrium relationship between the variables and write the model in a way that can easily be interpreted.

The approach described above sounds simple. However, it still requires judgment and skill on the part of the modeler. What it does do is ensure that the final model will be statistically well specified and this in turn will mean that tests of economic restrictions based on the final model will be more reliable than tests based on misspecified models.

Example: As an example of the general to specific approach to modeling, we will re-examine our model for US exports. Since our model is estimated using quarterly data we take, as our most general model, an equation that includes the current and four lagged values of the world trade variable and the real exchange rate, as well as four lags of the endogenous variable. When we estimate this general model, we obtain the results reported in Table 9.2.

TABLE 9.2 General Dynamic Equation for US Exports 1976q1–2008q2.

	Coefficient	Standard error	*T*-Ratio
Constant	−0.5054	0.3483	-1.4512
$\ln(W_t)$	0.2737	0.0583	4.6971
$\ln(W_{t-1})$	-0.0353	0.0567	-0.6233
$\ln(W_{t-2})$	-0.0211	0.0542	-0.3902
$\ln(W_{t-3})$	-0.0288	0.0550	-0.5245
$\ln(W_{t-4})$	-0.0719	0.0565	-1.2729
$\ln(E_t)$	0.1469	0.0844	1.7406
$\ln(E_{t-1})$	-0.0983	0.1073	-0.9165
$\ln(E_{t-2})$	-0.0697	0.1054	-0.6609
$\ln(E_{t-3})$	0.0235	0.1063	0.2214
$\ln(E_{t-4})$	-0.0534	0.0784	-0.6805
$\ln(X_{t-1})$	1.0515	0.0984	10.6836
$\ln(X_{t-2})$	-0.3635	0.1337	-2.7187
$\ln(X_{t-3})$	0.2339	0.1341	1.7440
$\ln(X_{t-4})$	-0.0380	0.0862	-0.4406

$$R^2 = 0.9980 \qquad \hat{\sigma}_u = 0.0191 \qquad DW = 1.8949$$

$$ARCH = 0.1726 \qquad Q_4 = 9.1480 \qquad JB = 6.3615*$$

ARCH is the chi-square test for a first-order ARCH process in the residuals, distributed as χ_1^2 under the null. Q_4 is the Ljung-Box test for fourth-order autocorrelation, distributed as χ_4^2 under the null). *JB* is the Jarque-Bera test for normality of the residuals, distributed as χ_2^2 under the null. * indicates significance at the 5% level.

Table 9.2 indicates a model that is reasonably well specified in a statistical sense. The first-order serial correlation which was present even in the model with a lagged endogenous variable is no longer evident here. The diagnostic test statistics reported below do not indicate significant misspecification at the 5% level except for the normality test. The Jarque-Bera test does indicate significant non-normality of the residuals, but this can be shown to depend on a few outlying observations.

The problem with the equation reported in Table 9.2 is that it contains too many insignificant variables, that is, this is not a parsimonious specification. The next stage of the general to specific process is, therefore, to conduct a specification search in which we eliminate insignificant variables until all variables in the equation are significant at some predetermined level. In doing this it is dangerous to eliminate too many variables at one time because variables that are insignificant in the general model may become significant when other variables are eliminated. It is, therefore, good practice to eliminate only a few variables at any one time and proceed cautiously until the final specification is obtained. There is no set procedure for the order in which insignificant variables are eliminated but a reasonable rule of thumb is to eliminate the least significant variables first. Gilbert [Gilbert1986] argues that this is where the judgment and art of econometric model building is introduced. However, significance is not the only criterion, we also need to check if the elimination of variables introduces misspecification problems such as serial correlation.

Historical Note: Pagan [Pagan1987] argues that the general to specific approach developed from a long oral tradition on the correct way to practice econometrics which developed at the London School of Economics from the 1960s onwards. Hence the approach is often described as the "LSE approach."

Following a specification search using the general equation as a starting point, the final specification for the US export function reported in (9.21) was obtained. The stopping criterion for the search was that all variables included should be significant at the 5% level. Note that, although the lagged world trade and exchange rate variables are not significant at the 5% level, they are retained in the final specification because eliminating these variables produced significant serial correlation in the model residuals.

$$\begin{aligned}\ln(X_t) = &-\underset{(0.2322)}{0.6388} + \underset{(0.0262)}{0.1656}\ln(W_t) - \underset{(0.0317)}{0.0456}\ln(W_{t-1}) - \underset{(0.0185)}{0.0339}\ln(E_{t-2}) \\ &+ \underset{(0.0852)}{1.1153}\ln(X_{t-1}) - \underset{(0.0751)}{0.2297}\ln(X_{t-2}) + \hat{u}_t\end{aligned} \tag{9.21}$$

$R^2 = 0.9979$ $\quad \hat{\sigma}_u = 0.0192$ $\quad DW = 1.9442$

$ARCH = 1.1376$ $\quad Q_4 = 8.9008$ $\quad JB = 11.1745$

The final specification contains only six estimated coefficients rather than the fifteen in the general model. This is, therefore, considerably more parsimonious in terms of the variables included. However, the exclusion of nine of the original variables has not reduced the fit of the equation to any noticeable extent. This can be seen by the fact that the R^2 and the standard error of the regression are virtually unchanged. A formal test of all the restrictions involved in moving from the general model to the specific model can be conducted using an F test based on the residual sums of squares of the two equations reported. The test statistic can be calculated as

$$F = \frac{(0.045488 - 0.041846)}{0.041846} \times \frac{(130 - 15)}{9} = 1.1121.$$

The 5% critical value for an F-test with 9 and 115 degrees of freedom is $F^{5\%}_{9,115} = 1.963$. Therefore, the restrictions involved in moving from the general to the specific model are not rejected.

General to specific analysis has given us a final equation that fits the data well statistically. However, it is less easy to assess whether the equation makes sense from the perspective of economics because of the complex lag structure of the final specification. One solution to this problem is to rewrite the equation in a form in which the parameters can be given a meaningful economic interpretation. A natural format for this is the *error correction model*. This is essentially just a different way of parameterizing (or writing) an equation that combines differences and levels of variables so that the investigator can separate out long and short-run dynamic effects.

Let us begin by considering the final equation we have estimated. This can be written in the form

$$\ln X_t = \beta_1 + \beta_2 \ln W_t + \beta_3 \ln W_{t-1} + \beta_4 \ln E_{t-2} + \beta_5 \ln X_{t-1} + \beta_6 \ln X_{t-2} + u_t. \quad (9.22)$$

We can rewrite this equation as:

$$\Delta \ln X_t = \gamma_1 + \gamma_2 \Delta \ln W_t + \gamma_3 \Delta \ln X_{t-1} + \gamma_4 \ln X_{t-1} + \gamma_5 \ln W_{t-1} + \gamma_6 \ln E_{t-2} + u_t. \quad (9.23)$$

Equations (9.22) and (9.23) are formally identical. They are in fact just two different ways of writing the same linear combination of the set of variables that are included in the final specification of our model. This can be seen by the fact that there is a unique mapping from the coefficients of (9.22) to those of (9.23), that is, $\gamma_1 = \beta_1, \gamma_2 = \beta_2, \gamma_3 = -\beta_6$ $\gamma_4 = \beta_5 + \beta_6 - 1$, $\gamma_5 = \beta_2 + \beta_3, \gamma_6 = \beta_4$. More importantly, the coefficient of (9.23) have natural economic interpretations. In particular, we can interpret the coefficients on the different terms as representing short-run dynamics. For example, γ_2 describes the impact effect of an increase in world trade on US exports. The coefficients on the level terms describe the long-run relationship between the variables. The long-run effect of an increase in world trade is given by the ratio $-\gamma_5 / \gamma_4$. The coefficient on the lagged endogenous variable γ_4 measures the speed of adjustment when the relationship between the variables is different from the equilibrium relationship, that is, it measures the speed with which *errors* (deviations from equilibrium) are *corrected* (by adjustment of the endogenous variable). At the risk of over emphasizing this point, it is worth representing our equation diagrammatically as shown in Figure 9.3.

Historical Note: Error correction models have their origin in the paper by Sargan [Hart1964]. This paper emphasized the importance of retaining levels terms in a first difference regression to allow for the existence of a long-run relationship between the variables.

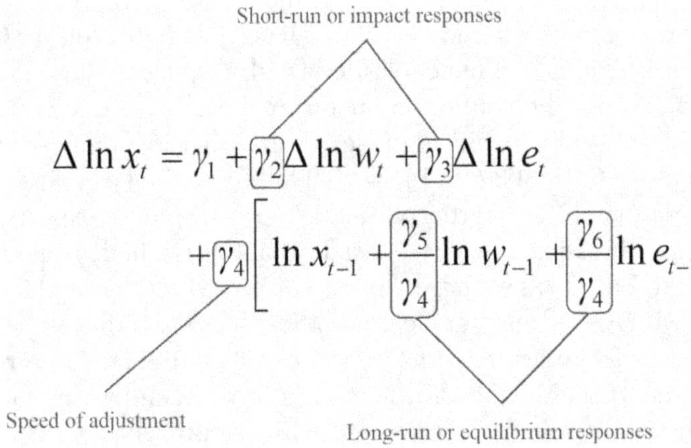

FIGURE 9.3 Interpretation of the Error-Correction Parameters.

Estimates of the final specification of the model in error-correction form are given in equation (9.24). Inspection of these results confirms that this is the same equation as presented in equation (9.21). This can be seen by the fact that the residual sums of squares are identical for the two equations, indicating that these are just two different ways of writing the same linear combination of variables. In fact, many of the equation statistics which are based on the residual sum of squares are identical, including the Durbin-Watson statistic, the standard error of the regression, and the Jarque-Bera statistic. However, there are some differences. In particular, the R^2 is much lower for the error-correction representation. This reflects the fact that the re-parameterization of the equation has changed the left-hand side variable from the (log) level of exports to its first difference. Thus the R^2 in (9.23) measures the fraction of the variance of the quarter on quarter growth rate of exports which is explained by the model rather than that of the level of exports. This gives a much more realistic guide as to the goodness of fit of the model since it is not artificially increased by the presence of a trend in the series.

$$\begin{aligned}\Delta \ln(X_t) = & -\underset{(0.2322)}{0.6388} + \underset{(0.0262)}{0.1656}\,\Delta \ln(W_t) + \underset{(0.0751)}{0.2297}\,\Delta \ln(X_{t-1}) - \underset{(0.0203)}{0.1144}\ln(X_{t-1}) \\ & + \underset{(0.0234)}{0.1200}\ln(X_{t-1}) - \underset{(0.0185)}{0.0339}\ln(E_{t-2}) + \hat{u}_t \end{aligned} \tag{9.24}$$

$R^2 = 0.4656$	$\hat{\sigma}_u = 0.0192$	$DW = 1.9442$
$ARCH = 1.1376$	$Q_4 = 8.9008$	$JB = 11.1745$

If we consider the parameter estimates in equation (9.24), then we see that it provides an economically plausible model of exports. The impact elasticity with respect to world trade is 0.165 which rises to $0.120011 / 0.1144 = 1.049$ in the long run. The impact elasticity with respect to the real exchange rate is zero because the contemporaneous real exchange rate variable was eliminated during the specification search but there is a long-run effect which is given by $-0.033868 / 0.1144 = -0.296$. The model explains just under half the variation of the quarter on quarter growth rate of exports, which is reasonably impressive when we consider that this is a highly variable series with no trend. Finally, as we have already confirmed for the model in levels, there is no evidence of misspecification other than a significant Jarque-Bera test statistic which indicates some non-normality in the equation residuals.

Pagan [Pagan1987] provides an interesting comparison of the general to specific methodology with two alternative approaches to econometric modeling. These are the "Extreme Bounds" approach of Leamer [Leamer1978] and the Vector Autoregression approach of Sims (1980). Although he is generally approving of the approach, Pagan expresses some concerns about the specification search approach of the general to specific methodology. In particular, he argues that there is a danger that the final specification of the model may be "path dependent." This arises because there is always the risk of making a Type I error during the search process and eliminating a variable that should be in the final specification. This means that other variables tend to be included in the final specification because they are correlated with the erroneously excluded variable. In more recent work, Krolzig and Hendry [Krolzig2001] have addressed this issue through the use of multiple search paths which are implemented in computer automated modeling software.

9.5 STATISTICAL PROBLEMS WITH LAGGED DEPENDENT VARIABLES

One problem encountered with dynamic econometric models is that least-squares estimates will be inconsistent when we have a combination of an equation including a lagged dependent variable and an autocorrelated error. To understand this, we will consider a simple example. Suppose we have a model in which Y is related to its own past value and the equation error u is also autocorrelated. This can be written

$$\begin{aligned} Y_t &= \beta Y_{t-1} + u_t \\ u_t &= \rho u_{t-1} + \varepsilon_t. \end{aligned} \tag{9.25}$$

Both processes are assumed to be stationary, that is, $|\beta|<1$ and $|\rho|<1$. $\varepsilon_t; t=1,\ldots,T$ are independent random errors, with mean zero and constant variance. The OLS estimator of β is consistent if $\operatorname{cov}(Y_{t-1}u_t)=0$. However, this condition is not satisfied in this case. We have

$$\begin{aligned} E(u_t Y_{t-1}) &= E\{(\rho u_{t-1}+\varepsilon_t)(\beta Y_{t-2}+u_{t-1})\} \\ &= E\{\beta\rho u_{t-1}Y_{t-2}+\rho u_{t-1}^2+\beta Y_{t-2}\varepsilon_t+u_{t-1}\varepsilon_t\}. \end{aligned} \tag{9.26}$$

By the assumption that the ε_t disturbances are serially uncorrelated, we have $E(\beta Y_{t-2}\varepsilon_t)=E(u_{t-1}\varepsilon_t)=0$ and, by the assumption of stationarity, we have $E(u_t Y_{t-1})=E(u_{t-1}Y_{t-2})=\operatorname{cov}(Y_{t-1}u_t)$. It follows that

$$\operatorname{cov}(Y_{t-1}u_t)=\frac{\rho}{1-\rho\beta}\sigma_u^2 \neq 0, \tag{9.27}$$

and therefore, OLS estimation of the autoregressive equation in Y will produce biased and inconsistent estimates. It is possible to derive the inconsistency in the OLS estimator explicitly and we can show that

$$\operatorname{plim}\hat{\beta}=\beta+\frac{\rho(1-\beta^2)}{1+\rho\beta}. \tag{9.28}$$

We can also show that the probability limit of the first-order sample autocorrelation is given by the expression

$$\operatorname{plim}\hat{\rho}=\rho-\frac{\rho(1-\beta^2)}{1+\rho\beta}.$$

The inconsistencies in these estimates are, therefore, offsetting in that $\text{plim}\left(\hat{\beta}+\hat{\rho}\right)=\beta+\rho$. Alternatively, if we over-estimate β then we under-estimate ρ by the same amount. Both these results are not intrinsically difficult to prove but require a lot of rather tedious algebra. They are, therefore, left as exercises for the interested student. One interesting property that follows from this is that, in cases where β and ρ are both positive, the Durbin-Watson statistic tends to be biased towards acceptance of the null hypothesis of no autocorrelation. To prove this note that $DW \approx 2\left(1-\hat{\rho}\right)$ in large samples and $\text{plim}\,\hat{\rho}<\rho$ in these circumstances. It is, therefore, better to use tests such as the Durbin h-test or the Breusch-Godfrey test if the regression contains a lagged endogenous variable.

EXERCISES

Excel files containing the data for these exercises are available as companion files for this book.

EXERCISE 9.1

Consider the following model

$$Y_t=\beta Y_{t-1}+u_t$$
$$u_t=\rho u_{t-1}+\varepsilon_t$$
$$|\beta|<1;\ |\rho|<1;\ \varepsilon_t \text{ are independent identically-distributed (iid) random variables with zero mean and constant variance.}$$

If $\hat{\beta}=\sum_{t=2}^{T}Y_{t-1}Y_t \,/\sum_{t=2}^{T}Y_{t-1}^2$, show that

$$\text{plim}\,\hat{\beta}=\beta+\frac{\rho\left(1-\beta^2\right)}{1+\rho\beta}$$

EXERCISE 9.2

Consider the following model

$$Y_t=\beta Y_{t-1}+u_t$$
$$u_t=\rho u_{t-1}+\varepsilon_t$$
$$|\beta|<1;\ |\rho|<1;\ \varepsilon_t \text{ are independent identically-distributed (iid) random variables with zero mean and constant variance.}$$

If $\hat{\rho} = \sum_{t=2}^{T} \hat{u}_{t-1}\hat{u}_t \, / \sum_{t=2}^{T} \hat{u}_t^2$, show that

$$\text{plim}\,\hat{\rho} = \rho - \frac{\rho\left(1-\beta^2\right)}{1+\rho\beta}$$

EXERCISE 9.3

An econometrician has estimated the following model which relates the demand for airline travel A to real personal disposable income Y and the relative price of airline travel P for the US economy. The data are annual from 1929 to 2003, the demand for airline travel is defined as revenue passenger miles per head of population and all variables are in natural logarithms.

$$\ln\left(A_t\right) = \underset{(1.5269)}{-24.3371} + \underset{(0.3197)}{1.7508}\ln\left(Y_t\right) - \underset{(0.5460)}{1.2691}\ln\left(P_t\right) + \hat{u}_t$$

$$R^2 = 0.9494 \qquad \hat{\sigma}_u = 0.5461 \qquad DW = 0.1013$$

a. Explain how the slope coefficients for this model can be interpreted as income and price elasticities of demand, respectively.

b. Explain why the standard errors of the coefficient estimates are probably too low and suggest an alternative specification that might be less susceptible to this problem.

Following your advice, the econometrician estimates a partial adjustment model for the demand for air travel. He obtains the following results:

$$\ln\left(A_t\right) = \underset{(0.8429)}{-3.3034} + \underset{(0.1067)}{0.3633}\ln\left(Y_t\right) + \underset{(0.1345)}{0.2623}\ln\left(P_t\right) + \underset{(0.0255)}{0.8749}\ln\left(Y_t\right) + \hat{u}_t$$

$$R^2 = 0.9980 \qquad \hat{\sigma}_u = 0.1044 \qquad DW = 1.5210$$

c. Calculate the long-run income and price elasticities of demand implied by this model. Are these consistent with economic theory?

d. Explain why the use of the Durbin-Watson test for serial correlation is problematic in this case. Using the data in the Excel workfile AIRLINE.XLSX, check the Ljung-Box tests for serial correlation and perform the Breusch-Godfrey test. Do the results change your conclusions?

EXERCISE 9.4

Another econometrician argues that the model should include a time trend to capture the long-term growth of this industry during the estimation period. Re-estimate the partial adjustment model and include a time trend as an additional variable. Does this affect the results noticeably?

REFERENCES

[Almon1965] Almon, S., "The Distributed Lag Between Capital Appropriations and Net Expenditures." *Econometrica*, 1965, 33, pp. 178–196.

[Gilbert1986] Gilbert, C. L., "Professor Hendry's Econometric Methodology." *Oxford Bulletin of Economics and Statistics*, 1986, 48, pp. 283–307.

[Hart1964] Sargan, J.D., "Wages and prices in the United Kingdom: A study in econometric methodology (with discussion)." In Hart, P.E., Mills, G. and Whitaker, J.K. (eds), *Econometric Analysis for National Economic Planning*, Vol 16 of *Colston Papers*, 1964, pp. 25–63, Butterworth Co.

[Hendry1978] Hendry, D. and Mizon, G., "Serial Correlation as a Convenient Simplification, Not a Nuisance: A Comment on a Study of the Demand for Money by the Bank of England." *Economic Journal*, 1978, 88, pp. 549–563.

[Jorgenson1966] Jorgenson, D., "Rational Distributed Lag Functions." *Econometrica*, 1966, 34(1), pp.135–149.

[Koyck1954] Koyck, L. M. *Distributed Lags and Investment Analysis*. 1954, North-Holland.

[Krolzig2001] Krolzig, H.-M., and Hendry, D.F., "Computer Automation of General-to-Specific Model Selection Procedures." *Journal of Economic Dynamics and Control*. 2001, 25, pp. 831–866.

[Leamer1978] Leamer, E. *Specification Searches*. 1978, Wiley.

[Muth1961] Muth, J.F., "Rational Expectations and the Theory of Price Movements." *Econometrica*. 1961, 29, pp. 315–335.

[Pagan1987] Pagan, A., "Three Econometric Methodologies: A Critical Appraisal." *Journal of Economic Surveys*. 1987, 1(1), pp. 3–24.

[Turner2001] Turner, P. and Benavides, G., "The Demand for Money and Inflation in Mexico 1980–1999: Implications for Stability and Real Seigniorage Revenues. *Applied Economics Letters*. 2001, 8, pp. 775–778.

CHAPTER 10

Time Series Analysis and ARIMA Modeling

So far, our analysis has been framed in terms of the relationships between random *variables*. Time series analysis marks a subtle difference in that the focus of interest now moves to random, or stochastic, *processes*. A stochastic process can be thought of as a random process that evolves over time. It consists of a variable or set of variables that move together over time subject to some degree of random variation. Our interest is now in the parameters which determine the process. For example, the random variable X observed at date t may depend on its own past value lagged one period plus a random disturbance

$$X_t = \theta X_{t-1} + \varepsilon_t. \tag{10.1}$$

This is an example of an *autoregressive process* because X depends on its own past value plus an additive random disturbance ε. For the purposes of this chapter, we assume that ε_t is a purely random disturbance in that it has no relationship to either its own past values or the past values of the X variable. We will also assume that the probability distribution of ε_t is independent of time so that we can think of $\varepsilon_t : t = 1,\ldots,T$ as a set of independent draws from a fixed distribution. These properties are often summarized by stating that ε is assumed to be a *white-noise* stochastic process. The parameters of interest in (10.1) are the autoregressive parameter θ and the variance of the disturbance σ_ε^2.

Equation (10.1) is an example of a discrete time stochastic process because we only observe X at fixed points in time $t = 1,\ldots,T$. Discrete time stochastic processes are characterized by a set of moments, that is, the mean or expected value, the variance, and the covariances of the random variable X. These moments depend on the nature of the random disturbance ε and the structure of the relationship between X and its own past values. As a trivial example consider a very simple stochastic process in which $X_t = \varepsilon_t$. In this case, the distribution of X is characterized by two parameters, the mean, and the variance. If $\varepsilon_t \sim N(0,\sigma_\varepsilon^2)$ then $X_t \sim N(0,\sigma_X^2)$ where $\sigma_X^2 = \sigma_\varepsilon^2$.

Let us return to a more interesting example in the form of our first-order autoregressive process defined by equation (10.1). As an alternative to this form, we can write this as a *moving average* process in which X depends on a weighted average of past random disturbances. By a process of backward substitution, (10.1) in moving average form,

$$X_t = \varepsilon_t + \theta\varepsilon_{t-1} + \theta^2\varepsilon_{t-2} + \theta^3\varepsilon_{t-3} = \sum_{i=0}^{\infty}\theta^i\varepsilon_{t-i}. \quad (10.2)$$

If $\varepsilon_t \sim N(0,\sigma_\varepsilon^2)$ then we can determine the mean of X very easily as

$$E(X_t) = \sum_{i=0}^{\infty} E(\varepsilon_t) = 0. \quad (10.3)$$

The variance is a little trickier but we note that $E(\varepsilon_t^2) = \sigma_\varepsilon^2$ and $E(\varepsilon_t\varepsilon_{t-k}) = 0$ for all $k \neq 0$ by assumption and therefore,

$$\begin{aligned} E(X_t^2) &= E(\varepsilon_t^2) + \theta^2 E(\varepsilon_{t-1}^2) + \theta^4 E(\varepsilon_{t-2}^2) + \ldots \\ &= \sum_{i=0}^{\infty}\theta^{2i}E(\varepsilon_{t-i}^2) = \frac{\sigma_\varepsilon^2}{1-\theta^2}. \end{aligned} \quad (10.4)$$

Note that, for autoregressive processes like this, the autocovariances are not equal to zero, that is, $E(X_tX_{t-k}) \neq 0$. For example

$$\begin{aligned} E(X_tX_{t-1}) &= \theta E(\varepsilon_{t-1}^2) + \theta^3 E(\varepsilon_{t-2}^2) + \theta^5 E(\varepsilon_{t-3}^2) + \ldots = \theta\sum_{i=0}^{\infty}\theta^{2i}E(\varepsilon_{t-1-i}^2) \\ &= \frac{\theta\sigma_\varepsilon^2}{1-\theta^2}, \end{aligned} \quad (10.5)$$

and, in general, we can show that

$$E(X_tX_{t-k}) = \frac{\theta^k\sigma_\varepsilon^2}{1-\theta^2} = \theta^k V(X_t). \quad (10.6)$$

For the variance to be a finite positive number we need $-1 < \theta < 1$. This is the assumption of stationarity which we will return to later. If this condition fails then the variance is either not defined, when $\theta = 1$, or negative, when $|\theta| > 1$.

Let us define $\gamma_k = E(X_tX_{t-k})$ to be the k'th order autocovariance for X. Next, we define $\rho_k = \gamma_k/\gamma_0$ to be the k'th order autocorrelation where γ_0 is the variance of X. This defines the *autocorrelation function*, or *correlogram*, for the random variable $X_t; t = 1,\ldots,T$, generated by the autoregressive process $X_t = \theta X_{t-1} + \varepsilon_t$, where ε_t are independent random disturbances with mean zero and constant variance. Note that $\rho_0 = 1$ by definition. In our example the autocorrelations are geometrically declining, that is, $\rho_k = \theta^k$, but this will not be true for more general stochastic processes. A sample plot of the correlogram for a first-order autoregressive process with positive θ is given in Figure 10.1. This shows a sequence of autocorrelations that decline exponentially to zero.

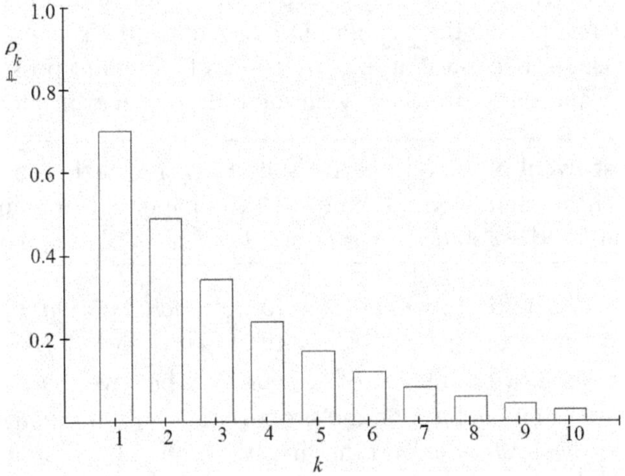

FIGURE 10.1 Correlogram for an AR(1) Process With Positive Autocorrelation Coefficient.

If the parameter $-1 < \theta < 0$, then the autocorrelations $\rho_k = \theta^k / \theta_0$ will still decline exponentially to zero. However, they will change sign depending on whether k is odd, which will generate negative ρ_k, or k is even, which will generate positive ρ_k. An example of the sort of correlogram we might observe for this case is given in Figure 10.2.

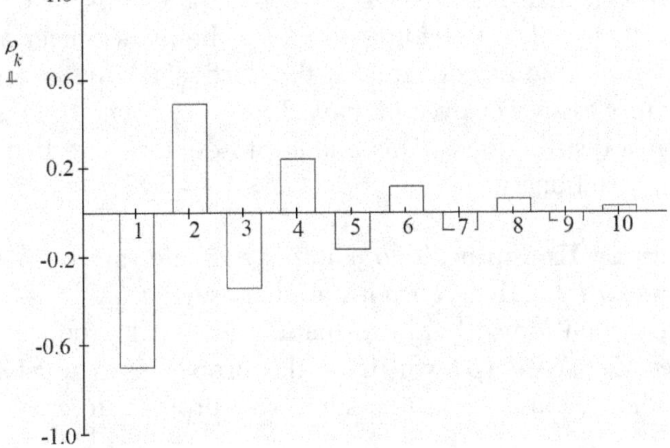

FIGURE 10.2 Correlogram for an AR(1) Process With Negative Autocorrelation Coefficient.

When we observe a time series, we are effectively observing only one possible realization of the process in question. For example, if we have $X_t = \theta X_{t-1} + \varepsilon_t$ then there is an infinite number of possible outcomes of this process depending on the particular values of ε observed. The question we must ask is whether it is possible to estimate the parameters of the process given that we observe only one realization. To do this, we must assume that the process is *ergodic*. This means that sample moments of a particular realization approach the population moments of the process as the length of the realization becomes large. Unfortunately, it is not possible to test for ergodicity and therefore we rely on the rather weaker property of *stationarity*.

> **Historical Note:** The term "stationary stochastic process" was first used by Khintchine [Khintchine1934] in paper written in German. It was translated as "stationary random process" by Wold [Wold1938].

A stochastic process is said to be *strictly stationary* if its moments are not affected by the choice of origin. Thus $E(X_t), E(X_t^2), E(X_t^3)\ldots$ are all unaffected by the choice of t. Even this, however, can be difficult to prove and we often fall back on the property of *weak stationarity*. A process is said to be *weakly stationary* if the first two moments of the distribution are unaffected by the choice of origin. This means that the mean and variance of the series are constants. This assumption is also referred to as the assumption of *second-order* or *covariance* stationarity. It is usually much easier to prove that a process is weakly stationary than to prove that it is strictly stationary. Finally, we note that, if the disturbances follow a normal distribution, then weak and strict stationarity are equivalent.

Example: Consider the AR(1) process $X_t = \theta X_{t-1} + \varepsilon_t$ where ε_t are white-noise Gaussian disturbances. We have already shown that $E(X_t) = 0$ and $\sigma_X^2 = \sigma_\varepsilon^2 / (1 - \theta^2)$. Providing $-1 < \theta < 1$, both the mean and the variance of X exist and do not depend on the particular time interval chosen. Hence, this process is weakly stationary. Moreover, since the disturbances follow a normal distribution, this is enough to demonstrate that the process is also strictly stationary.

Counter Examples: Two counter examples are of interest here. First, if $X_t = a + \beta t + \varepsilon_t$, then X is not stationary since $E(X_t) = a + \beta t$. However, there is a straightforward transformation that will produce a stationary series since $Z_t = X_t - a - \beta t$ will have the properties that $E(Z_t) = 0$ and $\sigma_Z^2 = \sigma_\varepsilon^2$. Another example of a non-stationary process that can be transformed to create a stationary series is the random walk $X_t = X_{t-1} + \varepsilon_t$. This fails the condition for weak stationarity because its variance is not defined. However, if we define $Z_t = X_t - X_{t-1}$, then we again have $E(Z_t) = 0$ and $\sigma_Z^2 = \sigma_\varepsilon^2$. Therefore, differencing is an appropriate transformation here to produce a stationary series.

10.1 IDENTIFICATION OF ARIMA PROCESSES

In the previous section, we defined the theoretical correlogram for a stochastic process. However, what we often wish to do is to identify the sort of process which might have generated an observed time series. To do this we can compute sample statistics corresponding to the unknown population parameters to obtain the *sample correlogram*. Suppose we have a set of observations $x_t; t = 1,\ldots,T$, the k'th order sample autocorrelation is defined as

$$\hat{\rho}_k = \frac{\sum (x_t - \bar{x})(x_{t-k} - \bar{x})}{\sum (x_t - \bar{x})^2}. \tag{10.7}$$

Examination of the sample correlogram is often very informative about the nature of the sort of stochastic process which might have generated the data.

Example: The data shown in Figure 10.3 have been artificially generated using a stochastic process of the form $X_t = 0.7X_{t-1} + \varepsilon_t$ where $\varepsilon_t, t = 1,\ldots,100$ are independent drawings from a normal distribution with mean zero and variance one.

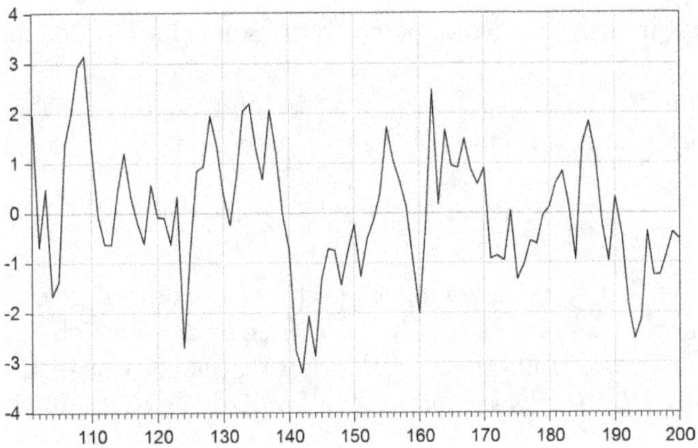

FIGURE 10.3 Realization of an AR(1) Stochastic Process.

The sample correlogram for this realization is shown in Figure 10.4. This shows the classic pattern for a stationary AR process with a positive autocorrelation coefficient in that the sample autocorrelations die down to zero exponentially. If the process was non-stationary, for example, if $\theta = 1$, then the autocorrelations would still die down to zero, but the rate of decline would be slower and be a linear, rather than an exponential, function of the lag length.

Sample: 101 200
Included observations: 100

	AC	PAC	Q-Stat	Prob
1	0.626	0.626	40.368	0.000
2	0.358	-0.055	53.718	0.000
3	0.163	-0.064	56.527	0.000
4	0.036	-0.044	56.667	0.000
5	0.033	0.086	56.784	0.000
6	0.013	-0.035	56.802	0.000
7	-0.029	-0.061	56.896	0.000
8	-0.065	-0.037	57.368	0.000
9	-0.136	-0.098	59.454	0.000
10	-0.215	-0.123	64.707	0.000
11	-0.219	-0.014	70.203	0.000
12	-0.156	0.046	73.031	0.000
13	-0.219	-0.223	78.667	0.000
14	-0.256	-0.098	86.455	0.000
15	-0.222	0.036	92.364	0.000

FIGURE 10.4 Sample Correlogram for the Time-Series Shown in Figure 10.3.

Up until now, we have only considered the sample autocorrelations defined in (10.7). These are not always helpful in determining the order of an autoregressive process, since the sample autocorrelations remain different from zero for lags longer than the order of the process itself. However, you will note that the correlogram output from EViews also includes a set of numbers labeled as PAC or *partial autocorrelations*. These are designed to capture the correlation between the series in question and its k'th lag while allowing for the intermediate effects of lags $1,2,\ldots,k-1$. This allows us to identify the order of the process of interest. We can think of the k'th order partial autocorrelation as the k'th coefficient in a relationship of the form

$$X_t = \theta_1 X_{t-1} + \theta_2 X_{t-2} + \ldots + \theta_k X_{t-k} + \varepsilon_t. \tag{10.8}$$

Note that the coefficients $\theta_i; i = 1,\ldots,k-1$ do not give the first $k-1$ partial autocorrelations. It is only the k'th coefficient which is relevant here. One method of estimating the k'th order partial autocorrelation would be to estimate (10.8) by least squares in order to obtain $\hat{\theta}_k$. In practice, however, this is not the method used to calculate the partial autocorrelations in most econometric packages. Instead, the method used is to solve for the partial autocorrelations using the *Yule-Walker equations*.

The Yule-Walker equations use a method of moments approach in that we derive expressions for the theoretical moments and then substitute the values of the sample moments into the resulting expression to create a set of equations that can be used to solve for the partial autocorrelations. The first-order partial autocorrelation is identical to the first-order sample autocorrelation because we begin with the assumption of a first-order autoregressive process $X_t = \theta_1 X_{t-1} + \varepsilon_t$. Multiplying by X_{t-1} and taking expectations we have $\gamma_1 = E(X_t X_{t-1})$ which gives the first-order autocovariance. The first-order partial autocorrelation is obtained by substituting the sample autocovariance in this expression and then dividing by the sample variance. This gives $\hat{\theta}_1 = \hat{\gamma}_1 / \hat{\gamma}_0$ which is the same formula we use to calculate the first-order sample autocorrelation. For the second-order partial autocorrelation, we begin by assuming a second-order autocorrelation process $X_t = \theta_1 X_{t-1} + \theta_2 X_{t-2} + \varepsilon_t$. Multiplying first by X_{t-1}, then by X_{t-2}, and taking expectations yields the relationships between the autocovariances given in (10.9)

$$\begin{aligned} \gamma_1 &= \theta_1 \gamma_0 + \theta_2 \gamma_1 \\ \gamma_2 &= \theta_1 \gamma_1 + \theta_2 \gamma_0 \end{aligned} \tag{10.9}$$

We can now substitute the sample autocorrelations $\hat{\gamma}_1$ and $\hat{\gamma}_2$ along with the sample variance into (10.9) and solve for the second-order partial autocorrelation $\hat{\theta}_2$. Note also that the estimates of $\hat{\theta}_i; i = 1,2$ obtained in this way will not be identical to OLS estimates but the two methods should converge as the sample size becomes large since both approaches provide consistent estimates. We can generalize this process to solve for the k'th order partial autocorrelation by assuming a k'th order autoregressive process which allows us to write down the relationships between the autocovariances and the θ parameters given in (10.10). We then replace the theoretical autocovariances with their sample equivalents and solve for $\hat{\theta}_k$

$$\begin{bmatrix} \gamma_1 \\ \gamma_2 \\ \vdots \\ \gamma_k \end{bmatrix} = \begin{bmatrix} \gamma_0 & \gamma_1 & \cdots & \gamma_{k-1} \\ \gamma_1 & \gamma_0 & \cdots & \gamma_{k-2} \\ \vdots & \vdots & \ddots & \vdots \\ \gamma_{k-1} & \gamma_{k-2} & \cdots & \gamma_0 \end{bmatrix} \begin{bmatrix} \theta_1 \\ \theta_2 \\ \vdots \\ \theta_k \end{bmatrix}, \tag{10.10}$$

> **Historical Note:** This method of solving for the partial autocorrelations derives from papers by George Udny Yule [Yule1927] and Gilbert Walker [Walker1931]. Yule uses the method to analyze the properties of time series data for sunspots.

The sample partial autocorrelations are useful when identifying the order of an autoregressive process. Consider the case of an AR(1) process $X_t = \theta_1 X_{t-1} + \varepsilon_t$ in which $\theta_2 = 0$ by definition. We have seen that $\rho_2 = \theta_1^2$ and the second-order sample autocorrelation will typically be non-zero. It is therefore difficult to distinguish a first-order autoregressive process from a second-order process by simply examining the regular autocorrelations because the regular autocorrelations do not "cut off" at the lag length of the process. However, the partial autocorrelations do cut off at this lag length and therefore values of $\hat{\theta}_k$ close to zero for $k \geq 2$ would indicate that a first-order process is appropriate.

Example: Consider the following stochastic process

$$X_t = 1.2X_{t-1} - 0.35X_{t-2} + \varepsilon_t.$$

where ε_t are independent identically distributed random variables with the standard normal distribution. A realization of this process was created using the EViews random number generator as shown in the graph below:

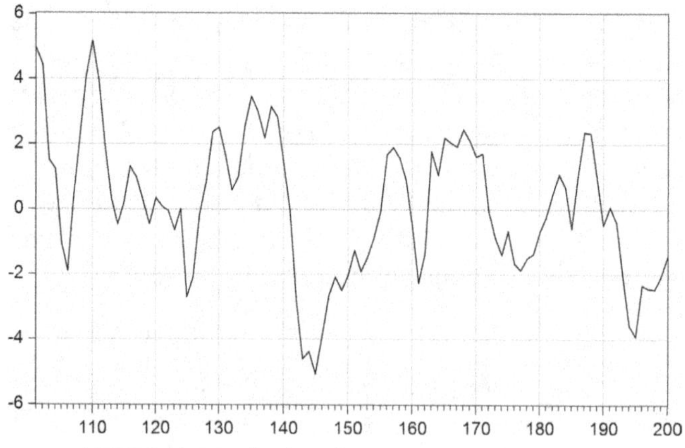

FIGURE 10.5 Realization of an AR(2) Stochastic Process.

Suppose we are given the data series shown in Figure 10.5 and asked to fit a model that captures its important features. The first question we need to ask is what sort of model should we estimate? To answer this, we turn to the sample correlogram as shown in Figure 10.6.

Sample: 101 200
Included observations: 100

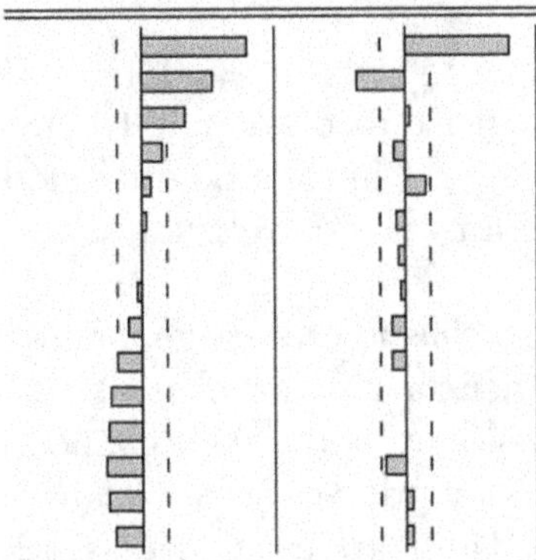

	AC	PAC	Q-Stat	Prob
1	0.827	0.827	70.548	0.000
2	0.565	-0.378	103.83	0.000
3	0.340	0.029	115.96	0.000
4	0.160	-0.083	118.69	0.000
5	0.077	0.156	119.33	0.000
6	0.041	-0.081	119.51	0.000
7	0.000	-0.064	119.51	0.000
8	-0.045	-0.049	119.73	0.000
9	-0.113	-0.109	121.15	0.000
10	-0.199	-0.101	125.62	0.000
11	-0.254	-0.001	133.00	0.000
12	-0.266	-0.008	141.21	0.000
13	-0.280	-0.156	150.41	0.000
14	-0.264	0.058	158.67	0.000
15	-0.209	0.042	163.92	0.000

FIGURE 10.6 Sample Correlogram for the Time-Series Shown in Figure 10.5.

This correlogram indicates that a second-order autoregressive process is appropriate. This is indicated by the fact that the sample autocorrelations approach zero exponentially while only the first two sample partial autocorrelations lie outside the standard error bands. The next stage in estimating the parameters of our model is to calculate the relevant sample moments. In this case, we have the following

$$\hat{\gamma}_0 = \sum_{t=1}^{100} (x_t - \overline{x})^2 = 632.193224$$

$$\hat{\gamma}_1 = \sum_{t=2}^{100} (x_t - \overline{x})(x_{t-1} - \overline{x}) = 522.823809$$

$$\hat{\gamma}_2 = \sum_{t=3}^{100} (x_t - \overline{x})(x_{t-2} - \overline{x}) = 357.189171.$$

The sample autocorrelations can then be calculated straightforwardly as

$$\hat{\rho}_1 = \hat{\gamma}_1 / \hat{\gamma}_0 = 0.8270$$

$$\hat{\rho}_2 = \hat{\gamma}_2 / \hat{\gamma}_0 = 0.5649.$$

Let θ_i be the partial autocorrelation of order i. The first-order sample partial autocorrelation can be calculated straightforwardly as $\hat{\theta}_1 = \hat{\rho}_1 = 0.8270$. We can then calculate the second-order partial autocorrelation using the Yule-Walker equations

$$\hat{\gamma}_1 = \hat{\theta}_1 \hat{\gamma}_0 + \hat{\theta}_2 \hat{\gamma}_1$$
$$\hat{\gamma}_2 = \hat{\theta}_1 \hat{\gamma}_1 + \hat{\theta}_2 \hat{\gamma}_0.$$

These equations can be solved to yield $\hat{\theta}_2 = -0.3778$ which gives our estimate of the second-order partial autocorrelation. Higher-order autocorrelations and partial autocorrelations can then be estimated using the method we have set out, thus generating the correlogram and partial correlogram shown in Figure 10.6.

To identify the type of stochastic process generating the data, we must form an assessment of whether the autocorrelations are significantly different from zero. The basic tool here is the 95% confidence interval for the autocorrelations under the assumption of no serial correlation shown by the broken lines in Figure 10.6. These are calculated as $\pm 2/\sqrt{T}$ which can be shown to generate a 95% confidence interval under the null hypothesis that the process is white noise. The standard error bands for both the regular and partial autocorrelations can be calculated in this way. This allows us to identify the order of the AR process generating the data by assessing how many of the partial autocorrelations lie outside the confidence interval.

The correlogram shown in Figure 10.6 indicates the presence of some form of autoregressive process since it shows sample autocorrelations that are consistently positive, but which decline exponentially to zero. However, the regular autocorrelations do not allow us to identify the order of this process since this pattern is consistent with either an AR(1) or a higher-order process. To identify the order of the process, we turn to the partial autocorrelations. Here we see two significant partial autocorrelations, which confirms that an AR(2) process is most appropriate in this case. We can therefore proceed on the basis that the process generating the data is likely to be an autoregressive process of order 2 and seek to estimate its parameters.

Examination of the correlogram also allows us to distinguish autoregressive processes from moving average processes. This is because the correlogram for a moving average process behaves differently from that of an autoregressive process. To see this, we will contrast an AR(1) process with an MA(1) process. We have seen that for a stationary AR(1) process the autocorrelations die down exponentially to zero, that is, $\rho_k = \rho^k$.

However, the partial autocorrelations cut out immediately after lag 1, that is, $\theta_k = 0; k = 2,\ldots,\infty$. Now consider an MA(1) process of the form

$$X_t = \varepsilon_t + a\varepsilon_{t-1}, \tag{10.11}$$

where ε_t are independent identically distributed random variables. We have

$$\begin{aligned} &E\left(X_t^2\right) = \sigma_\varepsilon^2\left(1 + a^2\right) \\ &E\left(X_t X_{t-1}\right) = a\sigma_\varepsilon^2 \\ &E\left(X_t X_{t-k}\right) = 0 \;\; \forall k \geq 2. \end{aligned} \tag{10.12}$$

It follows that the first-order autocorrelation $\rho_1 = a/\left(1 + a^2\right)$ is not zero, but all subsequent autocorrelations are equal to zero. Therefore, the regular correlogram cuts off abruptly after the first lag. The partial autocorrelations, however, do not cut off in this case. Consider the second-order partial autocorrelation. We can scale the Yule-Walker equations by dividing by the variance to obtain

$$\begin{aligned} \rho_1 &= \theta_1 + \rho_1\theta_2 \\ \rho_2 &= \theta_1\rho_1 + \theta_2. \end{aligned}$$

The second-order partial autocorrelation is given by θ_2 where θ_2 is the solution from this pair of equations for given ρ_1 and ρ_2. We have already shown that for an MA(1) process $\rho_1 = a/\left(1 + a^2\right)$ and $\rho_2 = 0$. Therefore, the second equation above gives

$$\theta_2 = -\theta_1\rho_1 \neq 0.$$

This result generalizes and we can show that, for a k'th order moving average process, the regular autocorrelations cut off after k lags. However, the partial autocorrelations continue to be non-zero for higher-order lags. This is a reversal of the pattern for an autoregressive process and allows us to distinguish between these alternatives by examination of the sample correlogram and the sample partial correlogram.

Example: Consider the following realization of a first-order moving average process $X_t = \varepsilon_t + 0.75\varepsilon_{t-1}$.

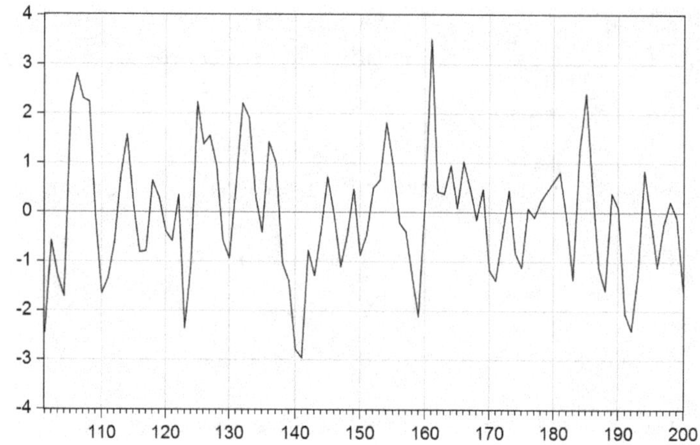

FIGURE 10.7 Realization of First-Order Moving Average Process.

The correlogram of this process takes the form shown in Figure 10.8. This has one significant first-order sample autocorrelation, but all the higher-order sample autocorrelations lie within the standard error bands. In contrast, both the first- and second-order partial autocorrelations lie outside the standard error bands. This is the characteristic pattern of a first-order moving average process and therefore we would fit a model of this type to the data.

Sample: 101 200
Included observations: 100

Autocorrelation	Partial Correlation		AC	PAC	Q-Stat	Prob
		1	0.451	0.451	20.999	0.000
		2	-0.051	-0.321	21.274	0.000
		3	-0.096	0.111	22.247	0.000
		4	-0.142	-0.214	24.384	0.000
		5	-0.083	0.109	25.131	0.000
		6	-0.037	-0.130	25.283	0.000
		7	-0.027	0.057	25.362	0.001
		8	0.019	-0.027	25.403	0.001
		9	-0.008	-0.035	25.411	0.003
		10	-0.137	-0.169	27.553	0.002

FIGURE 10.8 Correlogram of Time Series Shown in Figure 10.7.

10.2 ARIMA MODELING

So far, we have discussed the theory of stochastic processes and the differences between the correlograms of autoregressive and moving average processes. In this section, we will discuss how to use this knowledge to fit an ARIMA model to real-world data. Note that the acronym ARIMA means

Autoregressive Integrated Moving Average. The "autoregressive" and "moving average" parts of this acronym are self-explanatory, but we need to spend a little time discussing the meaning of "integrated" in this context before we use this model on actual data.

Historical Note: The method of ARIMA modeling was first set out systematically in the classic book by Box and Jenkins [Box1970] entitled *Time Series Analysis: Forecasting and Control*. This book has been immensely influential for both theorists and practitioners.

Let us suppose we have data $x_t : t = 1, \ldots, T$ that is a realization of some stochastic process. We wish to identify the nature of the stochastic process which has generated the data and to estimate the relevant parameters. The most important first stage of any investigation is to examine the data by plotting it against time. This will often reveal important features of the data generation process immediately.

Example: To illustrate the process of fitting an ARIMA model to data, we will use the example of United States unemployment data. The series shown below in Figure 10.9 has been downloaded from the Federal Reserve Board of St. Louis (FRED) database and consists of quarterly figures for the percentage of the workforce unemployed for the period 1948 to 2019.

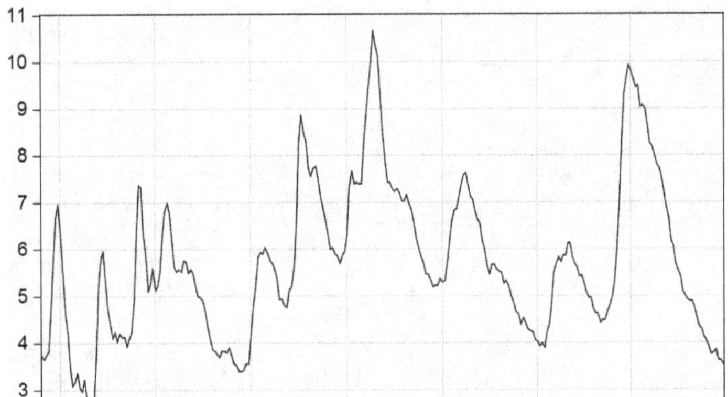

FIGURE 10.9 Percentage Unemployed US Economy.

The first thing we need to decide when constructing an ARIMA model is whether to difference the data. Formally, we need to decide whether the series in question is an *integrated* series, in the sense that it must be differenced in order to make it stationary. Recall that stationarity is an essential property if we are to be able to use standard statistical methods to estimate and interpret our model. Differencing the data in time-series modeling is often interpreted as "taking out the trend" in the data prior to estimation. However, the reasons for differencing are more general than this. A stochastic process may be non-stationary even when no trend is present. This applies in our example. Examination of Figure 10.9 would suggest that there is no long-run trend in unemployment. However, if we examine the correlogram of log unemployment,[1] as shown in Figure 10.10, then we see that the autocorrelations behave as we would expect from a series generated by a non-stationary stochastic process. In particular, we see that the autocorrelations that die down very slowly to zero according to a linear relationship with the lag length rather than the exponential relationship we would expect to see from a stationary process. This suggests that the data is generated by an integrated process and should be differenced prior to fitting a model.

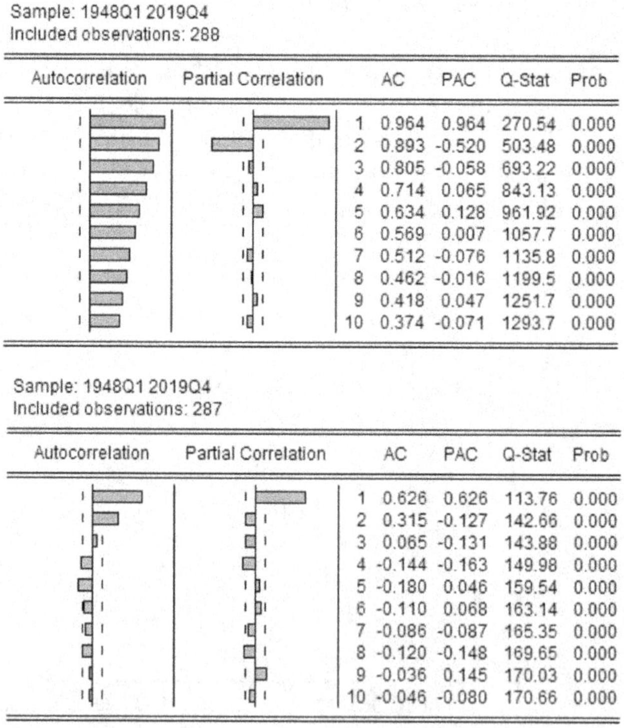

FIGURE 10.10 Correlograms for Log Unemployment (upper) and its First Difference (lower).

[1] Why take the log transformation here? The answer is that without this transformation it would be possible for the model to generate negative predictions for unemployment which is clearly inconsistent with the nature of the series.

Turning to the lower panel of Figure 10.10, we see that the first difference of log unemployment has a correlogram which suggests that it is plausible that this data has been generated by a stationary stochastic process. Unlike the case of the undifferenced series, the autocorrelations appear to decline exponentially rather than linearly. The first partial autocorrelation clearly lies outside the standard error bands and a number of subsequent partial autocorrelations are very close. To avoid overfitting, our first model includes just two autocorrelations. That is, we estimate an ARIMA(2,1,0) model where the numbers in parentheses represent the order of the autoregressive process for the variable itself, the number of time the data has been differenced and the order of the moving average process in the errors of the model. When this model is applied to the data, we obtain the results shown in equation (10.13)

$$\Delta\ln(u_t) = \underset{(0.0080)}{-0.0004} + \underset{(0.0411)}{0.7049}\Delta\ln(u_{t-1}) - \underset{(0.0464)}{0.1274}\Delta\ln(u_{t-2}) + \hat{\varepsilon}_t \qquad (10.13)$$

$$R^2 = 0.40 \qquad \hat{\sigma} = 0.0528 \qquad DW = 2.0271.$$

Both autoregressive coefficients are significant at the 5% level and the model is a reasonable fit, explaining 40% of the variation of the dependent variable. To assess whether any further autoregressive or moving average terms would improve the fit of the model, we examine the correlogram of the residuals from equation (10.13), which is shown in Figure 10.11. This correlogram is reasonably flat but some of the remaining autocorrelations are just about significant. However, to avoid overfitting the model, we chose not to make further adjustments.

Sample: 1948Q1 2019Q4
Included observations: 287
Q-statistic probabilities adjusted for 2 ARMA terms

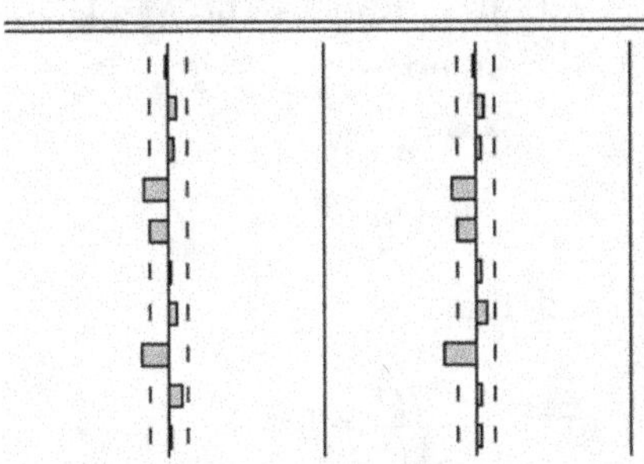

	AC	PAC	Q-Stat	Prob
1	-0.014	-0.014	0.0610	
2	0.060	0.060	1.1003	
3	0.034	0.036	1.4444	0.229
4	-0.168	-0.171	9.6794	0.008
5	-0.124	-0.137	14.175	0.003
6	0.020	0.037	14.290	0.006
7	0.042	0.078	14.813	0.011
8	-0.176	-0.209	24.051	0.001
9	0.091	0.028	26.505	0.000
10	0.010	0.040	26.535	0.001

FIGURE 10.11 Correlogram of Residuals from ARIMA(2,1,0) Model for Log US Unemployment Series.

So far, we have used an informal method to identify the nature of the stochastic process generating the data. This has involved examination of the correlogram to see if we can match the pattern we observe with one of the standard patterns for known stochastic processes. Basically, we look for linear damped autocorrelations as evidence for non-stationarity, which requires differencing, or for exponentially damped autocorrelations which indicates a stationary process. The partial autocorrelations allow us to determine the order of a stationary autoregressive process. If we observe sample autocorrelations that cut off abruptly, then this is interpreted as evidence of a moving average process. In many situations, this informal approach will allow us to quickly identify a reasonable model. However, it does suffer from the problem that different investigators might choose different models based on the same data.

Another, more systematic approach, to ARIMA modeling, is to estimate a comprehensive set of possible models and compare them using one of the model selection criteria available. The two most commonly used selection criteria are the Akaike and Schwartz criteria. These criteria balance the extra explanatory power provided by adding AR or MA terms against the loss of degrees of freedom this involves. Consider the model

$$X_t = \sum_{i=1}^{p} \hat{\rho}_i X_{t-i} + \sum_{j=1}^{q} \hat{\theta}_j \varepsilon_{t-j} + \hat{\varepsilon}_t. \tag{10.14}$$

The standard error of the regression is calculated as $\hat{\sigma} = \sqrt{\sum_{t=1}^{T} \hat{\varepsilon}_t^2 / (T - p - q)}$. Using this we can calculate several possible information criteria that can be used to compare alternative models. Each of these criteria represents a trade-off between the reduction in the residual sum of squares obtained by including extra autoregressive or moving average terms and the loss of degrees of freedom in doing so. We have a variety of possible information criteria that perform this function and there is no clear consensus on which performs the best. The two most often used are the Akaike and the Schwartz criteria. The Akaike criterion is defined as

$$AIC = \ln \hat{\sigma}^2 + \frac{2(p+q)}{T}. \tag{10.15}$$

while the Schwartz criterion is defined as

$$SC = \ln \hat{\sigma}^2 + \frac{(p+q)}{T} \ln T. \tag{10.16}$$

If we add either AR or MA terms to a model, then this will tend to reduce the standard error of the equation $\hat{\sigma}$. However, these criteria put a penalty of the loss of degrees of freedom from doing this in the form of the second terms in (10.15) and (10.16). In both cases, we would look for the "best" model as the one that contains the combination of AR and MA terms that minimize the relevant criterion. The Schwartz criterion can be shown to be more parsimonious in that it will typically include fewer AR and MA terms than the Akaike criterion.

Example: Using the EViews random number generator a realization of the stochastic process $X_t = 0.7X_{t-1} + \varepsilon_t$ was generated with random number seed 200. We estimated all possible models with p and q up to order 3. This generates the following Schwartz criteria:

		MA Lags			
		0	1	2	3
AR Lags	0	3.26322	2.84021	2.82190	2.86138
	1	2.86965	2.82798	2.86476	2.90074
	2	2.81503	2.86094	2.90705	2.94626
	3	2.85903	2.87860	2.92157	2.87150

Note that this process picks out the "wrong" model in that the lowest Schwartz criterion corresponds to an AR(2) model, rather than an AR(1) model. The model chosen here is

$$X_t = \underset{(0.1785)}{-0.9080} + \underset{(0.0959)}{0.7775}X_{t-1} - \underset{(0.0943)}{0.3023}X_{t-2} + \hat{\varepsilon}_t$$

$$R^2 = 0.4174 \qquad DW = 1.97 \qquad \hat{\sigma} = 0.9368.$$

The roots of this process are complex with modulus equal to 0.55.

10.3 FORECASTING WITH AN ARIMA MODEL

One of the big advantages of the ARIMA modeling approach is that it can be used to generate a forecasting model much more quickly and easily than other approaches. For example, if we compare the time and effort needed to construct even a small econometric model, then the ARIMA approach wins hands down. Even more encouragingly, studies have shown that ARIMA models outperform structural econometric models in forecasting exercises, at least over relatively short horizons. However, the factors that give ARIMA models the competitive edge in the short term can lead to problems when it comes to longer-term forecasting. By focusing on one variable at a time, the ARIMA approach neglects the interactions between variables which can become important over time. It is also possible to argue that forecasts are about more than providing a central estimate. Structural econometric models allow the forecaster to tell a story about how the variable(s) in question are likely to develop and to experiment with alternative scenarios. Having said that, however, there is a certainly a big role for ARIMA modeling as part of the toolbox of the professional forecaster. If nothing else, it can provide a useful benchmark against which more structural approaches can be judged. Therefore, in this section, we will examine the properties of forecasts generated by a time series approach.

Let us start with an example, consider the AR(1) model $X_t = \theta X_{t-1} + \varepsilon_t$ for which we have data for $t = 1,\ldots,T$. For the moment we will assume that θ is a known parameter. The one period ahead forecast is

$$E\left(X_{T+1} \middle| \Omega_T\right) = E\left(\theta X_T + \varepsilon_{T+1} \middle| X_T\right) = \theta X_T, \tag{10.17}$$

and the one period ahead forecast error is

$$X_{T+1} - E\left(X_{T+1} \middle| \Omega_T\right) = \varepsilon_{T+1}, \tag{10.18}$$

with the one step ahead variance being

$$E\left\{X_{T+1} - E\left(X_{T+1} \middle| \Omega_T\right)\right\}^2 = \sigma_\varepsilon^2. \tag{10.19}$$

Now consider the k step ahead forecast

$$\begin{aligned} X_{T+k} &= \varepsilon_{T+k} + \theta\varepsilon_{T+k-1} + \theta^2 \varepsilon_{T+k-2} + \ldots + \theta^{k-1}\varepsilon_{T+1} + \theta^k X_T \\ &= \theta^k X_T + \sum_{i=0}^{k-1} \theta^i \varepsilon_{T+k-i}. \end{aligned} \tag{10.20}$$

The k step ahead forecast is $E\left(X_{T+k} \middle| \Omega_T\right) = \theta^k X_T$ and the k step ahead forecast variance is

$$E\left(X_{T+k} - E\left(X_{T+k} \middle| \Omega_T\right)\right)^2 = \sum_{i=0}^{k-1} \theta^{2\,i} \sigma_\varepsilon^2 = \left(\frac{1-\theta^{2k}}{1-\theta^2}\right)\sigma_\varepsilon^2. \tag{10.21}$$

As $k \to \infty$, this converges to $\sigma_\varepsilon^2 / \left(1-\theta^2\right)$. This establishes an important principle of forecasting with a time series model – as the forecast horizon increases the forecast error variance increases. However, providing the process is stationary, the forecast error variance will converge to a constant value in the long run.

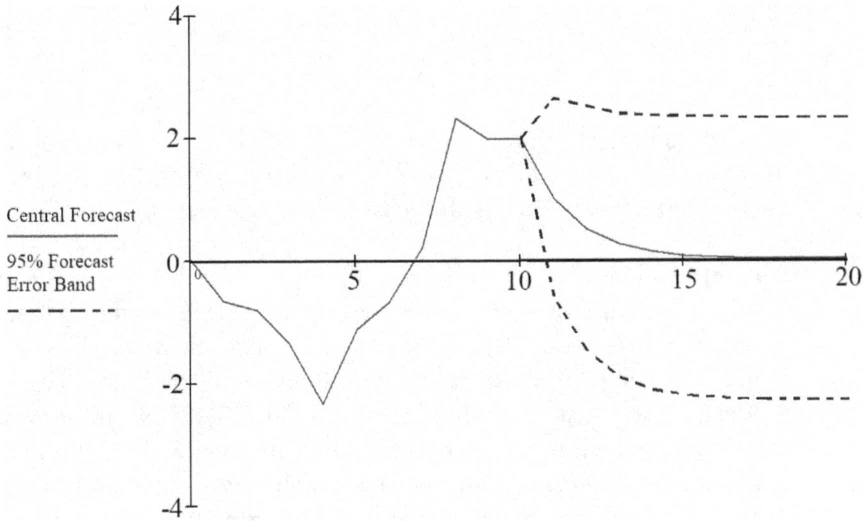

FIGURE 10.12 Forecasting with an ARIMA Model.

Example: For $t = 1,\ldots,10$ the data shown by the solid line in Figure 10.12 has been generated artificially using an equation of the form $X_t = 0.5X_{t-1} + \varepsilon_t$ where ε_t are Gaussian white-noise disturbances with mean zero and variance 1. For $t = 11,\ldots,20$ the solid line shows the central forecast calculated as $\theta^k X_{10}$. The broken lines show a 95% confidence interval for a k step ahead forecast. This is calculated as $\theta^k X_{10} \pm 2 \times SE$ where $SE = \sqrt{\left(1-\theta^{2k}\right)/\left(1-\theta^2\right)}$.

Our discussion of forecasting so far has assumed that the stochastic process is stationary. However, when constructing a forecasting model using the ARIMA approach, stationarity is often only achieved by differencing the data prior to estimation. For example, when constructing the model of unemployment given by equation (10.13), we found that the unemployment percentage was not stationary and we, therefore, differenced the data before estimating our final model. If our original series is not stationary, and the data is, therefore, differenced prior to estimation, it follows that the forecast confidence interval will not converge in the way described for stationary data. To demonstrate this result, consider a special case in which the random variable X is generated by a simple random walk, that is, $X_t = X_{t-1} + \varepsilon_t$. If we have data up to date T, then the forecast value of X for all future periods is simply X_T. It is easy to see that the forecast error is

$$X_{T+k} - E\left(X_{T+k} \middle| \Omega_T\right) = \sum_{i=1}^{k} \varepsilon_{T+i}, \tag{10.22}$$

and the forecast error variance is

$$E\left(\sum_{i=1}^{k} \varepsilon_{T+i}\right)^2 = k\sigma_\varepsilon^2, \tag{10.23}$$

and therefore, the forecast error variance does not converge as $k \to \infty$. Instead, the standard error bands around the central forecast will increase proportionally with k as the forecast horizon increases.

Example: The divergence of the forecast error bands can be using our model of unemployment. Using equation (10.13) to forecast unemployment over the period 2018.1 to 2019.4 we obtain the results shown in Figure 10.13. The solid line shows actual unemployment, and the broken line shows the central forecast generated by our model. The lines with circles give the 95% confidence interval around the central forecast. This shows the standard error bands increasing because of the differencing operator employed to transform the series to stationarity prior to estimation.

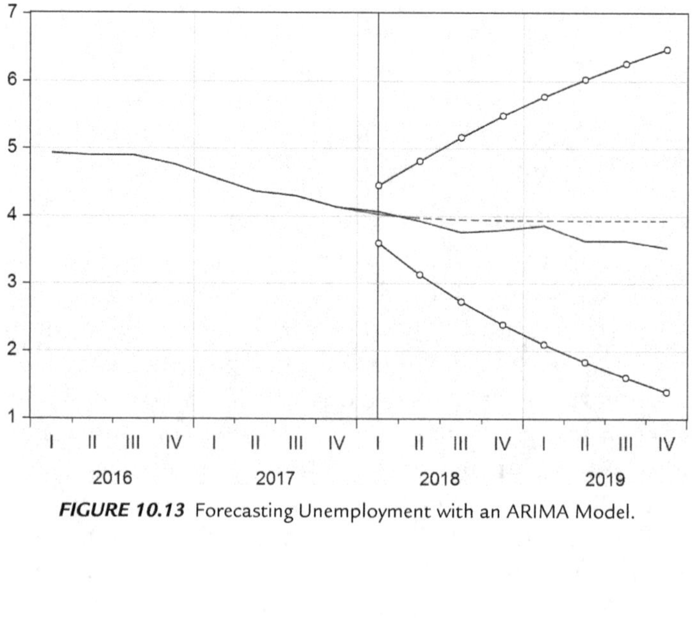

FIGURE 10.13 Forecasting Unemployment with an ARIMA Model.

In summary, we note that ARIMA models provide a very quick and convenient method for the construction of a forecasting model. These are especially useful for generating short horizon forecasts when we do not have any strong theoretical priors about the process generating the data. For longer forecast horizons, the forecast error variance will increase for all types of ARIMA model. However, if the original series is stationary, then this increase is limited, and the forecast error variance will eventually converge to a constant value. In contrast, if the original series is not stationary, and the data is differenced prior to estimation, then the forecast error variance will continue to increase indefinitely as the forecast horizon increases.

10.4 IMPULSE RESPONSES

We now turn to the analysis of the effects of shocks, or disturbances, in ARIMA models. These are measured in what we call the *impulse response*. The impulse response shows the effect on the variable X of a shock to the disturbance term ε. Note that sometimes the shock considered is a unit impulse and, in other cases, it is one standard deviation of the disturbance, but this is just a matter of scaling. Consider the example of a stationary AR(1) process of the form $X_t = \theta X_{t-1} + \varepsilon_t$. The impulse response can be derived from a process of backward substitution. This yields the moving average representation given by $X_t = \sum_{i=0}^{\infty} \theta^i \varepsilon_{t-i}$. To calculate the impulse response we assume a one-off innovation, for example, $\varepsilon_0 = 1$ and then assume all future ε's are zero. Setting $\theta = 0.7$ we can then calculate the impulse response as:

Time	Impulse	Effect on X
0	1	1
1	0	0.7
2	0	0.49
3	0	0.343
4	0	0.2401
5	0	0.16807
6	0	0.117649

Since the stochastic process here is stationary, it follows that the impulse response eventually converges on zero. If the stochastic process is not stationary, and therefore the data have been differenced prior to estimation, then the impulse response will not converge to zero.

Example: In this section, we consider a real-world example. We fit an ARIMA model to annual data for US consumption of gasoline and then use this to construct impulse responses. The series itself is shown in Figure 10.14.

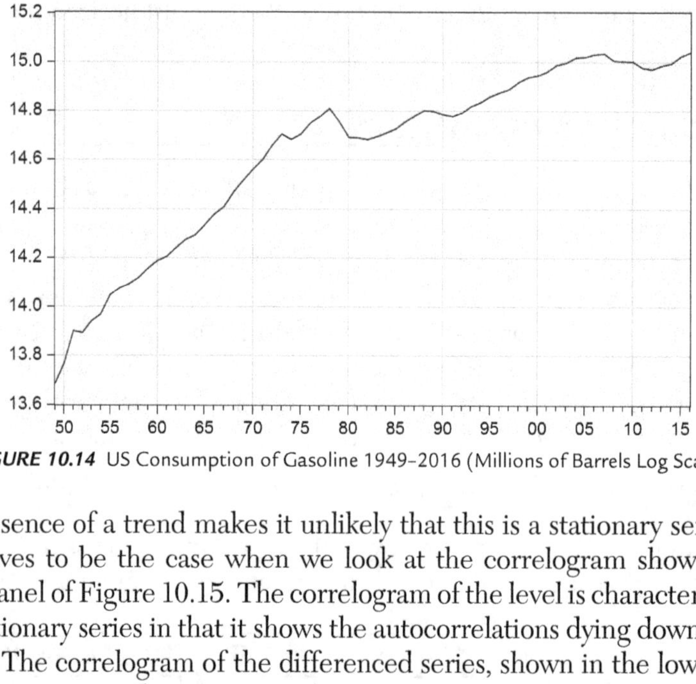

FIGURE 10.14 US Consumption of Gasoline 1949–2016 (Millions of Barrels Log Scale).

The presence of a trend makes it unlikely that this is a stationary series, and this proves to be the case when we look at the correlogram shown in the upper panel of Figure 10.15. The correlogram of the level is characteristic of a non-stationary series in that it shows the autocorrelations dying down linearly to zero. The correlogram of the differenced series, shown in the lower panel of Figure 10.15, is characteristic of an AR(1) process, in that it shows a significant first-order autocorrelation, but this dies down fairly quickly to zero. Although it is possible for this to arise from a higher-order autoregressive process, none of the higher-order partial autocorrelations are significant. The correlogram, therefore, suggests fitting an ARIMA(1,1,0) model to the data.

Sample: 1949 2016
Included observations: 68

Autocorrelation	Partial Correlation		AC	PAC	Q-Stat	Prob
		1	0.937	0.937	62.439	0.000
		2	0.879	-0.000	118.14	0.000
		3	0.831	0.060	168.74	0.000
		4	0.779	-0.056	213.92	0.000
		5	0.729	-0.008	254.12	0.000
		6	0.678	-0.045	289.42	0.000
		7	0.631	0.007	320.48	0.000
		8	0.584	-0.029	347.53	0.000
		9	0.536	-0.032	370.70	0.000
		10	0.486	-0.047	390.10	0.000

Sample: 1949 2016
Included observations: 67

Autocorrelation	Partial Correlation		AC	PAC	Q-Stat	Prob
		1	0.414	0.414	11.991	0.001
		2	0.229	0.070	15.734	0.000
		3	0.196	0.096	18.517	0.000
		4	0.229	0.132	22.375	0.000
		5	0.237	0.102	26.557	0.000
		6	0.100	-0.077	27.313	0.000
		7	0.022	-0.061	27.351	0.000
		8	0.057	0.028	27.603	0.001
		9	0.077	0.018	28.078	0.001
		10	0.087	0.040	28.697	0.001

FIGURE 10.15 Correlogram of Log Gasoline Consumption Series.

Based on an examination of the correlograms in Figure 10.15, an AR(1) model in first differences (i.e., an ARIMA(1,1,0) model), appears to be an appropriate model for this time series. Estimation of such a model yields the results shown in equation (10.24)

$$\Delta X_t = \underset{(0.0039)}{0.0109} + \underset{(0.1090)}{0.4138}\,\Delta X_{t-1} + \hat{\varepsilon}_t \qquad (10.24)$$

$$R^2 = 0.1839 \qquad DW = 2.14 \qquad \hat{\sigma} = 0.0258.$$

We have seen that, in models with unit roots, a one-off disturbance will have a temporary effect on the growth rate of the series but a permanent effect on its level. This can be demonstrated, in this case, by writing the model in levels rather than first differences

$$X_t = 0.0109 + 1.4138X_{t-1} - 0.4138X_{t-2} + \varepsilon_t. \tag{10.25}$$

Thus, a first-order equation in first differences becomes a second-order equation in the levels of the series. Moreover the difference equation defined in equation (10.25) contains a unit root. Consider a one standard deviation shock to the series, that is, an increase in ε of 0.0219 in the first period. This produces the dynamic response shown in Figure 10.16 which shows that demand continues to rise after the initial shock but eventually levels off at a new equilibrium value.

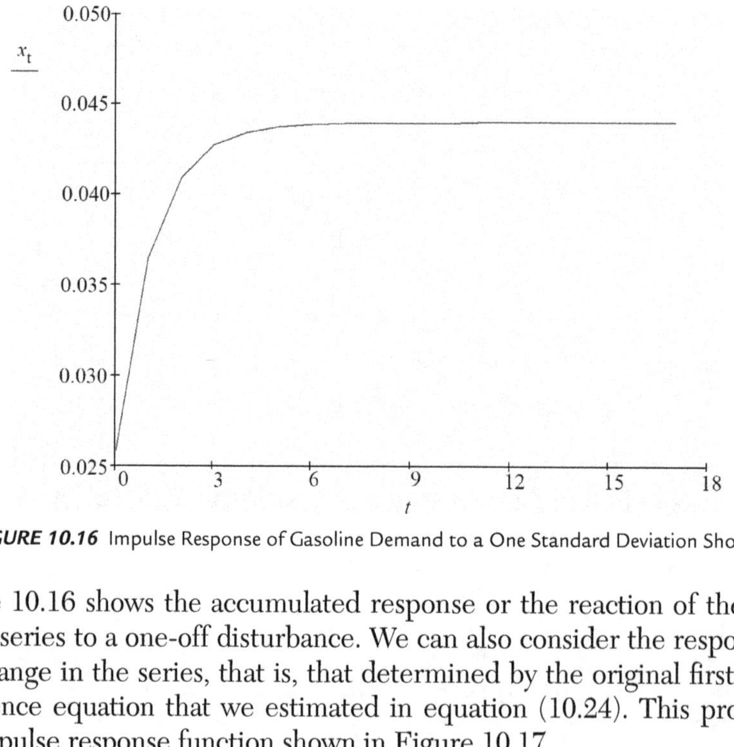

FIGURE 10.16 Impulse Response of Gasoline Demand to a One Standard Deviation Shock.

Figure 10.16 shows the accumulated response or the reaction of the level of the series to a one-off disturbance. We can also consider the response of the change in the series, that is, that determined by the original first-order difference equation that we estimated in equation (10.24). This produces the impulse response function shown in Figure 10.17.

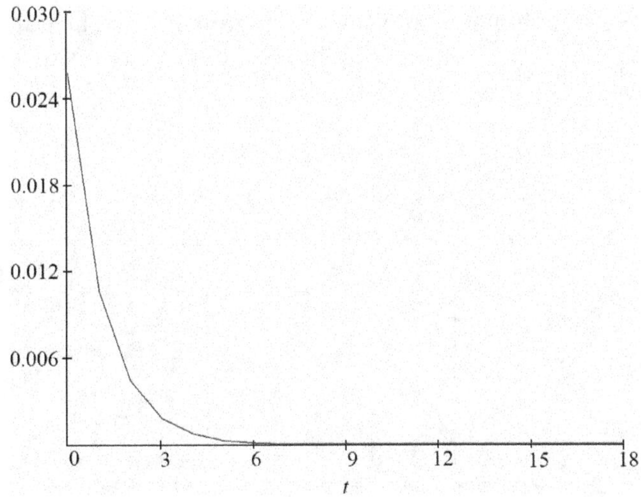

FIGURE 10.17 Impulse Response of the Change in Gasoline Demand to a One Standard Deviation Shock.

10.5 MOVING AVERAGE PROCESSES

We saw earlier through model (10.11) that the correlogram of a moving average process has very different characteristics from that of an autoregressive process. However, there are several other properties of moving average processes which it is important to consider when modeling time series. This first is that, except for one special case, the moving average process is not uniquely identified by the correlogram. We have $\rho_1 = a/\left(1+a^2\right)$, suppose we know ρ_1 and wish to solve for a. We have

$$\rho_1 a^2 - a + \rho_1 = 0. \tag{10.26}$$

It is easy to see that, for any a which is a solution of this equation, $1/a$ will also be a solution. Hence, the correlogram of an MA(1) process does not allow us to uniquely identify the parameter of that process. This result generalizes to higher-order moving average processes.

Example: A set of Gaussian white-noise disturbances was generated using the EViews random number generator. These were then used to construct two random series following moving average processes of the form $X_t = \varepsilon_t - 0.5\varepsilon_{t-1}$ and $X_t = \varepsilon_t - 2\varepsilon_{t-1}$. The correlograms of these series were then calculated and are shown in Figure 10.18. The correlograms for these two processes are virtually identical. There is no way in which we can choose between the processes on this basis.

Sample: 101 200
Included observations: 100

Autocorrelation	Partial Correlation		AC	PAC	Q-Stat	Prob
		1	-0.362	-0.362	13.497	0.000
		2	-0.057	-0.216	13.833	0.001
		3	0.059	-0.054	14.193	0.003
		4	-0.119	-0.153	15.705	0.003
		5	0.053	-0.059	16.007	0.007
		6	0.014	-0.023	16.027	0.014
		7	-0.051	-0.059	16.307	0.022
		8	0.028	-0.032	16.396	0.037
		9	0.057	0.057	16.764	0.053
		10	-0.049	0.003	17.042	0.073

Sample: 101 200
Included observations: 100

Autocorrelation	Partial Correlation		AC	PAC	Q-Stat	Prob
		1	-0.362	-0.362	13.466	0.000
		2	-0.059	-0.218	13.823	0.001
		3	0.065	-0.047	14.274	0.003
		4	-0.123	-0.151	15.882	0.003
		5	0.038	-0.076	16.037	0.007
		6	0.020	-0.033	16.082	0.013
		7	-0.059	-0.075	16.458	0.021
		8	0.015	-0.062	16.483	0.036
		9	0.047	0.013	16.736	0.053
		10	-0.052	-0.035	17.047	0.073

FIGURE 10.18 Correlograms for Alternative Moving Average Processes $X_t = \varepsilon_t - 0.5\varepsilon_{t-1}$ (upper) and $X_t = \varepsilon_t - 2\varepsilon_{t-1}$ (lower).

In practice, when faced with the choice illustrated by Figure 10.18, we normally choose the solution in which $\alpha < 1$. Why is this the case? The answer lies in an alternative representation of the moving average process. Using the lag operator we can write $X_t = \varepsilon_t - \alpha\varepsilon_{t-1}$ as $X_t = (1 - \alpha L)\varepsilon_t$. Dividing by the lag polynomial now allows us to write this process as

$$\frac{1}{1-\alpha L} X_t = \varepsilon_t. \tag{10.27}$$

Expanding the expression on the left-hand side allows us to write this as

$$\begin{aligned}&\left(1+aL+a^2L^2+a^3L^3+\ldots\right)X_t=\varepsilon_t\\&\sum_{i=0}^{\infty}a^i X_{t-i}=\varepsilon_t.\end{aligned} \tag{10.28}$$

This is a general result – any finite moving average process can be written as an infinite autoregressive process. Now, if we chose the solution in which $a>1$, the weights on past values of X would increase exponentially the further back into the past is the value of X. On the other hand, if $a<1$, then the weights on past values of X decline. It, therefore, makes more sense to choose this second option. Moving average processes in which $a<1$ are said to be invertible processes. Invertibility refers to the feature that the weights on past X values decline as the lag increases.

There is one special case that we have not yet considered. Suppose we have $|a|=1$. In the case the moving average parameter is uniquely identified by the correlogram. If $a=1$ then the first-order autocorrelation is equal to ½, while if $a=-1$ it is equal to −½. However, this process is non-invertible since the weights in the moving average process will either always be equal to one (in the case $a=-1$) or will alternate between +1 and −1 in the case $a=1$. In either of these situations, however, the weights do not decline to zero which is the required condition for invertibility. This special case can arise naturally when we difference a series that is already stationary. For example, consider a series that consists of Gaussian white-noise errors around a deterministic trend, that is, $X_t=a+\beta t+\varepsilon_t$. Now suppose we difference, this series to remove the trend. This is a common procedure in time-series analysis. The resulting series takes the form

$$\Delta X_t=\beta+\varepsilon_t-\varepsilon_{t-1}. \tag{10.29}$$

The effect of differencing this series is to create a new series with a disturbance which is a non-invertible moving average process. Such a situation is described as "over-differencing."

Example: Using the EViews random number generator, we generate a realization of the following stochastic process $X_t=0.05\,t+\varepsilon_t$ where ε_t are Gaussian white-noise disturbances. Note that this is a trend-stationary process. The sample correlograms of the level and the first difference of this series are shown in Figure 10.19. The upper panel shows the correlogram for the level and the lower panel shows that for the first difference.

Sample: 101 200
Included observations: 100

Autocorrelation	Partial Correlation		AC	PAC	Q-Stat	Prob
		1	0.620	0.620	39.639	0.000
		2	0.585	0.325	75.246	0.000
		3	0.557	0.201	107.89	0.000
		4	0.493	0.058	133.70	0.000
		5	0.539	0.196	164.94	0.000
		6	0.532	0.134	195.68	0.000
		7	0.507	0.061	223.83	0.000
		8	0.522	0.095	254.02	0.000
		9	0.505	0.069	282.61	0.000
		10	0.446	-0.051	305.19	0.000

Sample: 101 200
Included observations: 100

Autocorrelation	Partial Correlation		AC	PAC	Q-Stat	Prob
		1	-0.459	-0.459	21.684	0.000
		2	-0.058	-0.341	22.039	0.000
		3	0.081	-0.160	22.724	0.000
		4	-0.119	-0.240	24.218	0.000
		5	0.061	-0.168	24.622	0.000
		6	0.023	-0.107	24.680	0.000
		7	-0.058	-0.137	25.045	0.001
		8	0.022	-0.135	25.101	0.001
		9	0.060	-0.026	25.503	0.002
		10	-0.041	-0.013	25.689	0.004

FIGURE 10.19 Correlograms Generated by a Trend Stationary Stochastic Process.

The correlogram of the level of the series is consistent with a unit root process in that the autocorrelations die down linearly. However, when we difference the series, we obtain an MA(1) process, with a first-order autocorrelation which is close to –0.5. This is consistent with a series that has been over-differenced. Finally, the correlogram in Figure 10.20 is that of the residuals from a regression of the series on a time trend. This is consistent with a white-noise process which indicates that this is the preferred method in this case.

Sample: 101 200
Included observations: 100

Autocorrelation	Partial Correlation		AC	PAC	Q-Stat	Prob
		1	0.005	0.005	0.0030	0.956
		2	-0.061	-0.061	0.3877	0.824
		3	-0.026	-0.026	0.4613	0.927
		4	-0.140	-0.144	2.5421	0.637
		5	-0.011	-0.013	2.5542	0.768
		6	-0.017	-0.037	2.5865	0.859
		7	-0.040	-0.051	2.7635	0.906
		8	0.028	0.003	2.8528	0.943
		9	0.030	0.019	2.9547	0.966
		10	-0.090	-0.101	3.8741	0.953

FIGURE 10.20 Correlogram of Residuals from a Regression of a Trend Stationary Series on a Time Trend.

EXERCISES

Excel files containing the data for these exercises are available as companion files for this book.

EXERCISE 10.1

Consider the following moving average process

$$X_t = \varepsilon_t + a_1 \varepsilon_{t-1} + a_2 \varepsilon_{t-2}$$

where $\varepsilon_t; t = 1, \ldots, T$ are independent, identically distributed Gaussian disturbances with mean zero and variance σ_ε^2.

a. Derive the variance and the first two autocovariances of X.

b. Show that all autocovariances of order three and higher are zero.

EXERCISE 10.2

Consider the following autoregressive process

$$X_t = \theta_4 X_{t-4} + \varepsilon_t$$

where $\varepsilon_t; t = 1, \ldots, T$ are independent, identically distributed Gaussian disturbances with mean zero and variance σ_ε^2 and we assume $|\theta_4| < 1$.

a. Derive the variance and the fourth-order autocovariance of X.

b. Show the first, second, and third autocovariances of X are zero.

EXERCISE 10.3

This exercise uses the data in the Excel worksheet *EXERCISE 10.3.XLSX.*

a. Plot the data series and assess its most important features.

b. Using the correlogram approach, identify an appropriate ARIMA model for this data series.

c. Estimate the ARIMA model you have identified and assess how well it fits the data by checking the correlogram of the residuals.

EXERCISE 10.4

Consider the model for unemployment given by equation (10.13). This can be written as a second-order difference equation in first differences:

$$\Delta \ln u_t = -0.0066 + 0.4832\, \Delta \ln u_{t-1} + 0.3094\, \Delta \ln u_{t-2} + \varepsilon_t$$

a. Solve for the roots of this equation and show that they lie within the unit circle.

b. Show that this can be written as third-order difference equation in the log of unemployment.

c. Using a spreadsheet, solve for the impulse response function of unemployment to a one standard deviation shock.

REFERENCES

[Box1970] Box, G.E.P and Jenkins, G. M. *Time Series Analysis: Forecasting and Control.* 1970, Holden-Day.

[Khintchine1934] Khintchine, A., "Korrelationstheorie der Stationären Stochastischen Prozesse, *Mathemathische Annalen.*" 1934, 109, p. 604.

[Walker1931] Walker, G., "On Periodicity in Series of Related Terms." *Proceedings of the Royal Society of London, Ser.* A. 1931, 131, pp. 518–532.

[Wold1938] Wold, H. A *Study in the Analysis of Stationary Time Series.* 1938, Almqvist and Wiksell.

[Yule1927] Yule, G. U., "On a Method of Investigating Periodicities in Disturbed Series, With Special Reference to Wolfer's Sunspot Numbers." *Philosophical Transactions of the Royal Society of London, Ser.* A. 1927, 226, pp. 267–298.

CHAPTER 11

UNIT ROOTS AND SEASONALITY

The assumption of stationarity is important in ARIMA modeling. By stationarity, we mean that the moments (mean, variance and autocovariances) of the series in question both exist, and are constant, through time. This is not automatically the case. Consider, for example, the AR(1) process $X_t = \theta X_{t-1} + \varepsilon_t$. We have shown that the variance is equal to $\sigma_X^2 = \sigma_\varepsilon^2 / \left(1 - \theta^2\right)$. Therefore, we require the assumption $-1 < \theta < 1$ for the variance to exist and to be positive. This is clearly not satisfied when the series follows a random walk, that is, $\theta = 1$, and many real-world time series appear to either follow such a process or contain a random walk element as part of a more complex dynamic specification. In Chapter 10, we showed that differencing the data can often act to remove a random walk element from a stochastic process. In this chapter, we will explore the implications of the random walk in more detail. We will start with the most basic random walk specification, in which the random variable X is equal to its value in the previous period plus a white-noise disturbance. We can write this in moving average form as shown in equation (11.1)

$$X_t = X_{t-1} + \varepsilon_t = \sum_{i=0}^{\infty} \varepsilon_{t-i}. \tag{11.1}$$

Since $E\left(\varepsilon_{t-k}\right) = 0$ for all values of k, we have $E\left(X_t\right) = 0$. However, this is an infinite sum of random disturbances in which the weights on past errors do not decline. It follows that the variance of this process is not defined.

Non-stationary time series like the random walk can be transformed into stationary series by a process of differencing. In the case of the random walk, we have

$$\Delta X_t = X_t - X_{t-1} = \varepsilon_t, \tag{11.2}$$

which is stationary. This is obvious in this case but can also be applied to more complex models. For example, suppose we have

$X_t = 1.5X_{t-1} - 0.5X_{t-2} + \varepsilon_t$. Using the lag operator, we can write this as $(1 - 1.5L + 0.5L^2)X_t = (1 - L)(1 - 0.5L)X_t = \varepsilon_t$, which, in turn, can be written as $\Delta X_t = 0.5\Delta X_{t-1} + \varepsilon_t$. Taking first differences has therefore produced a stationary stochastic process. Note that the behavior of the original stochastic process is driven by the roots of the polynomial in the lag operator. A process that contains one or more roots equal to one is non-stationary and is often described as a unit root process. The process of differencing effectively transforms the series to remove the unit root. Stochastic processes that contain a unit root are also referred to as *integrated processes*. This refers to the property that they must be differenced to make them stationary. The processes we have discussed above are said to be integrated of order one since they must be differenced once to remove the unit root.

Historical Note: Karl Pearson [Pearson1905] coined the term "random walk" in a short letter in *Nature* (Vol 72 p. 294) in which he requests help with an interesting statistical problem. It is doubtful that he would have anticipated just how much ink would be spilled on the subject in the following century.

Stochastic processes may contain more than one unit root, or have a higher order of integration, than those we have considered above. Consider, for example, a process of the form $X_t = 2.5X_{t-1} - 2X_{t-2} + 0.5X_{t-3} + \varepsilon_t$. This can be written $(1 - 2.5L + 2L^2 - 0.5L^3)X_t = (1 - L)^2(1 - 0.5L)\varepsilon_t$ or $\Delta^2 X_t = 0.5\Delta^2 X_{t-1} + \varepsilon_t$. Therefore, this is an example of a stochastic process that is integrated of order two, that is, must be differenced twice to produce a transformed series with moments that are both finite and constant. Another way of describing this process is to say that it contains a double unit root.

Any process whose moments are not constant through time is non-stationary. In principle, this applies to all the moments of the process but, in practice, we normally restrict our attention to the first two moments (the mean and the variance). Processes for which the first two moments are constant through time are described as being covariance stationary. The reason why we confine our interest to such processes is that it is relatively easy to derive conditions for covariance stationarity. However, we note that, if the disturbances follow a normal distribution, then covariance stationarity is sufficient to guarantee that all higher-order moments are also constant.

There are many different reasons why a process might be non-stationary and we list some of the more common below:

1. *A deterministic trend process*

Suppose we have $X_t = a + \beta t + \varepsilon_t$ where ε_t is a Gaussian white noise process. This is non-stationary since $E(X_t) = a + \beta t$ which is a function of time. However, it is very easy to make this process stationary through the transformation $Z_t = X_t - a - \beta t$. We often describe this process as being stationary around a deterministic trend.

2. *A simple random walk*

This process takes the form $X_t = X_{t-1} + \varepsilon_t$. This is non-stationary because the unit root in this process means that the variance is not defined. It can be made stationary by applying the first difference operator.

3. *A random walk with drift*

Let $X_t = \beta + X_{t-1} + \varepsilon_t$. The presence of the β term in this equation produces a trend or drift in the series. In this case, the series is non-stationary, both because the variance is not defined and because $E(X_t) = X_0 + \beta t$, which demonstrates that the mean is not constant. In this case, differencing will again result in a stationary process since $\Delta X_t = \beta + \varepsilon_t$ Nelson and Plosser [Nelson1982] have shown that the random walk with drift provides a good fit to many macroeconomic time series.

4. *More general unit root processes*

Any autoregressive process can be written in the form $A(L)X_t = \varepsilon_t$ where $A(L)$ is the lag polynomial. If any of the roots of this lag polynomial are equal to one, then the process is not stationary. The number of unit roots determines the order of integration of the process or the number of times it must be differenced to make it stationary.

11.1 TESTING FOR UNIT ROOTS

How do we determine if a stochastic process contains a unit root based on a given realization? One method is to examine the sample correlogram. If a process generating the data is stationary, then the autocorrelations should decline to zero exponentially. For example, the AR(1) process has autocorrelations $\rho_k = \theta^k$. Therefore, for a stationary AR(1) process, we should expect to see sample autocorrelations that decline to zero exponentially. For a non-stationary function, it is harder to derive a theoretical correlogram, since the theoretical moments, derived on the basis of an infinite moving average representation, are not defined. However, we do not have infinite samples in practice, and it is possible to derive a theoretical correlogram for a finite sample. Consider, for example, the random walk process $X_t = X_{t-1} + \varepsilon_t$. If the sample size T is large, but finite, then the variance of X will be approximately equal $E\left(\sum_{t=1}^{T} X_t^2\right) = T\sigma_\varepsilon^2$ and the covariance of X and its k'th lag will be approximately equal to $E\left(\sum_{t=k+1}^{T} X_t X_{t-k}\right) = (T-k)\sigma_\varepsilon^2$. Therefore, the k'th order autocorrelation will be $\rho_k = 1 - k/T$. The autocorrelations will still decline to zero, but now according to a linear function of the lag, rather than an exponential function. The rate of decline depends on the sample size with larger samples being associated with a slower decline. Thus, a sample correlogram in which the autocorrelations decline linearly to zero is an indicator that the underlying stochastic process contains a unit root. The difference between the two is illustrated in the correlograms shown in Figure 11.1 which are based on simulated data created using the EViews random number generator.

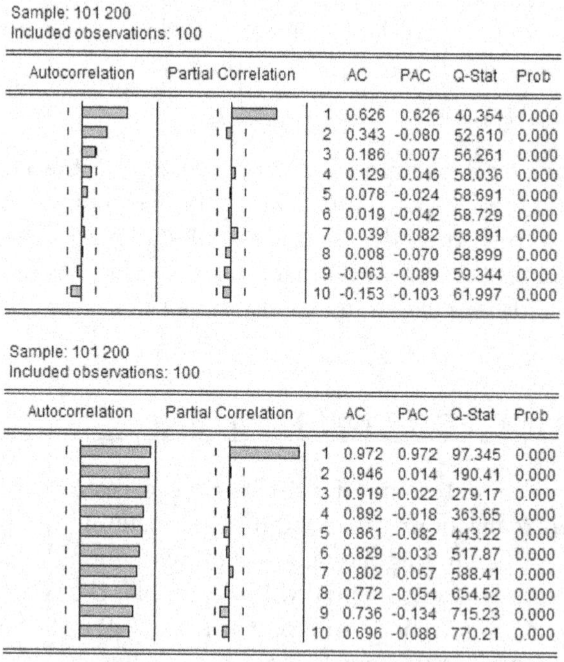

Sample: 101 200
Included observations: 100

	AC	PAC	Q-Stat	Prob
1	0.626	0.626	40.354	0.000
2	0.343	-0.080	52.610	0.000
3	0.186	0.007	56.261	0.000
4	0.129	0.046	58.036	0.000
5	0.078	-0.024	58.691	0.000
6	0.019	-0.042	58.729	0.000
7	0.039	0.082	58.891	0.000
8	0.008	-0.070	58.899	0.000
9	-0.063	-0.089	59.344	0.000
10	-0.153	-0.103	61.997	0.000

Sample: 101 200
Included observations: 100

	AC	PAC	Q-Stat	Prob
1	0.972	0.972	97.345	0.000
2	0.946	0.014	190.41	0.000
3	0.919	-0.022	279.17	0.000
4	0.892	-0.018	363.65	0.000
5	0.861	-0.082	443.22	0.000
6	0.829	-0.033	517.87	0.000
7	0.802	0.057	588.41	0.000
8	0.772	-0.054	654.52	0.000
9	0.736	-0.134	715.23	0.000
10	0.696	-0.088	770.21	0.000

FIGURE 11.1 Correlograms for Stationary (Upper) and Unit Root (Lower) Processes.

In cases like those shown in Figure 11.1, the distinction between a stationary and a unit root process is obvious. However, the difference may not be so clear in practice, and it may be hard to make a definite choice based simply on visual inspection of the correlogram. It, therefore, becomes necessary to develop formal tests for the presence of a unit root. Consider, for example, the general AR(2) process

$$X_t = \theta_0 + \theta_1 X_{t-1} + \theta_2 X_{t-2} + \varepsilon_t. \tag{11.3}$$

This can be parameterized as

$$\Delta X_t = \beta_0 + \beta_1 X_{t-1} + \beta_2 \Delta X_{t-1} + \varepsilon_t, \tag{11.4}$$

where $\beta_0 = \theta_0$, $\beta_1 = \theta_1 + \theta_2 - 1$ and $\beta_2 = -\theta_2$. Alternatively, (11.3) can be factorized as $(1-\varphi_1 L)(1-\varphi_2 L)X_t = \varepsilon_t$. If $\varphi_1 = 1$, then the process can be written as $\Delta X_t = \varphi_2 \Delta X_{t-1} + \varepsilon_t$. It follows that the restriction of a unit root in the stochastic process is equivalent to the restriction that $\beta_1 = 0$ in (11.4).

We can therefore test for the unit root hypothesis by estimating (11.4) and testing $H_0 : \beta_1 = 0$ against the alternative that $H_1 : \beta_1 < 0$.[1] If we cannot reject the null hypothesis, then we conclude that the process contains a unit root. A natural method for testing this hypothesis would be to estimate the model above and then use the t-statistic $\hat{\beta}_1 / \mathrm{se}(\hat{\beta}_1)$ as the basis for our test. The problem is that, if the null hypothesis is true, then this statistic does not follow the t distribution. This is because the distribution theory underlying the derivation of the t-statistic depends on the assumption that the process generating the data is stationary. This does not mean that we cannot use the statistic, but it does mean that the critical values from the Student's t tables are not appropriate. Instead, we need to use critical values determined by Monte Carlo methods. These were first calculated in a pioneering paper by Dickey and Fuller [Dickey1979] and are shown in Table 11.1.

TABLE 11.1 Dickey-Fuller Critical Values for Unit Root Test.

	Without trend		With trend	
Sample size	**1%**	**5%**	**1%**	**5%**
$T = 25$	−3.75	−3.00	−4.38	−3.60
$T = 50$	−3.58	−2.93	−4.15	−3.50
$T = 100$	−3.51	−2.89	−4.04	−3.45
$T = 250$	−3.46	−2.88	−3.99	−3.43
$T = 500$	−3.44	−2.87	−3.98	−3.42
$T = \infty$	−3.43	−2.86	−3.96	−3.41

[1] The alternative hypothesis provides a necessary but not a sufficient condition for stationarity in a second-order process – see Exercise 1 for a proof of this statement.

It should be noted that the critical values for the Dickey-Fuller unit root test are sensitive to the choice of deterministic variables (constant and trend) to be included in the test equation. For example, if we add a deterministic trend to equation (11.4) then this changes the critical values. From Table 11.1, we note that the critical values for the cases when a deterministic trend is included are noticeably higher than those calculated when it is not present. Therefore, if we estimate a model of the form

$$\Delta X_t = a + \beta t + \beta_1 X_{t-1} + \beta_2 \Delta X_{t-1} + \varepsilon_t, \tag{11.5}$$

rather than (11.4), then we must adjust the decision rule for the test to take account of the change in the distribution of the test statistic. We note, however, that the critical values for these tests are not sensitive to the inclusion of lagged differences on the right-hand side of our test equation. For example, equation (11.5) includes ΔX_{t-1} as an extra regressor but the critical values remain unchanged by this. This is important because it is often necessary to include lagged differences in the test equation so that the residuals are serially uncorrelated. If the residuals are serially correlated, then the test becomes unreliable. Unit root tests which include lagged differences for this purpose are referred to as *augmented Dickey-Fuller* (ADF) tests. The number of lagged difference terms that are included on the right-hand side is chosen to ensure that serial correlation is not present in the test equation. Econometric software such as EViews will automatically choose the number of lagged difference terms to include according to one of the information criteria which were discussed in Chapter 10.

Historical Note: Tables of critical values for unit root tests were first set out in a paper by David Dickey and Wayne Fuller in the *Journal of the American Statistical Association* in 1979.

Although the Dickey-Fuller (or more generally ADF) critical values were originally presented in table form as shown in Table 11.1, they are now more generally calculated using the *response surfaces* calculated in MacKinnon et al [MacKinnon1999]. These are functional relationships that allow the calculation of critical values for different test sizes and numbers of observations and take the form

$$C(p,T) = \hat{\beta}_\infty + \frac{\hat{\beta}_1}{T} + \frac{\hat{\beta}_2}{T^2}, \tag{11.6}$$

where is p is the size of the test and T is the sample size. MacKinnon *et al* give values of the $\hat{\beta}$ coefficients for these equations based on Monte Carlo studies using artificially generated data. For example, the 5% critical value for a unit root test using a test equation that included a constant and a deterministic trend is given by the following equation

$$C(0.05,T) = -3.4126 - \frac{4.039}{T} - \frac{17.83}{T^2}.$$

These response surfaces provide a convenient way of calculating accurate critical values and are built into many econometric packages as a standard feature.

Example: Suppose we wish to test for the presence of a unit root in the unemployment series shown we analyzed in Chapter 10. There is no obvious trend in the series and so we estimate an augmented Dickey-Fuller equation which includes a constant but no trend. This produces the following results

$$\Delta u_t = \underset{(0.0449)}{0.0782} - \underset{(0.006)}{0.0118}\, u_{t-1} + \underset{(0.080)}{0.6066}\, \Delta u_{t-1} + \underset{(0.080)}{0.2085}\, \Delta u_{t-2} + \hat{\varepsilon}_t$$

$$R^2 = 0.56 \qquad DW = 1.83.$$

The choice of two lagged differenced terms in this equation was made automatically by EViews as the lag length which minimizes the Schwartz criterion. The test statistic here is:

$$\tau = -\frac{0.0118}{0.006} = -1.97.$$

Using the standard t-critical values we would reject the null hypothesis that the coefficient on lagged unemployment was equal to zero in favor of the alternative that it is less than zero, because, with $T = 182$, the 5% critical value for a one-tailed t-test is -1.645. However, this is not the appropriate critical value here. Using the MacKinnon response surfaces we can calculate the correct 5% critical value as

$$C(0.05,T) = -2.8621 - \frac{2.738}{T} - \frac{8.36}{T^2} = -2.87.$$

Therefore, since the test statistic is less than the critical value in absolute terms, we are unable to reject the null hypothesis and we conclude that the unemployment series does contain a unit root.

11.2 FORECASTING WITH UNIT ROOT PROCESSES

We often wish to forecast future values of a trending variable and the random walk with drift model often provides a good model of such processes. In this section, we wish to consider the properties of forecasts based on such a model. Suppose we have a model of the form $X_t = \beta + X_{t-1} + \varepsilon_t$. It is easy to show that

$$X_{T+k} = X_T + k\beta + \sum_{i=1}^{k} \varepsilon_{T+i}. \tag{11.7}$$

The expected value of X_{T+k} based on information up to date T is given by

$$E\left(X_{T+k} \middle| \Omega_T\right) = X_T + k\beta, \tag{11.8}$$

and the variance is given by

$$E\left\{\left(X_{T+k} - E\left(X_{T+k} \middle| \Omega_T\right)\right)^2 \middle| \Omega_T\right\} = k\sigma_\varepsilon^2. \tag{11.9}$$

Forecasts based on this model, therefore, have the property that the variance increases the further in the future we wish to forecast. This is intuitively plausible in the sense that it indicates that our confidence in our forecast declines for values further into the future.

Example: An artificial data set was generated with the following properties $X_t = 0.05 + X_{t-1} + \varepsilon_t$ where ε_t are iid disturbances with $\varepsilon_t \sim N(0, 0.01)$. Data was generated for the period $t = 1,\ldots,100$ and then forecasts were calculated for the period $t = 101,\ldots,110$. The results are shown in Figure 11.2 along with the 95% confidence interval for the forecast. As you can see, the forecast error bands widen very quickly.

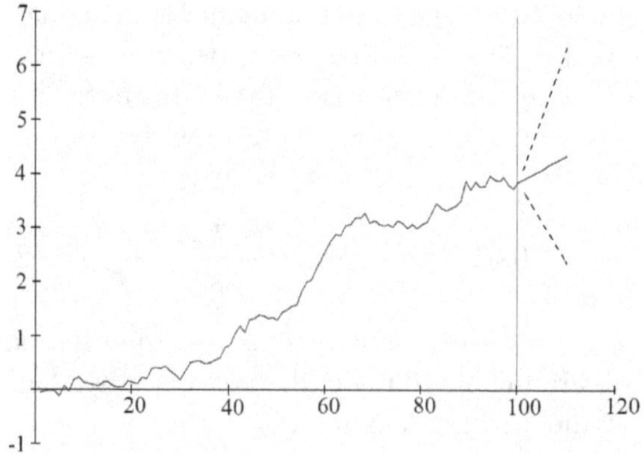

FIGURE 11.2 Forecasting Using a Random Walk with Drift Model.

Now let us contrast this case with one in which we can model the series as a trend stationary process. Let the process generating the data be $X_t = a + \beta t + \varepsilon_t$ where ε_t are Gaussian white-noise disturbances. The central forecast for X_{T+k}, on the basis of information up to date T, is given by

$$E\left(X_{T+k}\left|\Omega_T\right.\right) = a + \beta(T+k) = \left(X_T - \varepsilon_T\right) + \beta k. \qquad (11.10)$$

This is very similar to the central forecast produced by the difference stationary model in that the expected value of the variable increases by β in each time period. In this case, however, the current disturbance ε_T is not built into future forecasts.

Now consider the forecast variance. This is given by

$$E\left\{\left(X_{T+k} - a - \beta(T+k)\right)^2 \left|\Omega_T\right.\right\} = \sigma_\varepsilon^2. \qquad (11.11)$$

In this case, the forecast variance does not increase further into the future we wish to forecast. To illustrate the effects of this we again generate a random series, this time assuming a trend stationary process, and plot the central forecast plus the 95% confidence interval. The results are shown below. Note that the standard error bands do not widen as they do for the difference stationary process. It follows that there are considerable advantages if we can legitimately model a series as being trend, rather than difference, stationary.

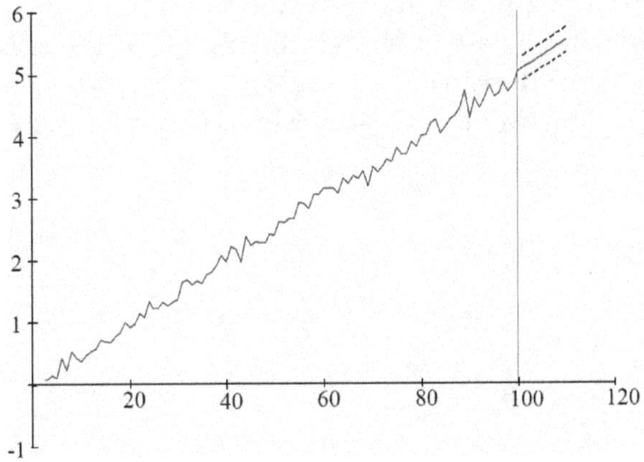

FIGURE 11.3 Forecasting with a Trend-Stationary Model.

11.3 SEASONALITY

So far, we have identified unit root behavior as being associated with the first lag of the stochastic process under consideration. However, there is an alternative form of non-stationarity which is relevant for economic time series. This arises because of the highly seasonal nature of many such series. Many economic variables are associated with a regular interval of publication, for example, annual, quarterly or monthly. For such series, we often observe a regular pattern of fluctuations within any given year. For example, household consumption expenditures regularly increase sharply in the fourth quarter of the year and subsequently fall back in the next quarter. Government statistical agencies frequently treat such within year movements as a nuisance and seasonally adjust data prior to publication. In some situations, however, this variation is of interest in itself and it is useful to capture it during the modeling process, rather than eliminate it prior to estimation.

For modeling purposes, the presence of seasonality means that correlations between observations made during the same period in the previous year are more important than correlations with the immediately preceding observations. A simple stochastic process with this property is provided by the fourth order autocorrelation process in equation (11.12)

$$X_t = \theta_4 X_{t-4} + \varepsilon_t. \tag{11.12}$$

This model is appropriate for quarterly data. If $|\theta_4| < 1$, then this is a stationary process in which the autocorrelations decline exponentially to zero as the lag length increases. The difference between this and the correlogram for a first-order autocorrelation process, is that the autocorrelations only differ from zero at lags corresponding to the seasonal frequency. For the case of quarterly data, we would observe non-zero autocorrelations at the fourth, eighth, twelfth, etc., lags. This is illustrated in Figure 11.4 for the case $\theta_4 = 0.7$.

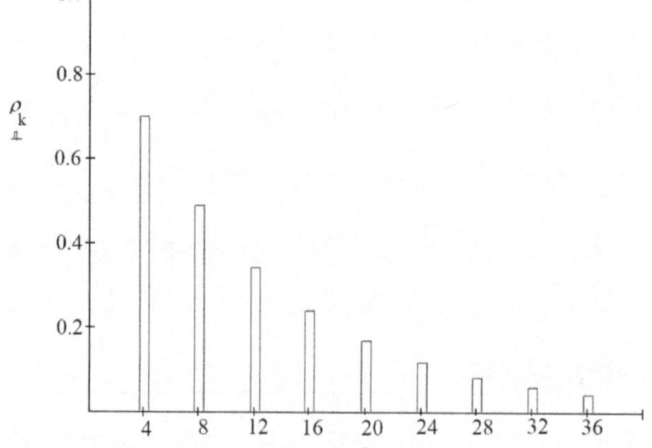

FIGURE 11.4 Correlogram for $X_t = 0.7X_{t-4} + \varepsilon_t$.

If $\theta_4 = 1$ then the process is no longer stationary and is characterized by a *seasonal unit root*. In such cases, the correlogram will still have the property that only the autocorrelations corresponding to the seasonal frequency differ from zero. In this case, however, they will decline linearly to zero rather than following the exponential path shown in Figure 11.4. This is similar to the patterns for the first-order autocorrelation process and the simple random walk processes which we derived earlier. In practice, the correlogram of most seasonal processes will reflect both seasonal and non-seasonal effects. The problem facing the modeler is to disentangle these effects to formulate a model that captures the stochastic process generating the data.

Testing for a seasonal unit root is less straightforward than testing for a nonseasonal unit root. Hylleberg, Engle, Granger and Yoo (HEGY) [Hylleberg1990] have set out a testing procedure that allows for simultaneous testing for either or both types of non-stationarity. Let us assume a stochastic process of the form $A(L)X_t = \varepsilon_t$, where $A(L)$ is a fourth-order polynomial function. This polynomial function can be factorized as shown in equation (11.13)

$$A(L) = (1-\gamma_1 L)(1+\gamma_2 L)(1-\gamma_3 iL)(1-\gamma_4 iL), \tag{11.13}$$

where $i = \sqrt{-1}$. This function contains several special cases of interest. In particular, if $\gamma_1 = 1$ and $\gamma_2 = \gamma_3 = \gamma_4 = 0$, there is a nonseasonal unit root, while if $\gamma_1 = \gamma_2 = \gamma_3 = \gamma_4 = 1$, then there is a seasonal unit root. Before we can turn this into an operational procedure for testing, however, we need to deal with the presence of complex numbers in (11.13). HEGY do this by defining transformed variables that generate a testing equation that is free from complex numbers. These transformed variables are as follows

$$\begin{aligned} X_{1t-1} &= \sum_{k=1}^{4} X_{t-k} \\ X_{2t-1} &= \sum_{k=1}^{4} (-1)^{k-1} X_{t-k} \\ X_{3t-1} &= X_{t-1} - X_{t-3}. \end{aligned} \tag{11.14}$$

The testing equation then becomes

$$\Delta^4 X_t = \varphi_1 X_{1t-1} - \varphi_2 X_{2t-1} + \varphi_3 X_{3t-1} - \varphi_4 X_{3t-1} + \varepsilon_t. \tag{11.15}$$

This equation will also typically include deterministic regressors such as a constant, a time trend and/or seasonal dummy variables, depending on the nature of the process under consideration. The testing procedure then involves sequential tests for (1) $H_0 : \varphi_1 = 0$, which is equivalent to $\gamma_1 = 1$ in

(11.13), and therefore indicates the presence of a nonseasonal unit root; (2) $H_0 : \varphi_2 = 0$ which is equivalent to $\gamma_2 = 1$ in (11.13), and therefore indicates the presence of a unit root with a semi-annual frequency, and, finally (3) $H_0 : \varphi_3 = 0$ or $\varphi_4 = 0$, which is equivalent to $\gamma_1 = \gamma_2 = \gamma_3 = \gamma_4 = 1$ in (11.13), and therefore indicates the presence of a seasonal unit root. The test statistics for the first two tests are the t-ratios, while that for the third test, is the F-test statistic. The distributions, however, are non-standard and HEGY report Monte Carlo critical values for a 5% test based on one hundred observations in each case. These are given in Table 11.2. Note that, as with most unit root tests, the critical values differ according to the deterministic variables included in the testing equation.

TABLE 11.2 Monte Carlo 5% Critical Values for the HEGY Test (100 Observations).

	Nonseasonal unit root	**Semi-annual unit root**	**Seasonal unit root**
Intercept	-2.88	-1.95	3.08
Intercept and trend	-2.95	-2.94	6.57
Intercept, trend and seasonal dummy variables	-3.53	-2.94	6.60

As an example of testing for a seasonal unit root, we will consider seasonally unadjusted UK Household Consumers' Expenditure data for the period 1955q1–2019q4. This is a highly seasonal time series as illustrated in Figure 11.5 which shows the logarithm of the original series. There is a strong upward trend throughout the period with strong visual evidence of a regular seasonal pattern.

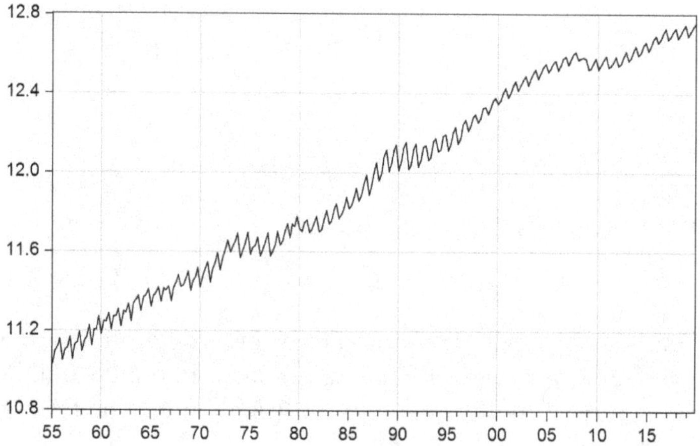

FIGURE 11.5 UK Household Consumer's Expenditure (Log Scale) 1955q1 to 2019q4.

To perform the HEGY tests, we must first make a decision as to what deterministic regressors to include in the equation. From the results in Table 11.3, we see that this is important for our conclusions. When we include just a constant, or a constant and a linear trend, then we are unable to reject any of the null hypotheses, and would therefore conclude that the series contains a simple unit root, a seasonal unit root and a unit root at the semi-annual frequency. If we include seasonal dummy variables, then we can reject the null of a seasonal unit root. This indicates that the inclusion of seasonal dummy variables is an adequate method for dealing with the seasonality in this series and there is no need to take seasonal differences prior to modeling this series.

TABLE 11.3 HEGY Test Results UK Household Consumers' Expenditure.

	Nonseasonal unit root	Semi-annual unit root	Seasonal unit root
Intercept	1.68	−1.28	2.58
Intercept and trend	1.49	1.33	3.48
Intercept, trend and seasonal dummy variables	1.59	1.41	12.10*

**indicates rejection of the null hypothesis at the 5% level.*

11.4 STRUCTURAL BREAKS AND UNIT ROOTS

We saw earlier that the distinction between difference and trend stationary processes can have important implications for how we interpret the behavior of economic time series. In particular, the fact that trend stationary processes exhibit to return to a stable trend means that the standard error bands around long term forecasts remain relatively narrow. In contrast, difference stationary processes generate forecasts in which the standard error bands rapidly become very wide. This means that Nelson and Plosser's [Nelson1982] claim, that most economic time series are difference stationary, has important implications for how we interpret and think about long term trends in economics. However, as we have also already noted, tests for unit roots often have low power against alternatives close to the null and, as Perron [Perron1989] has noted, allowing for a limited number of structural breaks in either the intercept or trend often produces outcomes which are more likely to be trend stationary.

Testing for trend stationarity when we have structural breaks follows the same procedure as standard unit root tests. The null hypothesis is that the process is difference stationary, and the alternative is that it is trend stationary. In this case, however, the test equation includes dummy variables that capture structural shifts in either the intercept or the trend. Suppose, for example, we wish to test for trend stationarity of a stochastic process X while allowing for both an intercept and a trend change as some date t^* in the interval $t = 1,\dots,T$. The first stage, is to estimate an equation of the form

$$X_t = \beta_0 + \beta_1 t + \beta_3 D_L + \beta_4 D_T + \varepsilon_t, \tag{11.16}$$

where $D_L = 0$ for $t = 1,\dots,t^*$ and 1 for $t = t^*+1,\dots,T$; and $D_T = 0$ for $t = 1,\dots,t^*$ and $\left(t - t^*\right)$ for $t = t^*+1,\dots,T$. The residuals from this equation then provide the basis for a second stage regression of the form

$$\hat{\varepsilon}_t = a_0 + a_1 \hat{\varepsilon}_{t-1} + \sum_{j=1}^{p} a_{1+j} \Delta\varepsilon_{t-j} + v_t, \tag{11.17}$$

which we use to test $H_0 : a_1 = 1$ against the alternative $H_1 : a_1 < 1$. Rejection of the null leads to the conclusion that the process is trend stationary. Of course, the addition of extra deterministic regressors in (11.16) changes the critical values, meaning that they are higher than the standard Dickey-Fuller values and, as Perron shows, the timing of the structural break is also important. Let $\lambda = t^*/T$, Perron shows that values of λ close to 0.5, that is, a structural break in the middle of the sample period, will have the largest critical values. In contrast, values of λ close to the extremes, either zero or one, have relatively little effect and the critical values are close to the standard Dickey-Fuller values. His paper also provides formulae that allow us to generate appropriate critical values for such tests.

While Perron's paper is important for its methodological innovations, it is also important for the application which he provides. His conclusion is that many of the economic time series, which Nelson and Plosser had found to be difference stationary, could be shown to be trend stationary when allowance was made for a limited number of structural breaks. As an illustration of this, we will consider an application of his method to UK time series data for Gross Domestic Product. Our time series consists of an index of real UK GDP constructed by splicing Feinstein's [Feinstein1972] historical series for 1855 to 1965 to the Office for National Statistics data for later periods. When we apply the standard Dickey-Fuller test for trend stationarity, we are unable to reject the null of a difference stationary process. However, the results are different when we allow for a structural break in 1921. This is an important year in British economic history because, from then, the data no longer include what would later become the Irish Republic in national income calculations. Not surprisingly, there is an immediate fall in the level of GDP, which we can model using an intercept dummy variable. At the same time, we see a modest increase in the long-term growth rate which, in the longer term, has even more significant effects on the series. We, therefore, adjust our test equation to allow for a structural break in both the intercept and trend in 1921, yielding the results shown in equation (11.18)

$$\ln(y_t) = \underset{(0.0134)}{3.4926} + \underset{(0.0003)}{0.0189}t - \underset{(0.0162)}{0.3307}D_L + \underset{(0.0004)}{0.0058}D_T + \hat{\upsilon}_t$$
$$R^2 = 0.9972 \quad \hat{\sigma}_u = 0.0507 \quad DW = 0.3805. \tag{11.18}$$

Equation (11.18) shows a downward intercept shift in 1921 and an increase in the equilibrium growth rate. Not surprisingly, there is still a high degree of autocorrelation in the residuals, as indicated by the low value of the Durbin-Watson statistic. This, of course, means that the standard errors of the coefficient estimates are probably underestimated.

Historical Note: The inclusion of dummy variables to capture a structural break in the series can also provide an alternative method for the construction of the Chow [Chow1960] test for parameter constancy. Rather than estimating separate regressions for sub-periods and then comparing the residual sums of squares, we can base the Chow test on a test for the joint significance of the dummy variables in an equation of the form (11.18).

To test for trend stationarity around the equilibrium defined by (11.18), we next estimate the auxiliary regression

$$\hat{\upsilon}_t = -\underset{(0.0022)}{0.0003} + \underset{(0.0474)}{0.7520}\hat{\upsilon}_{t-1} + \underset{(0.0758)}{0.3175}\Delta\hat{\upsilon}_{t-1} + \hat{\eta}_t$$
$$R^2 = 0.681 \quad DW = 1.9612 \tag{11.19}$$

The test statistic for the null hypothesis that the coefficient on the lagged residual is equal to zero is $(0.7520 - 1)/0.0474 = -5.23$. Calculating Perron's critical value, we first calculate $\lambda = 68/165 = 0.4121$. This, in turn, gives a 5% critical value of –4.22 for the test, allowing for both an intercept and trend change. We, therefore, reject the null hypothesis and conclude that this process is stationary around a linear trend with a structural break in 1921.

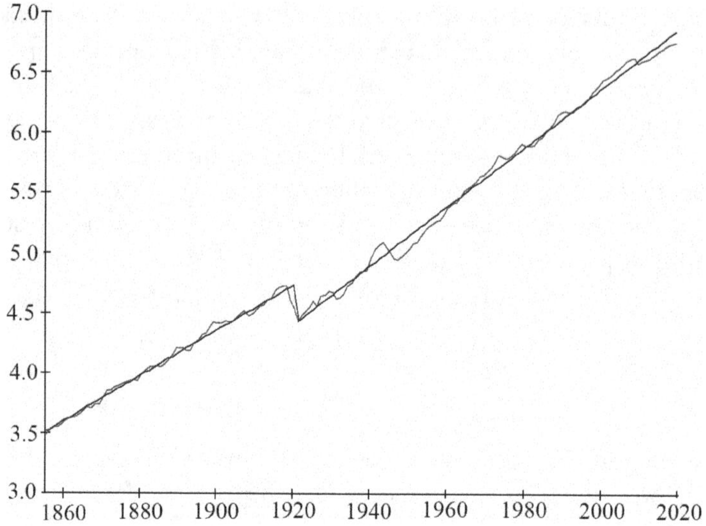

FIGURE 11.6 UK GDP (Log Scale) Plus Trend with Break in 1921.

Figure 11.6 illustrates why the conclusion of the Perron test has important implications here. Stationarity around a linear trend indicates a stable growth path, to which the economy returns in the long run following a disturbance. Without allowance for a structural break in 1921, we would have rejected this as a null hypothesis. The presence of such a structural break is clear from a simple visual inspection of the graph of the series. Allowance for this break changes the conclusion of the unit root test, producing an equilibrium trend growth path as shown by the trend line in Figure 11.6. A conclusion like this has the potential to radically change our interpretation of the economic history of the period.

EXERCISES

Excel files containing the data for these exercises are available as companion files for this book.

EXERCISE 11.1

Consider the second-order stochastic process $X_t = \theta_1 X_{t-1} + \theta_2 X_{t-2} + \varepsilon_t$ where $\varepsilon_t; t = 1,\ldots,T$ are independent, identically distributed random errors. Derive the roots of this process as a function of the parameters θ_1 and θ_2 and hence, show that the condition $\theta_1 + \theta_2 - 1 < 0$ is necessary, but not sufficient, for this process to be stationary.

EXERCISE 11.2

Consider the stochastic process $X_t = 0.85X_{t-1} + 0.25X_{t-2} - 0.1X_{t-3} + \varepsilon_t$, where ε_t; $t = 1,\ldots,T$ are independent, identically distributed random errors. Show that this process contains a single unit root but is stationary in first differences.

EXERCISE 11.3

Using the data in the file *UK EXCHANGE RATES.XLSX*, carry out tests for the null hypothesis that the stochastic process which determines the logarithm of the sterling-dollar exchange rate (EXRATE) is difference stationary.

EXERCISE 11.4

Quarterly data for new registrations of motor vehicles in the UK is given in the spreadsheet file *NEWREG.XLSX.* Use this data to fit an ARIMA model to the series and comment on your results.

REFERENCES

[Chow1960] Chow, G. C., "Tests of Equality Between Subsets of Coefficients in Two Linear Regression Models." *Econometrica*. 1960, 28, pp. 591–605.

[Dickey1979] Dickey, D. A. and Fuller, W. A., "Distribution of the Estimators for Autoregressive Time Series With a Unit Root." *Journal of the American Statistical Association*. 1979, 74, pp. 427–31.

[Feinstein1972] Feinstein, C. H., *Statistical Tables of National Income, Expenditure and Output of the U.K. 1855–1965*. 1972, Cambridge University Press.

[Hylleberg1990] Hylleberg, S. Engle, R. Granger, C. and Yoo, B., "Seasonal Integration and Cointegration." *Journal of Econometrics*. 1990, 44, pp. 215–38.

[MacKinnon1999] MacKinnon, J. G., Haug, A.A. and Michelis, L., "Numerical Distribution Functions of Likelihood Ratio Tests for Cointegration." *Journal of Applied Econometrics*. 1999, 14, pp. 563–577.

[Nelson1982] Nelson, C. R. and Plosser, C. I., "Trends and Random Walks in Macroeconomic Time Series: Some Evidence and Implications." *Journal of Monetary Economics*. 1982, 10, pp. 139–62.

[Pearson1905] Pearson, K., "The Problem of the Random Walk." *Nature*. 1905, 72, p. 294.

[Perron1989] Perron, P., "The Great Crash, the Oil Price Shock, and the Unit Root Hypothesis." *Econometrica*. 1989, 57(6), November, pp. 1361–1401.

CHAPTER 12

COINTEGRATION

In Chapter 11, we introduced the idea of unit root stochastic processes. Processes that contain one or more unit roots are often described as *integrated processes*. An integrated process is defined to be a process that must be differenced to create a stationary process. In such cases we write $X \sim I(d)$, that is, the stochastic process which determines the random variable X must be differenced d times to create a stationary process. We can describe such a process as being integrated of order d or as containing d unit roots. Since integrated processes are non-stationary it follows that the standard statistical distributions and hypothesis tests, which assume stationarity, do not apply. An implication of this is that the estimation of models using integrated data runs the risk of generating *spurious regressions*. That is, there is a danger that these regressions may appear to detect a significant relationship between variables, even when no such relationship exists.

Given the possibility of spurious regressions, the question arises as to whether it is ever sensible to estimate models involving integrated data. One argument is that the data should be differenced prior to estimation with the object of eliminating unit roots so as to avoid the spurious regression problem. There is a danger, however, that, by differencing the data, we eliminate valuable long-run information from the model. To understand this, we must introduce the idea of *cointegration*. In this chapter, we will show that there are circumstances in which it is possible, and indeed valuable, to estimate models using non-stationary data. In such circumstances, however, the econometrician needs to act with caution and to test the models estimated rigorously to avoid the danger of spurious regression.

> **Historical Note:** Statisticians have long been aware of the danger of spurious regressions. George Udny Yule provides an early, and very thorough, account of this problem in his 1926 paper [Yule1926]. Granger and Newbold [Granger1974] were the first to link the problem of spurious regressions with the presence of unit roots in the processes determining the regression variables.

In general, any linear combination of integrated processes will itself be integrated with the order of integration being equal to that of the highest order variable in the relationship. For example, if $X \sim I(1)$ and $Y \sim I(1)$ then $Z = Y - \beta X$ will also be $I(1)$. However, in some cases, it is possible to find a β such that $Z = Y - \beta X \sim I(0)$. In such cases, we say that there is a *cointegrating relationship* between Y and X, and that β is the cointegrating parameter. What this indicates is that, even if X and Y both contain unit roots, they are linked in such a way that they do not move too far from each other. An example from economics could be the relationship between consumption and disposable income. In both cases, a random walk with drift provides a good description of the behavior of the variable in question. It is plausible, however, to argue that consumption and income should not move too far from each other even in the very long run. Thus, it is possible, or even likely, that consumption and income will be cointegrated.

Example: Consider the joint process defined by (12.1) where ε_1 and ε_2 are independent Gaussian white-noise errors,

$$\begin{aligned} X_t &= 0.05 + X_{t-1} + \varepsilon_{1t} \\ Y_t &= 0.5X_t + 0.5Y_{t-1} + \varepsilon_{2t}. \end{aligned} \tag{12.1}$$

Now consider the realization of this process shown in Figure 12.1. The X variable is integrated of order one and, since the Y variable is a linear combination of its own lagged value and the current X and a stationary random variable ε_2, it follows that Y is also integrated of order one. However, the variables are linked so that they move together over time.

Let us define $Z_t = Y_t - X_t$, this can be shown to yield a stationary series by rewriting the second equation of our system as

$$\Delta Y_t = -0.5Z_{t-1} + \varepsilon_{2t}. \tag{12.2}$$

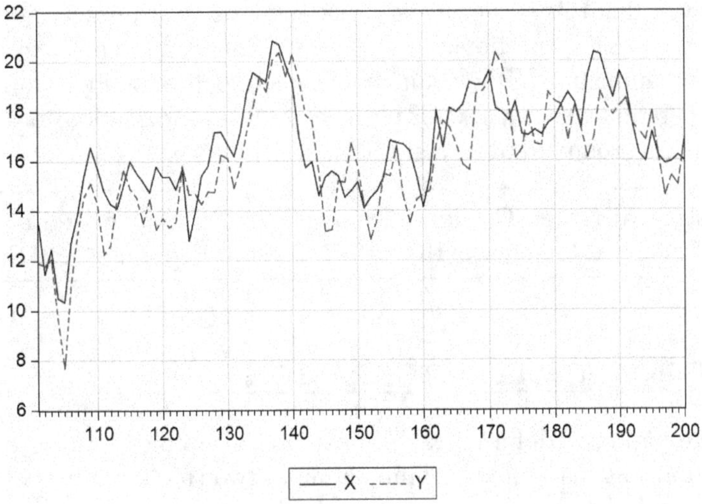

FIGURE 12.1 Linked Unit Root Processes.

Since Y is integrated of order one, it follows that ΔY is $I(0)$ and, by assumption, ε_2 is stationary. Since Z_{t-1} is a linear combination of $I(0)$ variables, it follows that Z is also $I(0)$. This result is illustrated in Figure 12.2 which shows that, unlike the Y and X series, the Z series is mean-reverting, that is deviations from the average value are "corrected" over time. Therefore, the Z series appears to be stationary in that there is no obvious trend and deviations from the average value appear to dissipate quickly. It should be noted,

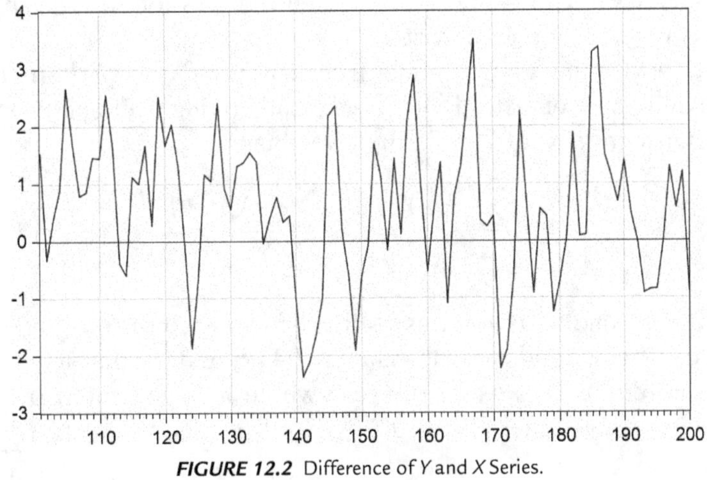

FIGURE 12.2 Difference of Y and X Series.

however, that this is the result of a somewhat arbitrary transformation and that our "test" for stationarity of the resulting variable involves a subjective visual examination of the data. We will need to develop more formal tests for cointegration which allow for more general transformations of the data and which are grounded in formal statistical inference.

> **Historical Note:** The term "cointegration" was first used in the paper by Engle and Granger in their 1987 paper in Econometrica [Engle1987].

12.1 TESTING FOR COINTEGRATION

Cointegration is defined as the existence of a cointegrating parameter β which means that a linear combination of two stochastic processes that are integrated of order d is integrated of a lower order. More generally, when there are more than two series β becomes a cointegrating vector rather than a cointegrating parameter. In our previous example, $Z_t = Y_t - \beta X_t$ is stationary for $\beta = 1$ even though the processes generating the Y and X series are non-stationary. if the value of the cointegrating parameter is unknown, then it can be estimated using a simple least squares regression of one variable on the other. If the variables Y and X are cointegrated, then we can show that the OLS estimator $\hat{\beta}$ is a *super-consistent* estimator of the cointegrating parameter β. This means that the probability limit of $\hat{\beta}$ converges on the true value β even in circumstances where the small sample estimator is biased and, moreover, converges more quickly than when the Y and X series are stationary. The residuals from such a regression $\hat{u}_t = Y_t - \hat{\beta} X_t$ form the basis for a test for cointegration.

Consider the simple model in which $Y_t = \beta X_t + u_t$ where Y and X are both integrated of order 1 and β is a cointegrating parameter. If we perform an OLS regression of Y on X, then we obtain

$$\hat{\beta} = \frac{\sum_{t=1}^{T} X_t Y_t}{\sum_{t=1}^{T} X_t^2} = \beta + \frac{\sum_{t=1}^{T} X_t u_t / T}{\sum_{t=1}^{T} X_t^2 / T}. \tag{12.3}$$

Sufficient conditions for this estimator to be consistent are (1) that the expected value tends to true value as $T \to \infty$ and (2) that its variance tends zero as zero as $T \to \infty$. In the case where X is generated by a stationary process, we assume $\text{plim} 1/T \sum_{t=1}^{T} X_t u_t = \sigma_{Xu}$ and $\text{plim} 1/T \sum_{t=1}^{T} X_t^2 = \sigma_X^2$.

That is, the sample moments converge on the fixed population moments in probability. We assume that the variance of X is positive $\sigma_X^2 > 0$ and therefore the first condition for consistency is met if the covariance of X and u is zero, that is, $\sigma_{Xu} = 0$. Turning to the second condition for consistency we have $V(\hat{\beta}) = \sigma_u^2 / \sum_{t=1}^{T} X_t^2$ and, from our second assumption, we have $\text{plim}\, V(\hat{\beta}) = \sigma_u^2 / T\sigma_X^2$. As $T \to \infty$, this expression clearly tends to zero, and therefore the estimator is consistent under these assumptions.

Now let us consider the rate at which the variance approaches zero as the sample size gets large. Consider the following expression

$$\sum_{t=1}^{T} \left(\frac{X_t}{\sqrt{T}} \right)^2 = \frac{1}{T} \sum_{t=1}^{T} X_t^2. \tag{12.4}$$

This converges on a finite limit by assumption. This demonstrates that the sum of squared values of X is stochastically bounded and that it is *root T convergent*. We can think of this as determining the rate at which the sample variance of X converges to its population value. Under the assumption that X is generated by a stationary process, this rate depends on the square root of T.

Now, if X is generated by a unit root process, we can show that $\text{plim} \sum_{t=1}^{T} X_t^2 / T$ does not converge to a finite positive value. Instead, it becomes infinitely large as $T \to \infty$. There are two implications of this. First, even if $\text{plim} \sum_{t=1}^{T} X_t u_t / T = \sigma_{Xu} \neq 0$, then the OLS estimator may still be consistent. Second, we can show that the rate at which the variance of the OLS estimator tends to zero depends on the level of T rather than its square root. This is because the variance of the OLS estimator tends to zero faster in probability limit when X data is generated by a unit root process. These two properties are summarized in the description of the OLS estimator as being *super-consistent*. Thus, OLS estimation, when the data is non-stationary, can still generate consistent estimates of the unknown β coefficient and the estimator will converge to the population value faster than would be the case with stationary data.

Historical Note: The property of super-consistency was first discussed in a paper by Stock in Econometrica in 1987 [Stock1987].

We have now established that, if the data series Y and X are cointegrated, then the OLS estimator is super-consistent, and that the linear combination $\hat{u}_t = Y_t - \hat{\beta} X_t$ is stationary. This forms the basis of a test for stationarity known as the *Engle-Granger Test*. This test is carried out using the following procedure:

1. If the model is $Y_t = \beta X_t + u_t$, estimate by OLS to obtain $\hat{\beta} = \sum_{t=1}^{T} X_t Y_t / \sum_{t=1}^{T} X_t^2$ and construct the regression residuals $\hat{u}_t = Y_t - \hat{\beta} X_t$.
2. Perform a second regression using the regression residuals. The model for the second regression takes the form $\Delta\hat{u}_t = \gamma_0 + \gamma_1 \hat{u}_{t-1} + \sum_{i=1}^{p} \Delta\hat{u}_{t-i} + \varepsilon_t$. The lag length p in this second regression should be chosen so that the residuals $\hat{\varepsilon}_t$ are free of serial correlation.
3. Test $H_0 : \gamma_1 = 0$ against the alternative $H_1 : \gamma_1 < 0$ using the test statistic $\tau = \hat{\gamma}_1 / SE(\hat{\gamma}_1)$. Note that this test statistic does not follow the standard t distribution under the null hypothesis, and the critical values used to perform this test are empirically determined values. The critical values we use for this test are given by the MacKinnon [MacKinnon1991] response surfaces.

It is also important to note that the critical values here will be higher, in absolute value, than those used when testing for the presence of a unit root in an individual time-series. The inclusion of more variables in the cointegrating equation will also increase the critical value for the unit root test. For example, the 5% critical value for a unit root test based on the residuals from a bivariate regression, which includes a constant but no time trend, is given by the following equation $C(0.05,T) = -3.3377 - 5.967 / T - 8.98 / T^2$. Therefore, with 100 observations, the 5% critical value for an Engle-Granger test for a cointegrating relationship between two variables is –3.39. In contrast, the 5% critical value for a unit root test for a single series with 100 observations is –2.89. If more variables are added to the cointegrating equation, or if a deterministic time trend is included, then the critical value will increase further.

Example: Using the stochastic process defined in (12.1) (but with different random shocks) we create the series shown in Figure 12.3.

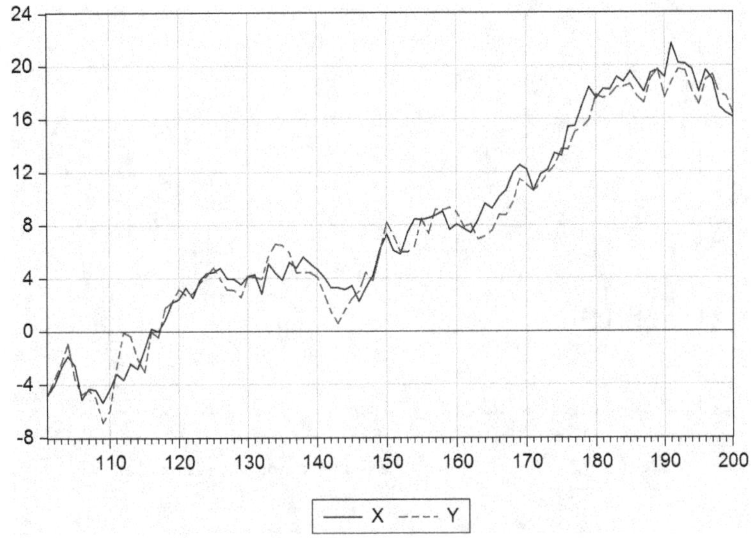

FIGURE 12.3 Simulated Cointegrated Variables.

The unit root test statistics for the X and Y variables are $\tau_X = -1.249$ and $\tau_Y = -1.332$, and the critical value is -2.89. We therefore cannot reject the null hypothesis of a unit root for either series. Next, we test for the presence of a cointegrating relationship linking the two series. In the first stage, we estimate the following bivariate regression equation

$$Y_t = 0.0071 + 0.9561X_t + \hat{u}_t. \tag{12.5}$$

We note that, if the variables are cointegrated, then the slope coefficient here is a super consistent estimator of the long-run or equilibrium parameter linking the two variables. Note that we do not report standard errors or other statistics for this equation because they will be biased and are irrelevant for our purposes anyway. The purpose of equation (12.5) is simply to allow us to calculate the residuals $\hat{u}_t$ so that we can perform a unit root test on this series. Since the data has been generated using the equation $Y_t = 0.5X_t + 0.5Y_{t-1} + u_t$, we can solve for the long-run relationship by removing the time subscripts and setting the equation error equal to its expected value of zero. This means that the long-run effect of an increase in X on Y is equal to one and our estimated value of 0.96 is reasonably close to this value.

Next, we construct the residuals for this equation and perform a unit root test on the resulting series. The residual series is shown in Figure 12.4.

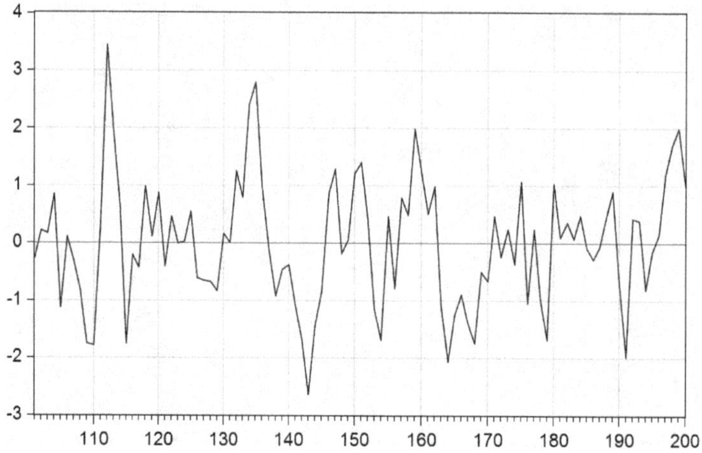

FIGURE 12.4 Residuals from the Cointegrating Regression (12.5).

Finally, we use the Dickey-Fuller approach to perform a unit root test on the residuals. The equation estimated to construct the test takes the form

$$\Delta\hat{u}_t = \underset{(0.0775)}{0.0076} - \underset{(0.0889)}{0.5096}\,\hat{u}_{t-1} + \hat{\varepsilon}_t, \tag{12.6}$$

which gives us a test statistic $\tau = -0.5096 / 0.0889 = -5.73$. This is greater than the 5% critical value of -3.39 in absolute value and so we reject the null hypothesis that the residual series contains a unit root at the 5% level. In other words, there is evidence here that, although the series in question are individually non-stationary, there exists a cointegrating relationship that links them.

In practice, the simple Dickey-Fuller test is rarely appropriate when dealing with economic data. This is because the process generating the data will often involve higher-order autocorrelations than the first-order process assumed in the example above. Therefore, the most usual way to apply this test is to use the Augmented Dickey Fuller (ADF) test on the residuals, where the second stage regression is augmented by the addition of lagged difference terms to capture the higher-order autocorrelations. Another alternative is to use the Phillips-Perron [Phillips1987] procedure for the second stage regression, in which a standard Dickey-Fuller regression is estimated, but the standard errors are adjusted for the presence of higher-order autocorrelations.

Sargan and Bhargava [Sargan1983] suggest an alternative test for cointegration in the form of the *cointegrating regression Durbin-Watson* (CRDW) test. Consider, the standard regression model $Y_t = \beta X_t + u_t$. We have already seen that there is an approximate relationship between the DW statistic and the first-order autocorrelation coefficient of the form $DW \approx 2(1-\hat{\rho})$. Therefore, if we wish to test the null hypothesis that the first-order autocorrelation coefficient is equal to one, then we can use DW as a test statistic since $H_0 : \rho = 1$ should imply a value of the DW statistic close to zero. It is again necessary to generate critical values for this test using Monte Carlo methods because the process generating the data is non-stationary under the null. For a bivariate regression and 100 observations, Sargan and Bhargava report a 5% critical value for the CRDW test equal to 0.386.

In practice, the Engle-Granger two step test is still very commonly used in econometric analysis, while the CRDW is not often applied. Normally, the choice between competing tests is based on relative power. However, this is not necessarily the case here. Engle and Granger (1987) provide a comparison of the relative power of a variety of different cointegration tests which shows that the CRDW does relatively well in terms of its ability to reject a false null hypothesis. In a simple comparison with the two-step test based on an ADF test of the residuals from a cointegrating regression, it has marginally higher power. However, the difference is very slight. Where the CRDW test does badly, is that it is very closely tied to a particular data generation process. The Monte Carlo DGP used to compare relative power assumes that the series in the bivariate regression are independent random walks. The CRDW test does well in this context, but less well when the DGP involves more complex autoregressive processes, which is likely to be the case for economic data. In contrast, the Engle-Granger test handles these cases rather better. For this reason, the Engle-Granger test is now the preferred option.

12.2 COINTEGRATION WITH MULTIPLE VARIABLES

So far, we have only considered bivariate cointegration. However, it is straightforward to extend this framework to deal with cases in which there are more than two variables in the cointegrating relationship. In such cases, the cointegrating parameter becomes a cointegrating vector.

Example: Consider the case of the aggregate production function which relates output to inputs of labor and capital. The Cobb-Douglas production function takes the form

$$\ln(Y_t) = \beta_1 + \beta_2 \ln(N_t) + \beta_3 \ln(K_t) + u_t, \tag{12.7}$$

where Y, N, and K are output, labor input and capital input respectively. When we estimate relationships like this using time series data, these series often contain unit roots. However, it is possible that there is a cointegrating relationship such that the equation error (or total factor productivity in this case) is stationary. Having established that the individual series contain unit roots using the Dickey-Fuller test, we can test for a possible cointegrating relationship using the Engle-Granger approach. The equation reported below uses a data set for US GDP and factor inputs given in Maddala [Maddala1989] which contains annual data for the period 1929–1967,

$$\ln(Y_t) = -3.9377 + 1.4508 \ln(N_t) + 0.3838 \ln(K_t) + \hat{u}_t. \tag{12.8}$$

We then take the residuals from this equation and estimate the auxiliary regression shown in equation (12.9)

$$\Delta\hat{u}_t = \underset{(0.0041)}{0.0013} - \underset{(0.135)}{0.6838}\,\hat{u}_{t-1} + \underset{(0.144)}{0.5411}\,\Delta\hat{u}_{t-1} + \hat{\varepsilon}_t. \tag{12.9}$$

The test statistic is $\tau = -0.6838/0.135 = -5.07$. We can then calculate the appropriate critical value using the MacKinnon response surface as

$$\tau^{5\%} = -3.7429 - \frac{8.352}{36} - \frac{13.41}{36^2} = -3.98. \tag{12.10}$$

Since the test statistic is greater than the critical value at the 5% level, this means that we can reject the null hypothesis at the 5% level and conclude that there is a cointegrating vector linking these three variables. In this case, it has a natural interpretation as a measure of total factor productivity, that is, the part of output that is unexplained by measured factor inputs.

In adopting the Engle-Granger approach, we start with the null hypothesis that the variables in our relationship are not cointegrated. We then test this against the alternative that there is a cointegrating relationship. Therefore, we must reject the null hypothesis if we are to conclude that a long-run (cointegrating) relationship exists. A problem with this approach is that the tests we use, including the Engle-Granger test, often have low power. This means that it is difficult to reject the null hypothesis even when it is false, and we, therefore, run the risk of making a Type II error. The Engle-Granger test is based on the idea that the residuals for the cointegrating regression should be stationary. Effectively we are looking for evidence that the coefficient on the lagged residual is less than one in a regression of the form

$$\hat{u}_t = \delta_0 + \delta_1 \hat{u}_{t-1} + \sum_{i=1}^{p} \delta_{1+i} \Delta \hat{u}_{t-i} + \varepsilon_t. \tag{12.11}$$

The problem is that the alternative hypothesis includes values that are very close to the null hypothesis. In practice, it may be difficult to reject $H_0 : \delta_1 = 1$ if the true value of δ_1 is equal to 0.95 for example, even though this would correspond to a stationary process (albeit one in which the return to the equilibrium relationship was quite slow). Simulation studies have shown that Engle-Granger test suffers from low power in these circumstances. Therefore, although the Engle-Granger approach provides a very intuitive way of testing for cointegration, it becomes necessary to look for alternative tests that may be more powerful in identifying cointegrating relationships.

12.3 COINTEGRATION AND ERROR CORRECTION

There is a close relationship between the ideas of cointegration and error correction which we can illustrate using the following example. Consider a general autoregressive distributed lag model of the form

$$Y_t = \beta_1 X_t + \beta_2 X_{t-1} + \beta_3 Y_{t-1} + u_t. \tag{12.12}$$

This could be a straightforward dynamic regression model in which both Y and X are $I(0)$ variables. However, it could also be a dynamic cointegrating relationship in which Y and X are $I(1)$ but u is $I(0)$. We can transform this equation and write it in the form

$$\Delta Y_t = \gamma_1 \Delta X_t + \gamma_2 X_{t-1} + \gamma_3 Y_{t-1} + u_t, \tag{12.13}$$

where $\gamma_1 = \beta_1; \gamma_2 = \beta_2 + \gamma_1$ and $\gamma_3 = \beta_3 - 1$. Note that this does not change the equation in any way. It simply reflects a reparameterization of the original equation. If, however, the variables are integrated of order one, then this transformation means that the new form of the equation contains variables which are $I(0)$ (the Δ terms) and variables which are $I(1)$ (the lagged X and Y terms). In the absence of cointegration, mixing orders of integration in a single equation will lead to very poor results. For example, regressing an $I(0)$ variable on an $I(1)$ variable will normally produce an insignificant result.

However, if the variables are cointegrated then there is a linear combination which is $I(0)$ and therefore the $I(1)$ variables on the RHS will be significant. Therefore, testing for the significance of the levels variables in an error correction equation provides us with an alternative test for cointegration.

Historical Note: The error correction model was first introduced by Sargan [Hart1964] and became popular in applied econometric work after its use by Davidson et al [Davidson1978]. Its link to the cointegration problem was first noted by Engle and Granger [Engle1987].

Another reparameterization of this equation can be obtained by writing it as

$$\Delta Y_t = \gamma_1 \Delta X_t + \gamma_3 (Y_{t-1} - \delta_1 X_{t-1}) + u_t, \tag{12.14}$$

where $\delta_1 = -\gamma_2 / \gamma_3 = (\beta_1 + \beta_2)/(1 - \beta_3)$. Here, we can interpret δ_1 as the "long-run" or cointegrating parameter which links Y and X. Therefore, $Y_{t-1} - \delta_1 X_{t-1}$ represents the lagged disequilibrium or "error" in the relationship between Y and X. This is where the term "error correction" originates. However, as Hendry has pointed out, the term "equilibrium correction" is probably more appropriate, since the parameter γ_3 determines the extent to which Y changes in response to disequilibrium. This version of the model suggests an alternative approach to testing for cointegration in which we estimate the error correction model and test $H_0 : \gamma_3 = 0$ against the alternative, $H_1 : \gamma_3 < 0$. The test statistic here is the t-ratio or $\hat{\gamma}_3 / \mathrm{SE}(\hat{\gamma}_3)$ but, as we would expect given the non-stationary nature of the levels of Y and X, this does not follow the t-distribution. Empirically determined critical values and response surfaces for this test are given in Ericsson and MacKinnon [Ericsson2002].

Example: Let us consider again our artificial example (12.1). We can combine the two equations of this example into a single error-correction equation which takes the form

$$\Delta Y_t = 0.025 - 0.5(Y_{t-1} - 1.0X_{t-1}) + u_t, \tag{12.15}$$

where $u_t = 0.5\varepsilon_{1t} + \varepsilon_{2t}$. We can again generate artificial data based on this model and then test for cointegration by estimating the error correction equation shown in equation (12.16)

$$\Delta Y_t = \underset{(0.406)}{-0.0213} + \underset{(0.121)}{0.5663}\,\Delta X_t - \underset{(0.081)}{0.4394}\,Y_{t-1} + \underset{(0.091)}{0.4536}X_{t-1} + \hat{u}_t \qquad (12.16)$$
$$R^2 = 0.39.$$

The Ericsson and MacKinnon test statistic can then be constructed as $\tau = -0.4394/0.081 = -5.42$. Using their response surfaces, we find a 5% critical value of -3.246. Therefore, we reject the null at the 5% level in this case. Note that the advantage of this approach is that the error correction model provides a better dynamic specification than the static regression used to generate the residuals in the Engle-Granger test. This means that this test tends to be more powerful and we are less likely to make a Type II error when we adopt this testing procedure. A variation on this theme is suggested by Turner [Turner2006], which proceeds by performing an F-test for the joint significance of the lags on the RHS of the test equation. In this case, we obtain a test statistic 16.37. Using Turner's response surfaces we find a 5% critical value of 5.88. Therefore, this also confirms the existence of a cointegrating relationship between the two variables in our example.

12.4 THE JOHANSEN TEST FOR COINTEGRATION

Consider again, our artificial example (12.1). We have seen that we can write this as a single ECM as in (12.15). However, a third option is to write it as a *Vector Error Correction Model* or VECM. We will discuss models of this type in more detail in Chapter 13. However, we introduce some of the ideas here so that we can develop a third testing procedure, known as the Johansen [Johansen1988] cointegration test, and compare it with the other cointegration tests. Solving the system of equations defined by (12.1) gives a matrix system of the form

$$\begin{bmatrix} \Delta Y_t \\ \Delta X_t \end{bmatrix} = \begin{bmatrix} 0.025 \\ 0.05 \end{bmatrix} + \begin{bmatrix} -0.5 & 0.5 \\ 0 & 0 \end{bmatrix} \begin{bmatrix} Y_{t-1} \\ X_{t-1} \end{bmatrix} + \begin{bmatrix} 0.5\varepsilon_{1t} + \varepsilon_{2t} \\ \varepsilon_{1t} \end{bmatrix}. \qquad (12.17)$$

Consider the matrix linking the changes in Y and X to their lagged levels. It can be shown that there is a relationship between the rank of this matrix and cointegration. This matrix has rank one because of the existence of the cointegrating vector linking Y and X. If there was no cointegrating vector, then all the elements of this matrix would be zero, and it would have rank zero. We could therefore test for cointegration by testing for the rank of this matrix. This forms the basis of the Johansen test for cointegration.

We can think of the Johansen test as a multivariate version of the error correction test. Essentially, this test is based on the idea that, if a cointegrating vector or vectors exist, then there should be detectable feedback from the lagged levels of the variables in the system to their rates of change. Let us consider a more general definition of the system under consideration. Let

$$\Delta \boldsymbol{y}_t = \boldsymbol{B}\boldsymbol{x}_t + \boldsymbol{\Pi}\boldsymbol{y}_{t-1} + \sum_{i=1}^{p} \boldsymbol{\Gamma}_i \Delta \boldsymbol{y}_{t-i} + \boldsymbol{v}_t, \tag{12.18}$$

where $\boldsymbol{y}_t$ is a vector of random variables, $\boldsymbol{x}_t$ is a vector of deterministic and exogenous variables and $\boldsymbol{v}_t$ is a vector of random errors. The matrices $\boldsymbol{B}$, $\boldsymbol{\Pi}$ and $\boldsymbol{\Gamma}_i$; $i =,\ldots,p$ are matrices of parameters. We are interested in the parameters of the $\boldsymbol{\Pi}$ matrix which capture the error-correction property of the system. The Johansen testing procedure is based on the properties of the $\boldsymbol{\Pi}$ matrix. However, it does not use this matrix directly. Instead, it works through a process of "concentrating out" the nuisance parameters contained in the $\boldsymbol{B}$ and $\boldsymbol{\Gamma}$ matrices, by regressing $\Delta \boldsymbol{y}_t$ and $\boldsymbol{y}_{t-1}$ on the $\boldsymbol{x}_t$ and $\Delta \boldsymbol{y}_{t-i}$ variables first and then working with the residuals from these regressions. A full description of the procedures used can be found in Davidson and MacKinnon [Davidson2004].

A rigorous derivation of the Johansen test is beyond the scope of this book. However, we provide a very brief account of the procedure here to give the reader an idea of how this works. Let $\hat{\boldsymbol{v}}_{1t}$ and $\hat{\boldsymbol{v}}_{2t}$ be the residuals from regressions of $\Delta \boldsymbol{y}_t$ and $\boldsymbol{y}_{t-1}$ on the $\boldsymbol{x}_t$ and $\Delta \boldsymbol{y}_{t-i}$ variables respectively. Next, define the matrices sample covariance matrices of these residuals as

$$\hat{\boldsymbol{\Sigma}}_{ij} = \frac{1}{T} \sum_{t=1}^{T} \hat{\boldsymbol{v}}_{jt}^{T} \hat{\boldsymbol{v}}_{lt}; \qquad j = 1,2, l = 1,2. \tag{12.19}$$

The Johansen testing procedure is based on the eigenvalues λ_i of the matrix

$$\mathbf{A} = \hat{\boldsymbol{\psi}}_{22}^{T} \hat{\boldsymbol{\Sigma}}_{21} \hat{\boldsymbol{\Sigma}}_{11}^{-1} \hat{\boldsymbol{\Sigma}}_{12} \hat{\boldsymbol{\psi}}_{22}, \tag{12.20}$$

where $\hat{\boldsymbol{\psi}}_{22} \hat{\boldsymbol{\psi}}_{22}^{T} = \hat{\boldsymbol{\Sigma}}_{22}^{-1}$. The interested reader is referred to Davidson and MacKinnon [Davidson2004] for a fuller treatment of this procedure.

The purpose of the transformations outlined in the previous paragraph is to calculate a set of eigenvalues $\lambda_i; i = 1,\ldots,k$ where k is the dimension of the $\boldsymbol{y}$ vector. These eigenvalues allow us to test for the rank of the $\boldsymbol{\Pi}$ matrix which captures the error correction properties of the system. If no error correction is present, that is, if there are no cointegrating vectors in the system, then this matrix will have rank zero. If there is a single cointegrating vector, then the matrix will have rank one, if there are two then it will have rank two, and so on. Thus, the Johansen procedure has the advantage of allowing for the possibility of multiple cointegrating relationships between the set of variables of interest. There are two test statistics used to assess the number of cointegrating vectors in the system. Suppose we wish to test the null hypothesis that there are r_1 cointegrating vectors against the alternative that there are r_2. The *trace test statistic* which is defined as

$$-T \sum_{i=r_1+1}^{r_2} \ln(1-\lambda_i), \tag{12.21}$$

and, if we wish to test the null hypothesis that there are r cointegrating vectors against the alternative that there are $r+1$, then we define the *maximum eigenvalue test statistic* as

$$-T \ln(1-\lambda_{max}), \tag{12.22}$$

where λ_{max} is the largest eigenvalue in the remaining set, after we have performed the first r tests.

Finally, we note that the distributions of both these test statistics are non-standard because of the assumption that the data generation process is not stationary under the null hypothesis. It, therefore, becomes necessary to use Monte Carlo methods to generate empirical critical values and response surfaces. As with most unit root testing procedures, the distributions, and critical values change according to the nature of the deterministic and/or exogenous variables included in the system. This is discussed in the paper by Pesaran, Shin and Smith [Pesaran2000].

It should have become obvious by now that the Johansen testing procedure is technically very demanding. However, it does offer some significant advantages over single equation testing procedures such as the Engle-Granger test or the error-correction test. The first of these is that it allows for the presence of multiple cointegrating vectors in the system. This is of obvious interest to economists who generally work with systems in which multiple equilibrium relationships are possible. Secondly, the Johansen procedure is not sensitive to the ordering of the variables in the system. One of the problems single equation methods, such as the Engle-Granger approach, is that the result of the test can differ depending on which variable is defined as the dependent variable and which is the regressor. This is not the case for the Johansen method. Finally, the Johansen method appears to offer a more powerful test that the single equation methods we have considered. This is important because cointegration tests notoriously lack power against alternatives close to the null.

It may be helpful at this stage to look at an example of the implementation of the Johansen procedure in practice. Figure 12.5 shows the EViews output for the Johansen test based on a sample realization of the stochastic process defined in (12.17). To interpret these results, we start with the null hypothesis that there are no cointegrating vectors and test this against the alternative that there is at least one. In this case, we reject this null hypothesis using both the trace and the maximum eigenvalue tests. Note that EViews automatically supplies the 5% critical values and p-values for these tests based on the response surfaces given in MacKinnon et al. (1999). Once we have established that there is at least one cointegrating vector, we redefine the null as the hypothesis that there is a single cointegrating vector and test this against the alternative that there is more than one. In this case, we cannot reject the second null hypothesis using either test. These conclusions are consistent with the DGP set out in (12.17) in which, by design, there is a single cointegrating vector.

Sample: 101 200
Included observations: 100
Trend assumption: Linear deterministic trend
Series: X Y
Lags interval (in first differences): 1 to 2

Unrestricted Cointegration Rank Test (Trace)

Hypothesized No. of CE(s)	Eigenvalue	Trace Statistic	0.05 Critical Value	Prob.**
None *	0.261652	32.00198	15.49471	0.0001
At most 1	0.016542	1.668010	3.841466	0.1965

Trace test indicates 1 cointegrating eqn(s) at the 0.05 level
* denotes rejection of the hypothesis at the 0.05 level
**MacKinnon-Haug-Michelis (1999) p-values

Unrestricted Cointegration Rank Test (Maximum Eigenvalue)

Hypothesized No. of CE(s)	Eigenvalue	Max-Eigen Statistic	0.05 Critical Value	Prob.**
None *	0.261652	30.33397	14.26460	0.0001
At most 1	0.016542	1.668010	3.841466	0.1965

Max-eigenvalue test indicates 1 cointegrating eqn(s) at the 0.05 level
* denotes rejection of the hypothesis at the 0.05 level
**MacKinnon-Haug-Michelis (1999) p-values

FIGURE 12.5 EViews Output for the Johansen Cointegration Test.

Example: To conclude our discussion of the Johansen method, let us consider an example using real-world data, rather than an artificial data set. Suppose we are interested in the relationship between interest rates on assets with different maturities. We have data on three such interest rates: Bankrate,[1] the Treasury Bill rate and the yield on government bonds. Figure 12.6 plots each of these series using monthly UK data for the period 1993.12–2006.12.

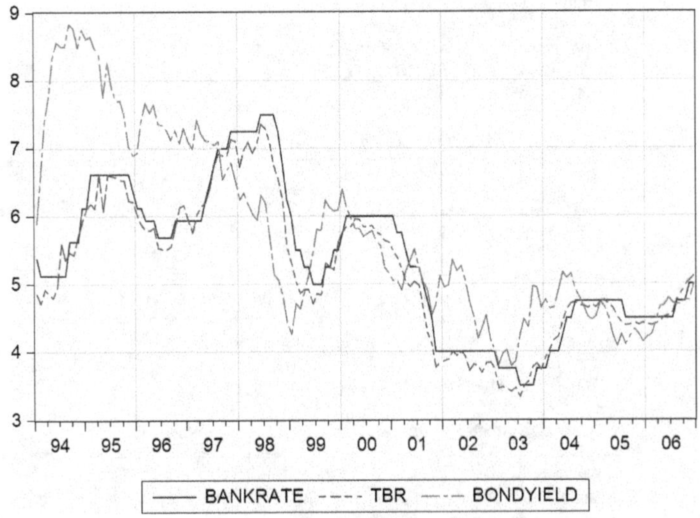

FIGURE 12.6 UK Interest Rate Data 1994.05-2006.12.

Applying augmented Dickey-Fuller tests indicates that each of these series contains a unit root. However, our interest is in whether these interest rates are cointegrated. We also note that it is possible that there may be multiple cointegrating vectors in this case. For example, in the long run, the Bankrate and Treasury Bill rate might move together, while there is a separate relationship between the Treasury Bill rate and the bond yield. Therefore, two separate cointegrating relationships might exist. The Johansen testing approach lends itself to situations like this as shown Figure 12.7 which gives the EViews output for this test. We begin with the null hypothesis that there are no cointegrating vectors and we test this against the alternative that there is at least one cointegrating vector. Both the trace test and the maximum eigenvalue test indicate that we should reject this null hypothesis at the 5%

[1] Bankrate is the rate at which the clearing banks can borrow from the Bank of England. It is used as one of the main tools of monetary policy.

level. We, therefore, move onto the second null hypothesis that there is at most one cointegrating vector which we test against the alternative that more than one cointegrating vector is present. Inspection of Figure 12.7 indicates that we cannot reject this null using either test, and therefore, we conclude that there is a single cointegrating vector present. Note that it is possible, but rare, for the trace and maximum eigenvalue tests to give contradictory results.

Sample (adjusted): 1994M05 2006M12
Included observations: 152 after adjustments
Trend assumption: Linear deterministic trend
Series: BANKRATE TBR BONDYIELD
Lags interval (in first differences): 1 to 4

Unrestricted Cointegration Rank Test (Trace)

Hypothesized No. of CE(s)	Eigenvalue	Trace Statistic	0.05 Critical Value	Prob.**
None *	0.133902	30.70037	29.79707	0.0393
At most 1	0.044815	8.849216	15.49471	0.3795
At most 2	0.012292	1.879914	3.841466	0.1703

Trace test indicates 1 cointegrating eqn(s) at the 0.05 level
* denotes rejection of the hypothesis at the 0.05 level
**MacKinnon-Haug-Michelis (1999) p-values

Unrestricted Cointegration Rank Test (Maximum Eigenvalue)

Hypothesized No. of CE(s)	Eigenvalue	Max-Eigen Statistic	0.05 Critical Value	Prob.**
None *	0.133902	21.85116	21.13162	0.0396
At most 1	0.044815	6.969302	14.26460	0.4928
At most 2	0.012292	1.879914	3.841466	0.1703

Max-eigenvalue test indicates 1 cointegrating eqn(s) at the 0.05 level
* denotes rejection of the hypothesis at the 0.05 level
**MacKinnon-Haug-Michelis (1999) p-values

FIGURE 12.7 Johansen Tests for Cointegration in a Model of Interest Rate Determination.

EXERCISES

Excel files containing the data for these exercises are available as companion files for this book.

EXERCISE 12.1

Consider the following pair of equations which describe the joint behavior of variables X and Y

$$X_t = 0.05 + X_{t-1} + \varepsilon_{1t}$$
$$Y_t = 0.25X_t + 0.75Y_{t-1} + \varepsilon_{2t}$$

where ε_1 and ε_2 are independent Gaussian white-noise variables.

a. Derive the stochastic process which determines the Y variable and shows that it contains a unit root.

b. Derive an error correction model for Y and use this to show that there is a cointegrating relationship linking Y and X.

c. Derive the cointegrating parameter which links the Y and X variables and find the coefficient which determines the speed of adjustment towards the equilibrium relationship.

EXERCISE 12.2

An econometrician estimates a model relating the logarithm of the exchange rate for the US dollar and the pound to the logarithms of the UK price level and the US price level. The results obtained are as follows:

```
Ordinary Least Squares Regression Results
Sample period: 1975 to 2019
Dependent Variable LOG(EX)
Sample Size 45
```

Variable	Coefficient	Std Err	T-Ratio
C	0.504781	0.288324	1.750744
LOG(PUK)	-1.436992	0.500604	-2.870513
LOG(PUS)	1.712601	0.655450	2.612863

R-squared	0.3761	F-statistic	12.6612
SEE	0.113053	RSS	0.536805
Durbin-Watson	0.5832	LogL	35.795363
ARCH(1) Test	8.7812	AIC	-1.457572
Jarque-Bera	0.2445	SIC	-1.337128

The econometrician claims that this model is a success! A rise in UK prices causes the pound to depreciate and a rise in US price causes the pound to appreciate. Moreover, the coefficients are reasonably close to being equal and opposite in sign which indicates that it is the relative price level that matters.

You are given the task of breaking it gently to our econometrician that his results may not be as good as he thinks they are. Explain carefully why this regression equation may suffer from the spurious regression problem and point out any evidence from the estimated equation which supports your argument.

EXERCISE 12.3

The Excel workfile UK INTEREST RATES.XLSX contains data for the yield on 20-year government bonds (R) and the Treasury Bill Rate (TBR).

a. Test each series individually to decide if they contain a unit root.

b. If the series are individually non-stationary, then use the Engle-Granger test to determine if there is a cointegrating relationship between the two series.

c. Estimate an error-correction model for the bond rate (with the Treasury Bill rate on the right-hand side) and use this to construct the Ericsson and MacKinnon *t*-test and the *F*-test for cointegration.

REFERENCES

[Davidson1978] Davidson, J. E. H., Hendry, D. F., Srba, F. and Yeo, J. S., "Econometric Modelling of the Aggregate Time-Series Relationship Between Consumers' Expenditure and Income in the United Kingdom." *Economic Journal*, 1978, 88 (352), pp. 661–692.

[Davidson2004] Davidson, R. and MacKinnon, J.G., *Econometric Theory and Methods.* 2004, Oxford University Press.

[Engle1987] Engle, R.F. and Granger, C.W.J., "Co-integration and Error Correction: Representation, Estimation and Testing." *Econometrica*, 1987, 55(2), pp. 251–276.

[Ericsson2002] Ericsson, N. and MacKinnon, J., "Distributions of Error Correction Tests for Cointegration." *Econometrics Journal*, 2002, 5, pp. 251–276.

[Granger1974] Granger, C.W.J. and Newbold, P., "Spurious Regressions in Econometrics." *Journal of Econometrics*, 1974, 2, pp. 111–20.

[Hart1964] Sargan, J.D., "Wages and Prices in the United Kingdom: A Study in Econometric Methodology (With Discussion)." In Hart, P.E., Mills, G. and Whitaker, J.K. (eds), *Econometric Analysis for National Economic Planning*, Vol 16 of *Colston Papers*, pp. 25–63 London: Butterworth Co.

[Johansen1988] Johansen, S., "Statistical Analysis of Cointegrating Vectors." *Journal of Economic Dynamics and Control*, 1988, 12, pp. 231–254.

[MacKinnon1991] MacKinnon, J.G. (1991) "Critical Values for Cointegration Tests," in Engle, R.F. and Granger, C.W.J. (eds) *Long-Run Economic Relationships*, Oxford: Oxford University Press.

[Maddala 1989] Maddala. G.S., *Introduction to Econometrics*. 1989, MacMillan Publishing Company.

[Pesaran2000] Pesaran, M.H., Shin, Y. and Smith, R.J., "Structural Analysis of Vector Error Correction Models with Exogenous I(1) Variables." *Journal of Econometrics*, 2000, 97, pp. 293–343.

[Phillips1988] Phillips, P.C.B. and Perron, P., "Testing for Unit Roots in Time Series Regression." *Biometrika*, 1988, 75, pp. 335–346.

[Sargan1983] Sargan, J.D. and Bhargava, A., "Testing Residuals from Least Squares Regression for Being Generated by the Gaussian Random Walk." *Econometrica*, 1983, 51(1), pp. 153–174.

[Stock1987] Stock, J., "Asymptotic Properties of Least Squares Estimators of Cointegrating Vectors." *Econometrica*, 1987, 55, pp. 381–386.

[Turner2006] Turner, P., "Response Surfaces for an F-Test for Cointegration." *Applied Economics Letters*, 2006, 13(8), pp. 279–282.

[Yule1926] Yule, G.U., "Why do We Sometimes Get Nonsense Correlations Between Time Series? A Study in Sampling and the Nature of Time Series." *Journal of the Royal Statistical Society*, 1926, 89, pp. 1–63.

CHAPTER 13

VECTOR AUTOREGRESSIONS

The stochastic processes discussed in Chapter 11 were designed to capture the important features of the behavior of a single time series. In many situations, however, we wish to describe the joint behavior of series which are linked in some way. One way of doing this is to build an econometric model that makes specific assumptions about the causal linkages between the variables in question. Models like this are, however, potentially open to misspecification, as the result of invalid theoretical restrictions. An alternative approach is to analyze the inter-relationships between variables using a vector autoregression (or VAR). This class of model was introduced by Sims [Sims1980] to capture the linear interdependencies between multiple time series while imposing as few theoretical restrictions as possible. VARs are particularly useful in capturing the dynamic linkages between variables. For example, the dynamic relationship between two variables X_1 and X_2 might be described by the following pair of equations, which we write in matrix form,

$$\begin{bmatrix} 1 & 0 \\ a_{21} & 1 \end{bmatrix}\begin{bmatrix} X_{1t} \\ X_{2t} \end{bmatrix} = \begin{bmatrix} b_{11} & b_{12} \\ b_{21} & b_{22} \end{bmatrix}\begin{bmatrix} X_{1t-1} \\ X_{2t-1} \end{bmatrix} + \begin{bmatrix} u_{1t} \\ u_{2t} \end{bmatrix}, \qquad (13.1)$$

where the u's are independent Gaussian white-noise disturbances. This form of the system is referred to as the *structural form of the VAR* because it allows interdependencies between the current values of the variables. The specification given in (13.1) does not place any restrictions on the dynamic relationship between the variables. However, it does make the assumption that the coefficient in the top right-hand corner of the matrix on the left-hand side of the system is equal to zero. This acts to identify the system. Without this assumption, or something similar, the equations in the system both consist of linear combinations of the same set of variables, and it would not be possible to separately identify them. The solution taken here is to restrict the contemporaneous relationship between the variables by assuming a particular causal ordering, that is, the variable X_2 has no immediate impact on X_1. This restriction is typical in VAR analysis. However, there are alternative identifying restrictions which we will discuss later.

The *reduced form of the VAR* is obtained by pre-multiplying by the inverse of the matrix of contemporaneous coefficients, so that each equation in the VAR contains only one variable dated in the current period. In this case, we obtain the following

$$\begin{bmatrix} X_{1t} \\ X_{2t} \end{bmatrix} = \begin{bmatrix} 1 & 0 \\ a_{21} & 1 \end{bmatrix}^{-1} \begin{bmatrix} b_{11} & b_{12} \\ b_{21} & b_{22} \end{bmatrix} \begin{bmatrix} X_{1t-1} \\ X_{2t-1} \end{bmatrix} + \begin{bmatrix} 1 & 0 \\ a_{21} & 1 \end{bmatrix}^{-1} \begin{bmatrix} u_{1t} \\ u_{2t} \end{bmatrix}$$

$$= \begin{bmatrix} c_{11} & c_{12} \\ c_{21} & c_{22} \end{bmatrix} \begin{bmatrix} X_{1t-1} \\ X_{2t-1} \end{bmatrix} + \begin{bmatrix} \varepsilon_{1t} \\ \varepsilon_{2t} \end{bmatrix}. \tag{13.2}$$

This provides a convenient way to estimate the VAR because each equation can be estimated individually by least squares. We can then combine the parameter estimates into the matrix system shown in (13.2). If necessary, we can use this system to recover estimates of the structural form parameters, but this is not always necessary. It should also be noted that, if the original structural disturbances u_1 and u_2 are independent, the reduced form disturbances will normally be correlated.

Example: Suppose we wish to describe the relationship between the Bank of England's base interest rate (*BRT*) and the yield on 3-month Treasury Bills (*TBR*). Using monthly data for the period 1994.1 to 2006.12, we obtain the following reduced form VAR estimates.

$$\begin{aligned} BRT_t &= -0.0248 + 0.5090\, BRT_{t-1} + 0.5098\, TBR_{t-1} + \hat{\varepsilon}_{1t} \\ TBR_t &= 0.0935 - 0.1597\, BRT_{t-1} + 1.1466\, TBR_{t-1} + \hat{\varepsilon}_{2t} \end{aligned} \tag{13.3}$$

The covariance matrix of the residuals is given by

	BRT	**TBR**
BRT	0.015476	0.013333
TBR	0.013333	0.033519

We can recover estimates of the structural parameters as follows. From (13.1), we can derive the relationship between the structural and reduced form disturbances as

$$\begin{bmatrix} \varepsilon_{1t} \\ \varepsilon_{2t} \end{bmatrix} = \begin{bmatrix} 1 & 0 \\ -a_{21} & 1 \end{bmatrix} \begin{bmatrix} u_{1t} \\ u_{2t} \end{bmatrix}, \tag{13.4}$$

and, since $\text{cov}(u_1, u_2) = 0$, it follows that

$$\begin{aligned} \sigma_{\varepsilon 1}^2 &= \sigma_{u1}^2 \\ \sigma_{\varepsilon 2}^2 &= a_{21}^2 \sigma_{u1}^2 + \sigma_{u2}^2 \\ \sigma_{\varepsilon 1, \varepsilon 2} &= -a_{21} \sigma_{u1}^2. \end{aligned} \tag{13.5}$$

We can now use a method of moments approach to solve for the unknown parameters. Substituting sample moments for the population moments, we have three equations in three unknown parameters which can be written

$$\begin{aligned} 0.015476 &= \hat{\sigma}_{u1}^2 \\ 0.033519 &= \hat{a}_{21}^2 \hat{\sigma}_{u1}^2 + \hat{\sigma}_{u2}^2 \\ 0.013333 &= -\hat{a}_{21} \hat{\sigma}_{u1}^2. \end{aligned} \tag{13.6}$$

Using the first equation in (13.6), we can solve for $\hat{\sigma}_{u1}^2$ as 0.015476. Next, using the third equation, we have $\hat{a}_{21} = -0.013333 / 0.015476 = -0.8615$. Finally, substituting these estimates into the second equation yields $\hat{\sigma}_{u2}^2 = 0.033519 - 0.8615^2 \times 0.015476 = 0.02203$. Once we have solved (13.6) to obtain the parameters of the contemporaneous relationships, it is straightforward to solve for the other structural parameters. To do this we would use the relationship $C = A^{-1}B$ where C is the matrix of reduced form coefficients on the lags, A is the matrix of contemporaneous structural coefficients and B is the matrix of structural coefficients on the lags. In practice, however, we rarely need to do this since these parameters are generally not of interest in themselves.

To identify this system, we made the assumption a particular causal ordering, that is, X_1 is not affected by X_2 in the current period, but we allow X_2 to be affected by X_1. This is a very common way of identifying the structure. It means that we assume that the matrix of contemporaneous coefficients has a lower triangular structure. The mathematical term for this assumption is that we apply a *Cholesky decomposition* to the system. This method generalizes to higher-order systems. Suppose, for example, we have three variables in the VAR. The VAR can be identified by the assumption that the matrix of contemporaneous coefficients has the triangular structure shown in equation (13.7).

$$A = \begin{bmatrix} 1 & 0 & 0 \\ a_{21} & 1 & 0 \\ a_{31} & a_{32} & 1 \end{bmatrix} \tag{13.7}$$

The variance–covariance matrix of the residuals from a VAR of order three will yield six sample moments, that is, three variances and three covariances. If the matrix of contemporaneous relationships has the structure shown in (13.7) then there are six unknown parameters that determine the relationship between the structural and reduced form disturbances. These are the three variances of the structural disturbances and the three a coefficients. Therefore, this satisfies the order condition for identification. More generally, if there are p variables in the VAR, then we have $p(p+1)/2$ sample moments corresponding to the reduced form parameters, and there is the same number of unknown structural parameters describing the relationship between the structural and reduced form disturbances in a lower triangular system. Hence, the assumption of a triangular structure for the matrix of contemporaneous relationships ensures that the order condition for identification is exactly satisfied. As with the two-variable system, the Cholesky decomposition in the general case can be thought of a causal ordering between the variables. For example, in our 3×3 example above, variable 1 is not affected by variable 2 or variable 3, variable 2 is affected by variable 1 but not variable 3 while variable 3 is affected by both the other variables.

It is also interesting to calculate the equilibrium of this system. First, we note that equation (13.3) can be written in matrix form as

$$\begin{pmatrix} BRT_t \\ TBR_t \end{pmatrix} = \begin{pmatrix} -0.0248 \\ 0.0935 \end{pmatrix} + \begin{pmatrix} 0.5090 & 0.5098 \\ -0.1597 & 1.1466 \end{pmatrix} \begin{pmatrix} BRT_{t-1} \\ TBR_{t-1} \end{pmatrix} + \begin{pmatrix} \hat{\varepsilon}_{1t} \\ \hat{\varepsilon}_{2t} \end{pmatrix} \tag{13.8}$$

The eigenvalues of the matrix on the right-hand side are 0.69 and 0.97 and therefore, since both lie within the unit circle, this is a stable system that will converge to an equilibrium, or steady-state, following a random disturbance. Setting the disturbance terms to zero and solving for the equilibrium yields

$$\begin{pmatrix} BRT \\ TBR \end{pmatrix} = \begin{pmatrix} 1-0.5090 & -0.5098 \\ 0.1597 & 1-1.1466 \end{pmatrix}^{-1} \begin{pmatrix} -0.0248 \\ 0.0935 \end{pmatrix} = \begin{pmatrix} 5.4377 \\ 5.2858 \end{pmatrix} \tag{13.9}$$

The system, therefore, has a sensible long-run solution with equilibrium interest rates between 5% and 6%. This is consistent with the average values of interest rates during the period used for estimation. However, since then, there has been a structural change which has meant that equilibrium interest rates have fallen quite dramatically. This does illustrate one of the weaknesses of the VAR approach in that shocks to the system can lead to changes in either the parameter values or the equilibrium properties of the system. Of course, the same criticism can be applied to standard econometric models. We should also note that we would normally expect the equilibrium Treasury Bill rate to be slightly higher than the equilibrium Bankrate but this ordering is reversed in equation (13.9).

13.1 SOME GENERAL RESULTS FOR VARS

We can write a very general form of a VAR as

$$\mathbf{A}_0\boldsymbol{x}_t = \mathbf{A}_1\boldsymbol{x}_{t-1} + \mathbf{A}_2\boldsymbol{x}_{t-2} + \ldots + \mathbf{A}_k\boldsymbol{x}_{t-k} + \boldsymbol{u}_t, \tag{13.10}$$

where $\boldsymbol{x}_t$ is a vector of p variables and therefore the $\mathbf{A}_i$ matrices have dimensions $p \times p$. $\boldsymbol{u}_t$ is a vector of p independent random disturbances. The reduced form of the VAR can be written as

$$\boldsymbol{x}_t = \sum_{i=1}^{k} \boldsymbol{B}_i\boldsymbol{x}_{t-i} + \boldsymbol{C}\boldsymbol{u}_t, \tag{13.11}$$

where $\boldsymbol{B}_i = \mathbf{A}_0^{-1}\mathbf{A}_i$ and $\mathbf{C} = \mathbf{A}_0^{-1}$.

We can also write any k'th order VAR in its companion form, that is, as a first-order VAR, by appropriate definitions of variables. For example, consider a single variable AR(2) process

$$X_t = a_1 X_{t-1} + a_2 X_{t-2} + \varepsilon_t. \tag{13.12}$$

Now define $Y_t = X_{t-1}$ and $Z_t = \begin{bmatrix} X_t & Y_t \end{bmatrix}^T$. We can write our equation as

$$\begin{bmatrix} X_t \\ Y_t \end{bmatrix} = \begin{bmatrix} a_1 & a_2 \\ 1 & 0 \end{bmatrix} \begin{bmatrix} X_{t-1} \\ Y_{t-1} \end{bmatrix} + \begin{bmatrix} \varepsilon_t \\ 0 \end{bmatrix}, \tag{13.13}$$

or $\boldsymbol{Z}_t = \boldsymbol{B}\boldsymbol{Z}_{t-1} + \boldsymbol{v}_t$ where $\boldsymbol{B}$ and $\boldsymbol{v}$ are defined appropriately. This has advantages because it is easier to analyze the properties of a first-order system than one with longer lags.

Consider the system $\boldsymbol{x}_t = \boldsymbol{B}\boldsymbol{x}_{t-1} + \boldsymbol{\varepsilon}_t$. Using the method of backward substitution, we can write any finite order VAR as an infinite moving average process.

$$\begin{aligned} \boldsymbol{x}_t &= \boldsymbol{\varepsilon}_t + \boldsymbol{B}\boldsymbol{\varepsilon}_{t-1} + \boldsymbol{B}^2\boldsymbol{\varepsilon}_{t-2} + \boldsymbol{B}^3\boldsymbol{\varepsilon}_{t-3} + \dots \\ &= \sum_{i=0}^{\infty} \boldsymbol{B}^i \boldsymbol{\varepsilon}_{t-i}. \end{aligned} \tag{13.14}$$

If the transition matrix $\boldsymbol{B}$ has unit roots, then $\boldsymbol{B}^i$ will not tend to zero as i tends to infinity. It follows that random disturbances will continue to affect the current value of X no matter how far back in the past they occurred. In addition, the variance of the elements of X will not be finite. It is therefore important that all the roots of the transition matrix B should lie within the unit circle.

Example: Monetary policy is often analyzed using VAR models. To replicate a famous model by Bernanke and Gertler [Bernanke1995], we estimate a VAR linking the log of GDP, the log of the price level, the Federal Funds Rate and the log of the price of crude oil. This was estimated using US quarterly data for the period 1965.1 to 1993.4 and the lag length was set at 4 because of the data frequency. The roots of the transition matrix are shown in Figure 13.1. Although all the roots fall within the unit circle, there is one root that is very close to one.

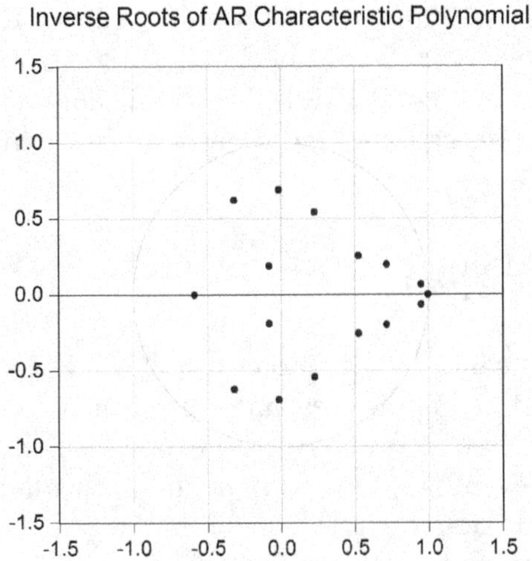

Roots of Characteristic Polynomial
Endogenous variables: LOG(GDP) LOG(PDEF) FFR/...
Exogenous variables: C
Lag specification: 1 4

Root	Modulus
0.995950	0.995950
0.950211 - 0.066479i	0.952534
0.950211 + 0.066479i	0.952534
0.716350 - 0.197974i	0.743203
0.716350 + 0.197974i	0.743203
-0.319291 - 0.621358i	0.698593
-0.319291 + 0.621358i	0.698593
-0.016196 - 0.692379i	0.692568
-0.016196 + 0.692379i	0.692568
-0.589574	0.589574
0.227055 - 0.542531i	0.588128
0.227055 + 0.542531i	0.588128
0.526072 - 0.254683i	0.584479
0.526072 + 0.254683i	0.584479
-0.081826 - 0.189131i	0.206073
-0.081826 + 0.189131i	0.206073

No root lies outside the unit circle.
VAR satisfies the stability condition.

FIGURE 13.1 Roots of a Monetary Policy VAR Estimated Using US data (EViews Output).

13.2 IMPULSE RESPONSES

So far, we have simply discussed how to estimate a VAR. In this and the following section, we show how the main outputs of the VAR methodology can be constructed. These outputs include impulse responses and variance decompositions. Using the interest rate example given in (13.3), we have already shown how we can calculate the matrix of contemporaneous coefficients from the variance-covariance matrix of the residuals of the reduced form VAR. In this case, this yields a matrix of contemporaneous coefficients of the form shown in equation (13.15)

$$\mathbf{A}_0 = \begin{bmatrix} 1 & 0 \\ -0.8615 & 1 \end{bmatrix}. \tag{13.15}$$

The relationship between the reduced and structural form disturbances is therefore given by

$$\begin{bmatrix} \varepsilon_{1t} \\ \varepsilon_{2t} \end{bmatrix} = \mathbf{A}_0^{-1} \begin{bmatrix} u_{1t} \\ u_{2t} \end{bmatrix} = \begin{bmatrix} 1 & 0 \\ 0.8615 & 1 \end{bmatrix} \begin{bmatrix} u_{1t} \\ u_{2t} \end{bmatrix}. \tag{13.16}$$

From equation (13.16) we see that a shock to the Bank Rate affects both the Bank Rate and the Treasury Bill Rate in the current period but a shock to the Treasury Bill Rate has no immediate effect on the Bank Rate. Over time, however, shocks to either variable will affect both variables as shown by the moving average representation of the VAR. The coefficients of the moving average representation give the impulse responses as shown in equation (13.17) where $\boldsymbol{B} = \mathbf{A}_0^1\mathbf{A}_1$.

$$\boldsymbol{x}_t = \sum_{i=0}^{\infty} \boldsymbol{B}^i \mathbf{A}_0^{-1} \boldsymbol{u}_{t-i} \tag{13.17}$$

The impulse responses for the interest rate model are shown in Figure 13.2. These are typical of the sort of impulse response functions we observe in VAR models. Each variable has its own associated shock which produces an initial impact effect that eventually dies down to zero, providing that the roots of the model lie within the unit circle. In addition, the shocks to each variable have dynamic effects on the other variables of the model. For example, an increase in bank rate feeds through into an increase in the Treasury Bill rate and vice versa. Note that the causal ordering we assume will affect the impulse responses. It is therefore important that the decision of what causal ordering to adopt is carefully considered, prior to the simulation of the model to create the impulse responses.

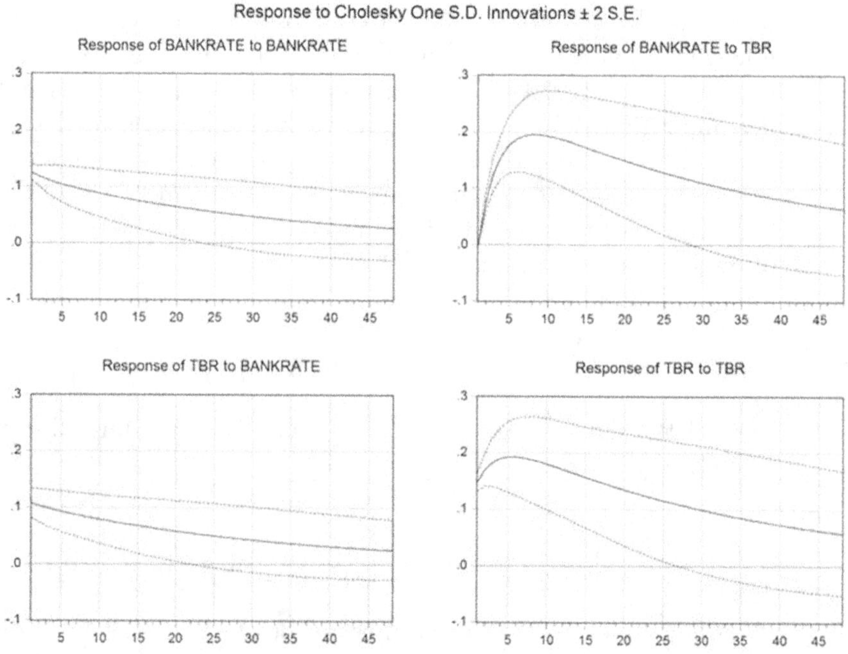

FIGURE 13.2 Impulse Responses for the Interest Rate Model (EViews Output).

The solid lines in Figure 13.2 show the impulse responses of the system to shocks to individual variables. These are equal to the coefficients of the moving average representation of the system. If the variables in the VAR are stationary, then the impulse responses will eventually converge to zero. The broken lines show the 95% confidence interval for the impulse responses.

13.3 VARIANCE DECOMPOSITIONS

Since the structural disturbances (the u's) are orthogonal, the moving average form of the model also allows us to perform a variance decomposition. Consider the moving average representation (13.17), we can use this to derive the k step ahead forecast variance matrix for the vector $\boldsymbol{x}$. We have

$$E\left(\boldsymbol{x}_k\boldsymbol{x}^T\right)=\sum_{i=0}^{k}\boldsymbol{B}^i\mathbf{A}_0^{-1}\boldsymbol{\Omega}\left(\mathbf{A}_0^{-1}\right)^T\left(\boldsymbol{B}^i\right)^T, \tag{13.18}$$

where $\boldsymbol{\Omega}=E\left(\boldsymbol{u}\boldsymbol{u}^T\right)$. This allows us to calculate the contribution of each of the orthogonal disturbance in the vector $\boldsymbol{u}$ to the variance of each of the variables in the vector $\boldsymbol{x}$ after k periods of time.

Example: An example might help make the concept of variance decomposition a little easier to understand. Consider our model of the relationship between Bankrate and the Treasury Bill rate, as set out in equation (13.8). From this model, we have the following

$$\mathbf{A}_0^{-1}=\begin{pmatrix}1 & 0\\ 0.8615 & 1\end{pmatrix}\qquad \boldsymbol{B}=\begin{pmatrix}0.5090 & 0.5098\\ -0.1597 & 1.1466\end{pmatrix}\qquad \boldsymbol{\Omega}=\begin{pmatrix}0.0155 & 0\\ 0 & 0.0220\end{pmatrix}.$$

Now consider the variances of Bankrate and the Treasury Bill rate in period 0. We can write this as

$$\mathbf{A}_0^{-1}\boldsymbol{\Omega}\left(\mathbf{A}_0^{-1}\right)^T=\begin{pmatrix}0.0155 & 0.0134\\ 0.0134 & 0.0335\end{pmatrix}.$$

Next, consider an alternative scenario in which the variance of u_2 is set equal to zero, that is, let

$$\boldsymbol{\Omega}^*=\begin{pmatrix}0.0155 & 0\\ 0 & 0\end{pmatrix},$$

and recalculate, this yields

$$\mathbf{A}_0^{-1}\boldsymbol{\Omega}^*\left(\mathbf{A}_0^{-1}\right)^T = \begin{pmatrix} 0.0155 & 0.0134 \\ 0.0134 & 0.0115 \end{pmatrix}.$$

This gives the variances and covariances of Bankrate and the Treasury Bill rate under the assumption that u_1 is the only source of variation. From this, we can calculate the contribution of u_1 to the variance of Bankrate (which in this case is 100%) and the Treasury Bill rate (which in this case is $100\% \times 0.0115/0.0335 = 34.3\%$). The contribution of u_2 is then given by the remainder in each case. For Bankrate, the contribution is zero, reflecting the causal ordering assumption made to estimate the VAR, while for the Treasury Bill rate it is 65.7%.

Now consider the one-step ahead forecast variance. We have

$$\mathbf{A}_0^{-1}\boldsymbol{\Omega}\left(\mathbf{A}_0^{-1}\right)^T + \boldsymbol{B}\mathbf{A}_0^{-1}\boldsymbol{\Omega}\left(\mathbf{A}_0^{-1}\right)^T\boldsymbol{B}^T = \begin{pmatrix} 0.0352 & 0.0384 \\ 0.0384 & 0.0731 \end{pmatrix}.$$

Again, setting the variance of u_2 to zero and recalculating yields:

$$\mathbf{A}_0^{-1}\boldsymbol{\Omega}^*\left(\mathbf{A}_0^{-1}\right)^T + \boldsymbol{B}\mathbf{A}_0^{-1}\boldsymbol{\Omega}^*\left(\mathbf{A}_0^{-1}\right)^T\boldsymbol{B}^T = \begin{pmatrix} 0.0294 & 0.0255 \\ 0.0255 & 0.0221 \end{pmatrix}.$$

Therefore the contribution of u_1 to the variance of Bankrate for a one-step ahead forecast is $100\% \times 0.0294/0.0352 = 83.5\%$ and that of u_2 is 16.5%. The contribution of u_1 to the variance of the Treasury Bill rate is $100\% \times 0.0221/0.0731 = 30.2\%$ and that of u_2 is 69.8%. Note that all these calculations are carried out to four decimal places, and the results calculated by the EViews package will be more accurate, as shown in Table 13.1 which gives variance decompositions up to a 12-step ahead forecast horizon.

TABLE 13.1 Variance Decomposition for the Interest Rate Model.

Variance Decomposition of BANKRATE:			
Period	S.E.	BANKRATE	TBR
1	0.124402	100.0000	0.000000
2	0.187397	83.69439	16.30561
3	0.251964	66.25688	33.74312
4	0.315933	53.81188	46.18812
5	0.376620	45.46596	54.53404
6	0.432764	39.78565	60.21435
7	0.484028	35.79606	64.20394
8	0.530542	32.90207	67.09793
9	0.572649	30.74079	69.25921
10	0.610762	29.08525	70.91475
11	0.645302	27.78897	72.21103
12	0.676663	26.75452	73.24548
Variance Decomposition of TBR:			
Period	S.E.	BANKRATE	TBR
1	0.183081	34.27072	65.72928
2	0.270368	30.23589	69.76411
3	0.341278	27.44256	72.55744
4	0.402151	25.44836	74.55164
5	0.455453	23.98410	76.01590
6	0.502606	22.88212	77.11788
7	0.544608	22.03476	77.96524
8	0.582226	21.37079	78.62921
9	0.616076	20.84180	79.15820
10	0.646660	20.41404	79.58596
11	0.674399	20.06351	79.93649
12	0.699642	19.77281	80.22719
Cholesky Ordering: BANKRATE TBR			

We can assess the variance decomposition shown in Table 13.1 as follows. First, the variance of Bankrate is largely driven by shocks to Bankrate in the short run. However, this is partly the result of the causal ordering we imposed on the model for estimation purposes which necessarily implies that 100% of the variance is due to the shock associated with the variable itself in period 1. In contrast, the causal ordering permits shocks to Bankrate to affect the variance of the Treasury Bill rate in the short run and, from Table 13.1, we see that this means that 34.3% of the variance of the Treasury Bill rate is estimated as being caused by shocks to Bankrate in period 1. Over time, we see that the impact of shocks to Bankrate on both itself, and the Treasury Bill rate, declines. By the end of 1 year, only 26.8% of the variance of Bankrate is caused by its own associated shock. A similar effect can be seen in the variance decomposition for the Treasury Bill rate which shows the effect of shocks to Bankrate declining to 19.8% by the end of one year.

13.4 STRUCTURAL VARS

We have seen that it is necessary to impose some restrictions on the joint relationship between variables to identify and estimate a VAR. The approach taken in the previous sections was to assume a particular causal ordering or Cholesky decomposition. However, we have seen that this can have implications for the impulse responses and the variance decompositions which the VAR generates. Thus, untested, and somewhat arbitrary restrictions, can have significant effects on the results generated by VAR analysis. An alternative is to use restrictions derived from economic theory to identify the VAR. This at least has the benefit of ensuring that the results can be justified and interpreted in the context of theory. At the simplest level, this might involve choosing the ordering of the Cholesky decomposition according to economic theory. However, it potentially involves more interesting restrictions such as those involving the long-run relationship between the variables. For example, suppose we start with the VAR given in equation (13.19)

$$\begin{bmatrix} 1 & a_{12} \\ a_{21} & 1 \end{bmatrix}\begin{bmatrix} X_{1t} \\ X_{2t} \end{bmatrix} = \begin{bmatrix} b_{11} & b_{12} \\ b_{21} & b_{22} \end{bmatrix}\begin{bmatrix} X_{1t-1} \\ X_{2t-1} \end{bmatrix} + \begin{bmatrix} u_{1t} \\ u_{2t} \end{bmatrix}. \tag{13.19}$$

This cannot be estimated as it stands because there are more parameters than there are sample moments – hence the need for restrictions prior to estimation. In the case of the Cholesky decomposition this is achieved by either setting $a_{12} = 0$ or setting $a_{21} = 0$.

Suppose instead we specify that the variable X_2 has no long-run effect on X_1. From the first equation in the VAR we can derive the equilibrium relationship

$$X_1 = \frac{b_{12} - a_{12}}{1 - b_{11}} X_2 \tag{13.20}$$

Therefore, an alternative restriction is to impose $b_{12} = a_{12}$, which will be enough to identify the VAR and permit estimation. This is not a standard procedure in EViews (or other regression packages) but can be done with a certain amount of manipulation.

13.5 VECTOR ERROR CORRECTION MODELS (VECMS)

When we estimate a VAR, the series included should be stationary so that the eigenvalues of the system lie within the unit circle. If they are not stationary, then the impulse responses will not converge, and the variances will not be defined. The normal recommendation is, therefore, that the data should be differenced prior to estimation, unless there exists a cointegrating relationship, in which case it is possible to estimate a vector error-correction model or VECM.

Consider a simple 2×2 VAR of the form given by equation (13.21). Note that we have assumed no contemporaneous interactions to simplify the notation. This means that, if such interactions were important, then the errors in this system would be correlated.

$$\begin{bmatrix} X_{1t} \\ X_{2t} \end{bmatrix} = \begin{bmatrix} a_{11} & a_{12} \\ a_{21} & a_{22} \end{bmatrix} \begin{bmatrix} X_{1t-1} \\ X_{2t-1} \end{bmatrix} + \begin{bmatrix} v_{1t} \\ v_{2t} \end{bmatrix} \tag{13.21}$$

It is always possible to write this in the following form

$$\begin{bmatrix} \Delta X_{1t} \\ \Delta X_{2t} \end{bmatrix} = \begin{bmatrix} d_{11} & d_{12} \\ d_{21} & d_{22} \end{bmatrix} \begin{bmatrix} X_{1t-1} \\ X_{2t-1} \end{bmatrix} + \begin{bmatrix} v_{1t} \\ v_{2t} \end{bmatrix}. \tag{13.22}$$

Consider the matrix on the right-hand side of equation (13.22). This links the changes in X and Y to their lagged values. It is therefore analogous to the error-correction coefficients in a single equation model. Let us assume that there is a single cointegrating vector linking X and Y which has parameters $\begin{bmatrix} \beta_1 & \beta_2 \end{bmatrix}$. We can then decompose the matrix on the right-hand side of our equation as shown in (13.23)

$$\begin{bmatrix} d_{11} & d_{12} \\ d_{21} & d_{22} \end{bmatrix} = \begin{bmatrix} a_1 \\ a_2 \end{bmatrix} \begin{bmatrix} \beta_1 & \beta_2 \end{bmatrix} = \begin{bmatrix} a_1 \beta_1 & a_1 \beta_2 \\ a_2 \beta_1 & a_2 \beta_2 \end{bmatrix}, \tag{13.23}$$

where the β coefficients are the cointegrating parameters, and the a coefficients are adjustment parameters that determine how the system responds to disequilibrium. For example, a_1 captures the speed at which X_1 changes when there is a deviation from the equilibrium relationship between X_1 and X_2. Another way of writing our model is

$$\begin{bmatrix} \Delta X_{1t} \\ \Delta X_{2t} \end{bmatrix} = \begin{bmatrix} a_1\left(\beta_1 X_{1t-1} + \beta_2 X_{2t-1}\right) \\ a_2\left(\beta_1 X_{1t-1} + \beta_2 X_{2t-1}\right) \end{bmatrix} + \begin{bmatrix} v_{1t} \\ v_{2t} \end{bmatrix}. \tag{13.24}$$

Note that the linear combinations $a_1\left(\beta_1 X_{1t-1} + \beta_2 X_{2t-1}\right)$ and $a_2\left(\beta_1 X_{1t-1} + \beta_2 X_{2t-1}\right)$ are both stationary because of the existence of a cointegrating vector linking X and Y. Estimation of systems like (13.24) can be tricky because it is non-linear in its parameters. However, standard estimation routines are built into programs like EViews. Once we have estimated the VECM, then it is straightforward to calculate impulse responses and variance decompositions in the same way that we did for the simple vector autoregression. The main difference is that the impulse responses will not converge to zero for a VECM because of the presence of a unit root in the system. This means that shocks to variables have permanent effects on their levels in VECM models.

Example: Suppose we estimate a VECM linking the bank rate to the Treasury Bill rate. The results are shown below.

$$\begin{bmatrix} \Delta BTR_t \\ \Delta TBR_t \end{bmatrix} = \begin{bmatrix} \underset{(0.0099)}{-0.0024} - \underset{(0.0473)}{0.4865}\left(0.1117 + BTR_{t-1} - \underset{(0.0180)}{1.0511} TBR_{t-1}\right) \\ \underset{(0.0147)}{0.0014} - \underset{(0.0700)}{0.1448}\left(0.1117 + BTR_{t-1} - \underset{(0.0180)}{1.0511} TRB_{t-1}\right) \end{bmatrix} + \begin{bmatrix} \hat{v}_{1t} \\ \hat{v}_{2t} \end{bmatrix}$$

The cointegrating vector, or equilibrium relationship, therefore, takes the form:

$$BTR = -0.1117 + 1.0511\, TBR$$

Note that shocks in this system will have permanent effects on the variables in the system. This is because there is a unit root in the system. This is evident from the impulse responses shown in Figure 13.3.

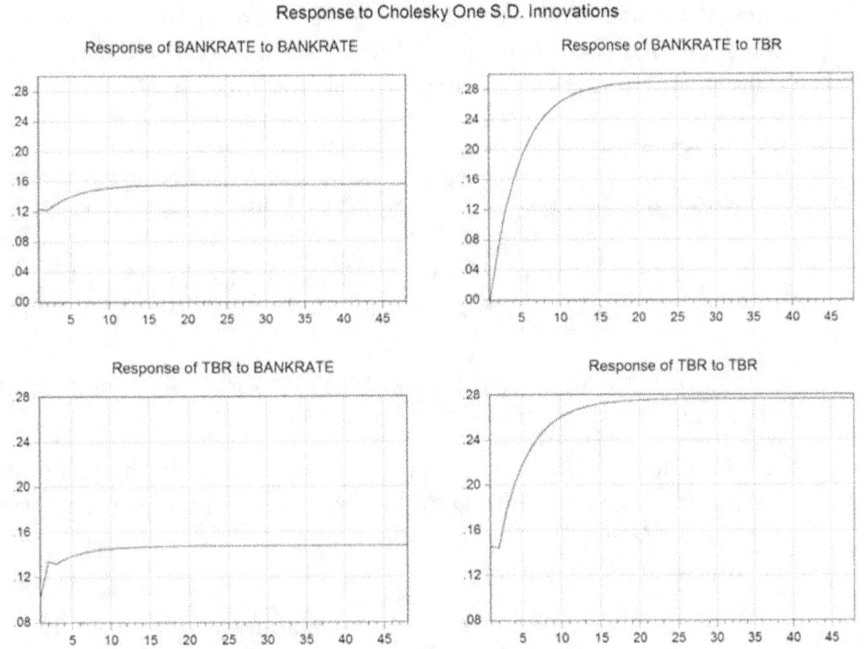

FIGURE 13.3 Impulse Response from VAR Model of Interest Rates.

Note also that the coefficient on the Treasury Bill rate is close to one (but significantly different from one if we accept the estimates of the standard error). We can impose the restriction that this coefficient is equal to one if we wish. This gives us the following results.

$$\begin{bmatrix} \Delta BTR_t \\ \Delta TBR_t \end{bmatrix} = \begin{bmatrix} -\underset{(0.010)}{0.0024} - \underset{(0.0480)}{0.4811}\left(-0.1532 + BTR_{t-1} - TBR_{t-1}\right) \\ \underset{(0.0147)}{0.0014} - \underset{(0.0700)}{0.1664}\left(-0.1532 + BTR_{t-1} - TBR_{t-1}\right) \end{bmatrix} + \begin{bmatrix} \hat{v}_{1t} \\ \hat{v}_{2t} \end{bmatrix}$$

EXERCISES

Excel files containing the data for these exercises are available as companion files for this book.

EXERCISE 13.1

Consider the third-order difference stochastic difference equation

$$X_t = a_1 X_{t-1} + a_2 X_{t-2} + a_3 X_{t-3} + u_t.$$

Show that this can be written as a vector autoregression in companion form and show that the eigenvalues of the transition matrix have the same solutions as the characteristic equation of the original equation.

EXERCISE 13.2

An econometrician has estimated a VECM model with the following results

$$\begin{pmatrix} \Delta X_{1t} \\ \Delta X_{2t} \end{pmatrix} = \begin{pmatrix} -0.5 & 0.5 \\ 0.25 & -0.25 \end{pmatrix} \begin{pmatrix} X_{1t-1} \\ X_{2t-1} \end{pmatrix} + \sum_{i=1}^{p} \hat{\Gamma}_i \begin{pmatrix} \Delta X_{1t-i} \\ \Delta X_{2t-i} \end{pmatrix} + \begin{pmatrix} \hat{\upsilon}_{1t} \\ \hat{\upsilon}_{2t} \end{pmatrix}$$

a. Show that this is consistent with the presence of a single cointegrating vector.

b. Solve for the decomposition of the matrix linking the difference to the levels terms into cointegrating parameters and adjustment parameters, that is, $\boldsymbol{\Pi} = \boldsymbol{\alpha\beta}^T$ where $\boldsymbol{\alpha}$ is the vector of adjustment parameters and $\boldsymbol{\beta}$ is the vector of cointegrating parameters.

For Exercises 3 and 4 you will need an econometrics package such as EViews, which allows for the analysis of vector autoregression models.

EXERCISE 13.3

The Excel workfile *POTATO.XLSX* contains Henry Ludwell Moore's [Moore1914] data on prices and output for the market for potatoes in the United States between 1866 and 1911. The data are expressed as annual percentage changes. Using this data

a. Estimate a vector autoregression model linking these two variables.

b. Calculate, and interpret, the impulse response functions.

c. Calculate, and interpret, the variance decomposition for this model.

EXERCISE 13.4

The Excel workfile *US EXPORTS.XLSX* contains quarterly data on exports for the United States USEXP and exports of other industrialized economies IEXP2 for the period 1975 to 2003. Using this data

a. Estimate a Vector Error Correction Model linking US exports to those of other industrialized economies.

b. Calculate, and interpret, the impulse response function.

c. Calculate, and interpret, the variance decompositions.

REFERENCES

[Bernanke1995] Bernanke, B.S., and Gertler, M., "Inside the Black Box: The Credit Channel of Monetary Policy Transmission." *Journal of Economic Perspectives*. 1995, 9(4), pp. 27–48.

[Moore1914] Moore, H. L. *Economic Cycles: Their Law and Cause*. 1914, The MacMillan Company.

[Sims1980] Sims, C.A., "Macroeconomics and Reality." *Econometrica*. 1980, 48, pp. 1–47.

ANSWERS TO ODD NUMBERED EXERCISES

CHAPTER 1: PROBABILITY AND THE STATISTICAL FOUNDATIONS OF ECONOMETRICS

EXERCISE 1.1

The purpose of this question is for students to become familiar with using sample data to calculate probabilities and to construct a contingency table. In order to answer the question let us first define the following notation M denotes a state in which an individual is male, F denotes a state in which the individual is female, E denotes a state in which they are employed, SE a state in which they are self-employed, U in which they are unemployed and finally, NE a state in which they are not economically active. Hence, $p(M)$ gives the (marginal) probability that an individual is male, $p(F \cap SE)$ is the joint probability that any individual is both female and self-employed and $p(NE|M)$ is the conditional probability that an individual is not economically active given that he is male.

Part (a) First, we need to calculate the total population as the sum of males plus females this gives us $29916 + 31059 = 60975$. Note that the unit of measurement here is thousands of people and so this gives an estimate of the 2007 population of just under 61 million people.

We can now use this figure to calculate the joint probabilities as the ratio of the number falling into each category to the total population. For example, the probability that an individual is both male and employed is given by $12950 / 60975 = 0.2124$. The complete table is given below:

	Male	Female
Employed	0.2124	0.2010
Self-Employed	0.0453	0.0173
Unemployed	0.0155	0.0116
Not Economically Active	0.2175	0.2795

Part (b) Next, we can calculate the marginal probabilities by taking the sums of the rows and the columns. For example, the probability that an individual is employed is found by taking the sum of the probability that they are male and employed plus the probability that they are female and employed. We, therefore, have

$$\begin{aligned} p(E) &= p(M \cap E) + p(F \cap E) \\ &= 0.2124 + 0.2010 = 0.4134. \end{aligned}$$

We can use this to calculate the marginal probabilities as shown in the table below:

	Male	Female	
Employed	0.2124	0.2010	*0.4134*
Self-Employed	0.0453	0.0173	*0.0626*
Unemployed	0.0155	0.0116	*0.0271*
Not Economically Active	0.2175	0.2795	*0.4970*
	0.4907	*0.5094*	

The numbers in italics are the sums of the rows and columns and give the marginal probabilities of the different events. It is straightforward to confirm that these probabilities sum to one when we add the row/column sums apart from rounding errors in the fourth decimal place, that is,

$$\begin{aligned} &p(M) + p(F) = 0.4907 + 0.5094 = 1.001 \\ &p(E) + p(SE) + p(UE) + p(NE) = 0.4134 + 0.0626 + 0.0271 + 0.4970 = 1.001. \end{aligned}$$

Part (c) To answer this question, we need to make use of Bayes' formula. This gives us the relationship between the joint, conditional and marginal probabilities of the relevant events. In this case, we have

$$p(M \cap SE) = p(M|SE)p(SE),$$

that is, the joint probability that an individual is both male and self-employed is the product of the conditional probability that the individual is male given that he/she is self-employed multiplied by the marginal probability that he/she is self-employed. From the numbers in the table we, therefore, have

$$0.0453 = p(M|SE) \times 0.0626 \Rightarrow p(M|SE) = 0.7236.$$

Part (d) We again use Bayes' formula to calculate the conditional probability that an individual is unemployed given that they are male. We have

$$p(UE|M) = \frac{p(UE \cap M)}{p(M)} = \frac{0.0155}{0.4906} = 0.0316.$$

EXERCISE 1.3

The purpose of this question is for students to become familiar with some important properties of the normal distribution. We have $X \sim N(\mu_X, \sigma_X^2)$ and $Y \sim N(\mu_Y, \sigma_Y^2)$. By a standard result any linear combination of normally distributed random variables is itself normally distributed. The mean of $X+Y$ is given by

$$\mu_{X+Y} = E(X+Y) = \mu_X + \mu_Y.$$

The variance is given by

$$\begin{aligned}\sigma_{X+Y}^2 &= E(X+Y-E(X+Y))^2 = E\left((X-E(X))^2 + (Y-E(Y))^2\right)\\ &= E(X-E(X))^2 + E(Y-E(Y))^2 + 2E\{(X-E(X))(Y-E(Y))\},\end{aligned}$$

we are told that these variables are independent and therefore

$$E\{(X-E(X))(Y-E(Y))\} = 0,$$

which leaves us with

$$\sigma_{X+Y}^2 = E(X - E(X))^2 + E(Y - E(Y))^2 = \sigma_X^2 + \sigma_Y^2.$$

For the case of $X - Y$ we have

$$\mu_{X-Y} = E(X - Y) = \mu_X - \mu_Y$$

and

$$\begin{aligned}\sigma_{X-Y}^2 &= E(X - Y - E(X - Y))^2 = E\left((X - E(X))^2 - (Y - E(Y))^2\right) \\ &= E(X - E(X))^2 + E(Y - E(Y))^2 - 2E\{(X - E(X))(Y - E(Y))\}.\end{aligned}$$

Using the assumption that the two variables are independent then allows us to write

$$\sigma_{X-Y}^2 = \sigma_X^2 + \sigma_Y^2.$$

CHAPTER 2: STATISTICAL INFERENCE

These questions are concerned with hypothesis testing. This requires the following:

1. A hypothesis to be tested (the "null hypothesis") and an alternative hypothesis with which is to be compared.
2. A test statistic whose distribution is known under the assumption that the null hypothesis is true.
3. A decision rule, that is, a statement of the circumstances under which the null hypothesis will be "accepted" and those under which it will be rejected.

We will consider each of the exercises set within this framework.

EXERCISE 2.1

For this question we are asked to test the null hypothesis that the population mean is 55 against the alternative that it is greater than 55. Thus, we can state the hypothesis to be tested formally as

$$H_0 : \mu = 55$$
$$H_1 : \mu > 55.$$

We are not told the distribution of the variable itself but, under the central limit theorem, we know that the distribution of the sample mean will converge onto the normal distribution for a sufficiently large sample. Here, we have 50 observations, so it is reasonable to assume that this will be the case and we can, therefore, write

$$\bar{X} \sim N\left(\mu, \frac{\sigma^2}{N}\right) \Rightarrow \frac{\bar{X} - \mu}{\sigma / \sqrt{N}} \sim N(0,1),$$

where σ^2 is the population variance and N is the sample size.

We do not know the population variance but we can substitute the following unbiased estimator

$$\hat{\sigma}^2 = \frac{\sum (X_i - \bar{X})^2}{N - 1}.$$

This produces a test statistic that has a t distribution with $N - 1$ degrees of freedom. That is

$$\frac{\bar{X} - \mu}{\hat{\sigma} / \sqrt{N}} \sim t_{N-1}.$$

We are now in a position to calculate the test statistic for this particular problem. We have

$$\frac{\bar{X} - \mu}{\hat{\sigma} / \sqrt{N}} = \frac{60.9253 - 55}{20.9229 / \sqrt{50}} = 2.0025.$$

Should we accept or reject the null hypothesis in this case? The decision depends on what we consider to be an acceptable probability of making a Type 1 error. A Type 1 error is the case in which we reject a null hypothesis that is true. This is a case in which we have a one-sided alternative – that is, a "one-tailed" test – and, therefore, we can find the probability of making a Type 1 error by looking at the area in the tail of the probability density function to the right of the value given by the test statistic.

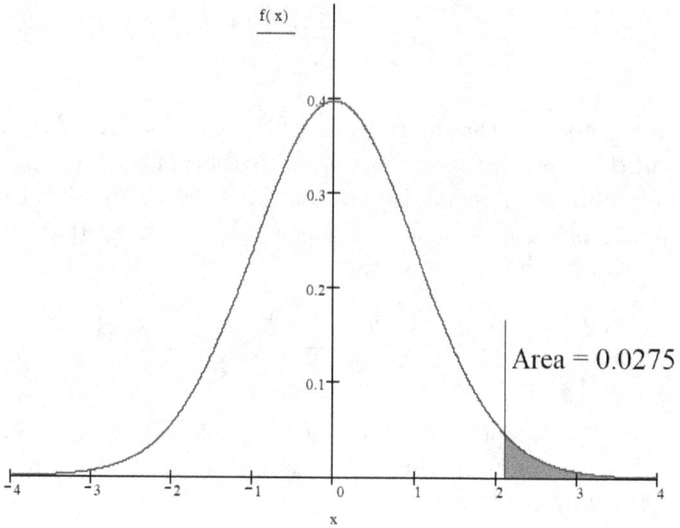

This area gives the p-value for the test statistic. Assuming that the null hypothesis is true, it gives us the probability of obtaining a test statistic at least as extreme as the value observed. In this case, the probability is quite low.

The most usual decision rule is to fix an acceptably low value p-value and then to reject the null hypothesis only if the actual p-value falls below this level. This defines the "significance level" for the test. Usually, we set the significance level at either 10% (0.1) or 5% (0.05). In this case, the p-value for the test falls below either of these levels and so we would choose to reject the null hypothesis.

Another way of looking at the testing procedure is to ask what value of the test statistic would be consistent with our chosen significance level. This defines the critical value for the test statistic, and we reject the null if the actual test statistic is more extreme than the critical value. In this case, the 10% critical value is 1.299 and the 5% critical value is 1.677. The actual test statistic is more extreme than either of these values and, therefore, we reject the null at both the 10% and 5% significance levels. (Note that if we reject the null at the 5% level it follows immediately that we reject it at the 10% and all lower levels of significance levels).

EXERCISE 2.3

The test statistic for normality is the Jarque-Bera test statistic, which is calculated using the skewness and kurtosis coefficients. It is defined as

$$JB = \frac{N}{6}\left[\hat{\gamma}_1^2 + \frac{1}{4}(\hat{\gamma}_2 - 3)^2\right],$$

where N is the sample size, $\hat{\gamma}_1$ is the coefficient of skewness and $\hat{\gamma}_2$ is the coefficient of skewness. Under the null hypothesis of normality this follows a chi-square distribution with two degrees of freedom. In this case, we have

$$JB = \frac{50}{6}\left[(-0.027204)^2 + \frac{1}{4}(2.394301 - 3)^2\right] = 0.7705.$$

The 5% critical value for the chi-square distribution with two degrees of freedom is 5.99. Therefore, we cannot reject the null hypothesis at the 5% level using this test. Alternatively, we note that the p-value for a test statistic of 0.7705 for this distribution is equal to 0.68. This confirms that we cannot reject the null hypothesis at any reasonable level of significance.

CHAPTER 3: THE BIVARIATE REGRESSION MODEL

EXERCISE 3.1

The purpose of this exercise is for students to develop a deeper understanding of the relationship between the OLS regression estimates and the residual sum of squares. It also demonstrates a result that proves useful in other derivations and proofs. We are asked to prove that the residual sum of squares is equal to $(N-1)$ multiplied by the difference between the sample variance of Y minus the ratio of the squared sample covariance of X and Y divided by the sample variance of X. The algebra of the proof is given below.

$$RSS = \sum_{i=1}^{N}\left(Y_i - \hat{a} - \hat{\beta}X_i\right)^2$$

$$= \sum_{i=1}^{N}\left(Y_i - \overline{Y} - \hat{\beta}\left(X_i - \overline{X}\right)\right)^2 \quad \text{since } \overline{Y} = \hat{a} + \hat{\beta}\overline{X}$$

$$= \sum_{i=1}^{N}\left(Y_i - \overline{Y}\right)^2 + \hat{\beta}^2\sum_{i=1}^{N}\left(X_i - \overline{X}\right)^2 - 2\hat{\beta}\sum_{i=1}^{N}\left(Y_i - \overline{Y}\right)\left(X_i - \overline{X}\right)$$

$$= \sum_{i=1}^{N}\left(Y_i - \overline{Y}\right)^2 - \frac{\left(\sum_{i=1}^{N}\left(Y_i - \overline{Y}\right)\left(X_i - \overline{X}\right)\right)^2}{\sum_{i=1}^{N}\left(X_i - \overline{X}\right)^2} \quad \text{since } \hat{\beta} = \frac{\sum_{i=1}^{N}\left(Y_i - \overline{Y}\right)\left(X_i - \overline{X}\right)}{\sum_{i=1}^{N}\left(X_i - \overline{X}\right)^2}$$

$$= (N-1)\left\{\hat{\sigma}_Y^2 - \frac{\hat{\sigma}_{XY}^2}{\hat{\sigma}_X^2}\right\}$$

since $\hat{\sigma}_Y^2 = \frac{1}{N-1}\sum_{i=1}^{N}\left(Y_i - \overline{Y}\right)^2, \hat{\sigma}_X^2 = \frac{1}{N-1}\sum_{i=1}^{N}\left(X_i - \overline{X}\right)^2$

and $\hat{\sigma}_{XY} = \frac{1}{N-1}\sum_{i=1}^{N}\left(Y_i - \overline{Y}\right)\left(X_i - \overline{X}\right)$

QED

EXERCISE 3.3

The purpose of this exercise is to familiarize students with the relationship between correlation, OLS regression and the sample moments of the data.

Part (a) The correlation coefficient can be calculated as

$$\hat{\rho}_{XY} = \frac{\hat{\sigma}_{XY}}{\sqrt{\hat{\sigma}_X^2 \hat{\sigma}_Y^2}} = \frac{-145.3071}{\sqrt{211.5094 \times 246.5624}} = -0.6363$$

Part (b) The slope coefficient and the intercept for a regression of price changes on quantity changes can be calculated as

$$\hat{\beta} = \frac{\hat{\sigma}_{XY}}{\hat{\sigma}_X^2} = \frac{-145.3071}{246.5624} = -0.5893$$

$$\hat{\alpha} = \overline{Y} - \hat{\beta}\overline{X} = 0.44115 - (-0.5893) \times (-2.034617) = -0.7578$$

Part (c) The slope coefficient and the intercept for a regression of quantity changes on price changes can be calculated as:

$$\hat{\delta} = \frac{\hat{\sigma}_{XY}}{\hat{\sigma}_Y^2} = \frac{-145.3071}{211.5094} = -0.6870$$

$$\hat{\gamma} = \overline{X} - \hat{\delta}\overline{Y} = -2.034617 - (-0.6870) \times (0.44115) = -1.7315$$

Part (d) The correlation coefficient and the regression slope coefficients differ because, although the numerator, that is, the covariance of the two variables, is the same, the denominator differs. However, they are related in that the product of the two regression slope coefficients is equal to the square of the correlation coefficient.

$$\hat{\rho}_{XY}^2 = \hat{\beta} \times \hat{\delta} = \frac{\hat{\sigma}_{XY}^2}{\hat{\sigma}_X^2 \hat{\sigma}_Y^2}$$

CHAPTER 4: THE MULTIVARIATE REGRESSION MODEL

EXERCISE 4.1

The purpose of this exercise is to ensure that students understand the construction of the least-squares estimator and its relationship to the sample moments of the data.

Recall that the definition of the sample standard deviation of a variable is

$$\hat{\sigma}_X = \frac{1}{N-1}\sum_{i=1}^{N}\left(X_i - \overline{X}\right)^2.$$

For each of the variables, we can calculate the sum of squared deviations from the mean using the definition of the standard deviation. We have the following:

Change in Unemployment	$69 \times 1.051518^2 = 76.292617$
% Change in GDP	$69 \times 2.286201^2 = 360.643336$
Time Trend	$69 \times 20.351085^2 = 28577.499587$

Recall that the definition of the sample correlation of two variables is

$$\hat{\rho}_{XY} = \frac{\sum_{i=1}^{N}\left(X_i - \overline{X}\right)\left(Y_i - \overline{Y}\right)}{\sqrt{\sum_{i=1}^{N}\left(X_i - \overline{X}\right)^2 \sum_{i=1}^{N}\left(Y_i - \overline{Y}\right)^2}} = \frac{\hat{\sigma}_{XY}}{\hat{\sigma}_X \hat{\sigma}_Y}$$

and therefore

$$\sum_{i=1}^{N}\left(X_i - \overline{X}\right)\left(Y_i - \overline{Y}\right) = \hat{\rho}_{XY} \times (N-1) \times \hat{\sigma}_X \times \hat{\sigma}_Y.$$

Therefore, from the correlation matrix, we can calculate the sums of the cross-products of deviations from the mean:

ΔUN and ΔGDP	−117.072721
ΔGDP and t	−1092.029353
ΔUN and t	−50.2476

We can now calculate estimates of the slope coefficients using the matrix formulation

$$\begin{pmatrix} \hat{\beta}_2 \\ \hat{\beta}_3 \end{pmatrix} = \begin{pmatrix} 360.643336 & -1092.029353 \\ -1092.029353 & 28577.499587 \end{pmatrix}^{-1} \begin{pmatrix} -117.072721 \\ -50.2476 \end{pmatrix}$$

$$= \begin{pmatrix} -0.373119 \\ -0.016016 \end{pmatrix}$$

The intercept can then be calculated by using the property that the regression line passes through the sample means of the data. This yields

$$\hat{\beta}_1 = -0.044285 + 0.373119 \times 3.223957 + 0.016016 \times 37.5 = 1.759235$$

Please note that that regression estimates using the original data set may give slightly different results because of rounding errors in the above calculations.

EXERCISE 4.3

Part (a) The data are measured in \$bn at 2000 prices. It follows that we can think of the coefficient estimates as marginal effects, that is, a \$lbn increase in autonomous expenditures results in a \$0.57bn increase in consumer expenditure while a \$1bn increase in the money supply results in an increase of \$0.39bn. Given that the equation is expressed in difference form, it is best to think of this as a short-run or impact effect. That is, it tells us what the immediate effect of changes in the exogenous variables will be but does not really give us a guide as to their long-run impact.

We can think of the link between consumer spending, autonomous expenditure and the money supply using the IS-LM model. An increase in autonomous expenditure produces a rightwards shift of the IS curve, GDP and disposable income both increase and consumption rises because consumption and income are linked through the consumption function. An increase in the money stock produces a rightwards shift of the LM curve with similar effects on GDP, disposable income and consumption. The IS-LM model predicts positive effects of both changes in autonomous expenditure and the money stock and therefore the fact that both estimated coefficients are positive and significant is consistent with the theory.

The coefficients in our estimated equation are functions of the whole set of structural parameters that define the IS-LM system and therefore depend on the marginal propensity to consume, the marginal tax rate, the interest sensitivity of investment, etc. It is, therefore, highly likely that these parameters would change if the structure of the economy changed – for example, if there were changes to the marginal tax rate. The best way to think about these estimates is therefore as an average over the sample period rather than fixed values which would remain unchanged into the future.

Part (b) We can calculate the residual sum of squares (RSS) using the relationship between the standard error of the regression and the RSS. We have

$$SEE = \sqrt{\frac{RSS}{T-k}} \Rightarrow RSS = SEE^2 \times (T-k).$$

Therefore, in this case, we have

$$RSS = 41.676845^2 \times (48-3) = 78{,}163.17.$$

Part (c) We can use the relationship between the coefficient of determination and the F statistic to answer this question. We have

$$F = \frac{R^2}{1-R^2}\frac{T-k}{k-1} = \frac{0.7341}{1-0.7341} \times \frac{45}{2} = 62.12.$$

Under the null hypothesis that both slope coefficients are zero, this statistic is distributed as F with 2 and 45 degrees of freedom. The 5% critical value for this distribution is 3.204. Therefore, we can reject the null hypothesis, in this case, in favor of the alternative that one or both of the coefficients is not zero.

EXERCISE 4.5

Part (a) Estimation of the equation allowing for separate effects of the different autonomous expenditure categories yields the following results:

```
Ordinary Least Squares Regression Results
Sample period: 1960 to 2007
Dependent Variable DC1
Sample Size 48

Variable          Coefficient            Std Err            T-Ratio

C                  -27.008916          17.449870          -1.547800
DI                   0.600674           0.086876           6.914146
DG                   0.585015           0.240838           2.429079
DX                  -0.120358           0.211238          -0.569776
DRM1                 0.038503           0.005691           6.765150

R-squared              0.7758     F-statistic               37.1994
SEE                 39.150912     RSS                  65910.136739
Durbin-Watson          1.4573     LogL                  -241.505366
ARCH(1) Test           0.2433     AIC                     10.271057
Jarque-Bera            0.9363     SIC                     10.465974
```

Note that the investment and government consumption variables are both positive and significant whereas the exports variable has the wrong (negative) sign and is statistically insignificant.

Part (b) To test the restriction that the coefficients on the three expenditure categories are equal we first note that the restricted version of the equation is given in exercise 4.1. The residual sum of squares for the restricted regression is calculated in Exercise 1 part (b) as 78163.17. We can, therefore, calculate the F-statistic for a test of the null hypothesis as

$$F = \frac{78163.17 - 65910.14}{65910.14} \times \frac{48-5}{2} = 3.997.$$

Under the null hypothesis this statistic is distributed as $F_{2,43}$. The 5% critical value for an F-test with these degrees of freedom is 3.214. Therefore, we reject the null hypothesis, in this case, in favor of the alternative that the coefficients are not equal.

CHAPTER 5: SERIAL CORRELATION

EXERCISE 5.1

The purpose of this exercise is to demonstrate an important property of autoregressive models. This property is that the errors in such equations will be correlated with the lagged values of the variable concerned if the errors are themselves autocorrelated. This, in turn, means that OLS estimation of models with lagged dependent variables and autocorrelated errors will yield biased results. We have

$$Y_t = \beta Y_{t-1} + u_t$$
$$u_t = \rho u_{t-1} + \varepsilon_t.$$

From the second equation, we note that, in general $E(u_t u_{t-k}) = \rho^k \sigma_u^2$. Next, using backward substitution, we can write the first equation as

$$Y_t = u_t + \beta u_{t-1} + \beta^2 u_{t-1} + \beta^3 u_{t-3} + \ldots,$$

and therefore

$$Y_{t-1} = u_{t-1} + \beta u_{t-2} + \beta^2 u_{t-3} + \beta^3 u_{t-4} + \ldots$$

Multiplying this expression by u_t, and taking expectations, yields

$$
\begin{aligned}
E(u_t Y_{t-1}) &= E(u_t u_{t-1}) + \beta E(u_t u_{t-2}) + \beta^2 E(u_t u_{t-3}) + \ldots \\
&= \rho\sigma_u^2 + \beta\rho^2\sigma_u^2 + \beta^2\rho^3\sigma_u^2 + \ldots \\
&= \rho\sigma_u^2\left(1 + \beta\rho + \beta^2\rho^2 + \ldots\right) \\
&= \frac{\rho\sigma_u^2}{1-\beta\rho},
\end{aligned}
$$

which is the required result. Finally, we note that $E(u_t Y_{t-1}) = 0$ is a necessary condition for OLS to be an unbiased estimator. Therefore, OLS estimation of the first equation will yield a biased estimate of the β parameter.

EXERCISE 5.3

The purpose of this exercise is to reinforce students' understanding of the property that a regression with a high R^2 is not necessarily a good model. It also requires them to construct two formal tests for serial correlation.

Part (a) The problem with this regression is that there is evidence of significant serial correlation. Even without a formal test, we can see that the Durbin-Watson statistic is much lower than the expected value of two under the null hypothesis that there is no serial correlation. The fact that the Durbin-Watson statistic is less than two indicates that there is likely to be positive serial correlation. This means that we cannot rely on the t-test for significance of the right-hand side variable because the standard error of the estimate will typically be biased downwards in cases like this. The high value of the R^2 statistic may simply indicate that the variables each contain a trend rather than the existence of any genuine relationship between them.

Part (b) Suppose the error process is of the form:

$$u_t = \rho u_{t-1} + \varepsilon_t$$

where ε_t is a white-noise error process. Tests for first-order autocorrelation are tests for the null hypothesis $H_0 : \rho = 0$ against the alternative $H_1 : \rho \neq 0$. Two tests are directly available to us using the information given in the table. The first is the Durbin-Watson (DW) test. The DW test statistic is defined as:

$$DW = \frac{\sum(\hat{u}_t - \hat{u}_{t-1})^2}{\sum \hat{u}_t^2}$$

where $\hat{u}_t; t = 1, \ldots, T$ are the regression residuals. Under the null the expected value of DW is 2 and tables for critical bounds are available in most books of statistical tables. Here we have 58 observations and one right-hand side variable. The tables we use do not give critical bounds for the exact number of observations but we can interpolated from those for $T = 55$ and $T = 60$ to obtain $d_L = 1.54$ and $d_U = 1.61$. The test statistic is 0.28 which lies below the lower bound, indicating that there is significant positive autocorrelation.

The other test available to us is the Box-Ljung test. For first-order serial correlation, the test statistic is calculated as:

$$Q = \frac{T(T+2)}{T-1}\hat{\rho}^2 = \frac{58 \times 60}{57} \times 0.8425^2 = 43.34$$

Under the null hypothesis, this follows a chi-squared distribution with one degree of freedom. The 5% critical value for the χ_1^2 distribution is 3.841. Therefore, using this test, we again reject the null hypothesis that there is no serial correlation.

EXERCISE 5.5

The purpose of this exercise is to give students hands-on experience of estimating, and interpreting, a regression equation. The data will be provided in spreadsheet form as a download so that they can use the software provided by the course organizer. The results reported below have been calculated using the program I-REG.

Part (a) If we estimate the model in first differences, then we obtain the following results:

```
Ordinary Least Squares Regression Results
Sample period: 1949 to 2005
Dependent Variable D(I)
Sample Size 57

Variable          Coefficient          Std Err          T-Ratio

C                -1336.827410       679.375168        -1.967730
D(Y)                 0.279684         0.034312         8.151172

R-squared              0.5471       F-statistic          66.4416
SEE               3205.596423       RSS             5.651717 E8
Durbin-Watson          1.7610       LogL             -540.002777
ARCH(1) Test           0.1528       AIC                19.017641
Jarque-Bera           17.8359       SIC                19.089327

Autocorrelations

                     AR Coeff           Q-stat     5% crit val
Order 1                0.1176           0.8315           3.841
Order 2               -0.2064           3.4372           5.991
Order 3               -0.0940           3.9875           7.815
Order 4                0.0852           4.4491           9.488
```

The following is a graph of the residuals. There is no obvious visual evidence of serial correlation.

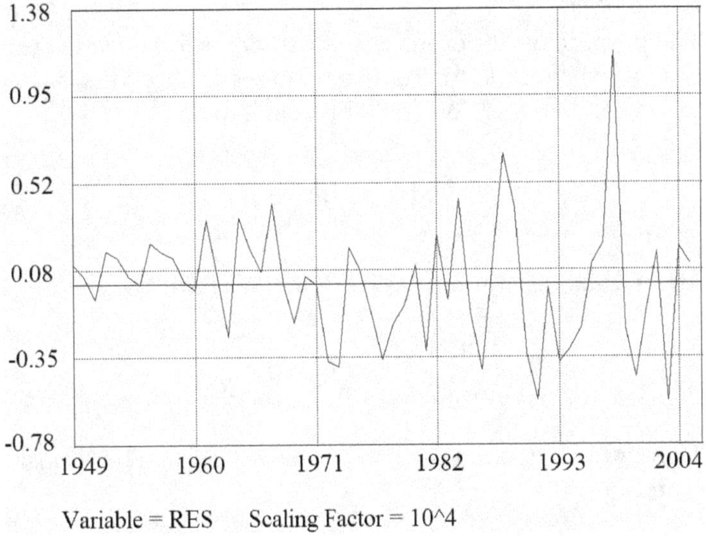

Part (b) The Durbin-Watson statistic is 1.761 and the critical bounds are 1.53 and 1.60. Since the *DW* statistic is above the upper bound (and less than 2) we conclude that there is no evidence of first-order autocorrelation. The Box-Ljung test for first-order serial correlation yields a test statistic of 0.83 which is less than the critical value of 3.841, therefore, again we find no evidence of first-order serial correlation. Calculation of the Breusch-Godfrey test yields a test statistic of 0.76 which is well below the critical value for the appropriate *F*-distribution. Hence, again we cannot reject the null hypothesis that there is no autocorrelation.

Part (c) The fact that there is no evidence of serial correlation, in this case, makes us more confident about our hypothesis tests on the coefficients than in the case of the model estimated in levels.

CHAPTER 6: HETEROSCEDASTICITY, FUNCTIONAL FORM, AND STRUCTURAL BREAKS

EXERCISE 6.1

The purpose of this exercise is for students to become aware of how models can be transformed to produce homoscedastic errors.

Let $v_i = u_i / X_i^h$. We look for h such that

$$E\left(v_i^2\right) = E\left(\frac{u_i}{X_i^h}\right)^2 = \sigma_u^2,$$

that is, the error variance is constant. We have

$$E\left(\frac{u_i}{X_i^h}\right)^2 = \frac{1}{X_i^{2h}} E\left(u_i^2\right) = \frac{1}{X_i^{2h}} \sigma_u^2 \sqrt{X_i}.$$

Therefore, we need $X_i^{2h} = X_i^{1/2}$ or $h = 1/4$. The transformed model is therefore

$$\frac{Y_i}{\sqrt[4]{X_i}} = \frac{a}{\sqrt[4]{X_i}} + \beta X_i^{3/4} + v_i.$$

EXERCISE 6.3

The purpose of this exercise is to give students practice in the construction and interpretation of tests for heteroscedasticity. The data set can be downloaded from the webpage for this book.

Part (a) The coefficient of determination for this regression is very close to one. By itself, this would indicate a good fit for the data. The coefficient for money growth is close to one which is consistent with the quantity theory of money.

Part (b) To apply the Goldfeld-Quandt test we first divide the sample into three groups. The first group contains the observations corresponding to the smallest values of the exogenous variable and should contain about 3/8 of the total sample, the third group contains the observations with the largest value of the exogenous variable and also contains 3/8 of the sample. In this case, 3/8 of the full sample is approximately equal to 31 observations. We then estimate separate regressions for each of these groups. The middle group, which contains 1/4 of the sample is discarded.

The two regressions estimated are shown as follows:

```
Ordinary Least Squares Regression Results
Sample period: 1 to 31
Dependent Variable INF
Sample Size 31

Variable          Coefficient          Std Err          T-Ratio

C                    0.846703         1.263791         0.669970
MG                   0.373297         0.168090         2.220807

R-squared              0.1453      F-statistic           4.9320
SEE                  2.408052      RSS               168.162752
Durbin-Watson          2.2790      LogL              -70.196743
ARCH(1) Test       0.8706 E-1      AIC                 4.657854
Jarque-Bera            6.2976      SIC                 4.750370

Ordinary Least Squares Regression Results
Sample period: 53 to 83
Dependent Variable INF
Sample Size 31

Variable          Coefficient          Std Err          T-Ratio

C                   -8.454817         1.761125        -4.800803
MG                   1.062356         0.017858        59.489085

R-squared              0.9919      F-statistic        3538.9513
SEE                  7.832127      RSS              1778.924011
Durbin-Watson          2.0161      LogL             -106.758634
ARCH(1) Test       0.4753 E-3      AIC                 7.016686
Jarque-Bera            8.2601      SIC                 7.109201
```

The test statistic for the null hypothesis that the error variance is constant throughout the sample is given by:

$$F = \frac{RSS_U}{RSS_L} = \frac{1778.9}{168.2} = 10.6$$

Under the null hypothesis, this statistic is distributed as $F_{31,31}$ and the 5% critical value for such as statistic is approximately 1.84. We, therefore, reject the null hypothesis at the 5% level in favor of the alternative that heteroscedasticity is present in this model.

Part (c) To perform the White test, we regress the squared residuals from the full sample regression on the original regressor plus its square. This gives us the following results:

```
Ordinary Least Squares Regression Results
Sample period: 1 to 83
Dependent Variable RSQ
Sample Size 83
```

Variable	Coefficient	Std Err	T-Ratio
C	-12.489376	10.009077	-1.247804
MG	2.187250	0.424890	5.147795
MGSQ	-0.005176	0.001322	-3.915132

R-squared	0.3309	F-statistic	19.7806
SEE	58.855564	RSS	277118.192244
Durbin-Watson	1.9077	LogL	-454.476287
ARCH(1) Test	0.5287	AIC	11.023525
Jarque-Bera	1038.1744	SIC	11.110953

The F-test for the joint-significance of the right-hand variables here is 19.8. Under the null hypothesis, this is distributed as F with 2 and 80 degrees of freedom. The 5% critical value here is 3.11. Therefore, we reject the null hypothesis in favor of the alternative that there is heteroscedasticity. This is consistent with the results of the Goldfeld-Quandt test which also indicated the presence of heteroscedasticity.

Note that a small sample test, such as the F-test, is not strictly appropriate here because the White test is an asymptotic test. However, it is often applied in practice. An alternative would have been to apply a Chi-squared test in which the test statistic is given by:

$$TR^2 = 83 \times 0.3309 = 27.5$$

Under the null hypothesis, this is distributed as chi-squared with two degrees of freedom in large samples. The 5% critical value for a chi-squared distribution with two degrees of freedom is 5.991 and therefore we would reject the null using this test statistic also.

Part (d) The presence of heteroscedasticity does not necessarily indicate bias in the coefficient estimates. However, it does indicate that the OLS estimator will be inefficient. In principle, there is an estimator that has the lower variance than the OLS estimator. Also, the estimates of the standard errors will be biased when there is heteroscedasticity which means that conventional *t*-tests are not reliable. It is more difficult to determine the direction of the bias in the case of heteroscedastic errors (in comparison with the case of serially correlated errors). However, if the variance of the error term is positively related to the value of the exogenous variable, then it is likely that the standard errors will be biased downwards. This is the case here as indicated in the regression we used to calculate the White test. Therefore, we should treat hypothesis tests based on this regression with caution as it is likely that we will tend to reject the null hypothesis too often on the basis of the t-statistics generated by an OLS regression.

An alternative would be to use the White variance-covariance matrix for the coefficient estimates as discussed in the main text. This procedure adjusts the standard errors of the coefficient estimates to allow for the presence of heteroscedasticity in the residuals. However, the coefficient estimates remain the same. The equation estimates with White standard errors are given in the table below:

```
Dependent Variable INF
Sample Size 83
White Heteroscedasticity Consistent Standard Errors
```

Variable	Coefficient	Std Err	T-Ratio
C	-5.681642	0.670408	-8.474893
MG	1.046654	0.024176	43.291358
R-squared	0.9902	F-statistic	8206.5392
SEE	5.657356	RSS	2592.460073
Durbin-Watson	1.7031	LogL	-260.595058
ARCH(1) Test	3.3717	AIC	6.327592
Jarque-Bera	56.2515	SIC	6.385877

Note that, in this case, we would not reject the null hypothesis that the slope coefficient is equal to one using a *t*-test. The *t*-test statistic is equal to 0.046654/0.024176 = 1.93. This is less than the 5% critical value of 1.99 for a two-tailed *t*-test with 80 degrees of freedom. The hypothesis that the coefficient for money growth is equal to one is generated by the quantity theory of money and therefore the model above generates results that are consistent with the quantity theory. If we had used the OLS regression results, then we would have rejected the null and we would have had a result that was inconsistent with the quantity theory.

CHAPTER 7: BINARY DEPENDENT VARIABLES

EXERCISE 7.1

The purpose of this exercise is for students to practice deriving the estimator for a parameter of a function using the method of maximum likelihood.

Part (a) The probability distribution function for the Poisson distribution is given by the following expression

$$f(\theta) = \theta^{x} \frac{\exp(-\theta)}{x!}; \; x = 1, 2, \ldots.$$

The likelihood function therefore takes the form

$$L(\theta | x_i; i = 1, \ldots, N) = \prod_{i=1}^{N} f(\theta) = \theta^{x_i} \frac{\exp(-\theta)}{x_i!},$$

which means that the log-likelihood takes the form

$$LL(\theta | x_i; i = 1, \ldots, N) = \sum_{i=1}^{N} \left\{ x_i \ln\theta - \theta - \sum_{i=1}^{N} \ln(x_i - i) \right\},$$

where $\ln(x_i!) = \sum_{i=1}^{x_i} \ln(x_i - i)$ is the trickiest part of finding the log-likelihood. However, since this latter term does not involve θ, it can safely be ignored since it will drop out when we take derivatives. To find the maximum likelihood estimator of θ, we find the derivative of the log-likelihood function, set it to zero and solve for $\hat{\theta}_{ML}$. We have

$$\frac{dLL(\theta)}{d\theta} = \frac{\sum_{i=1}^{N} x_i}{\theta} - N,$$

setting this equal to zero and solving yields

$$\hat{\theta}_{ML} = \frac{\sum_{i=1}^{N} x_i}{N}$$

Part (b) The maximum likelihood estimator of the variance is equal to minus the inverse of the information evaluated at the maximum likelihood estimator of the parameter. We have

$$I(\theta)=\frac{d^2LL}{d\theta^2}=-\frac{\sum_{i=1}^{N}x_i}{\theta},$$

and therefore,

$$V(\theta)=-\frac{1}{I(\theta)}=\frac{\theta^2}{\sum_{i=1}^{N}x_i}.$$

Substituting $\hat{\theta}_{ML}$ for θ yields

$$V\left(\hat{\theta}_{ML}\right)=\frac{\hat{\theta}_{ML}^2}{N\hat{\theta}_{ML}}=\frac{\hat{\theta}_{ML}}{N}.$$

EXERCISE 7.3

The purpose of this exercise is to make students aware of the different shapes of the probability function implied by the linear model and the logistic model.

Part (a) For part (a), we obtain the following results. These were calculated using Mathcad but they could easily be done using a spreadsheet.

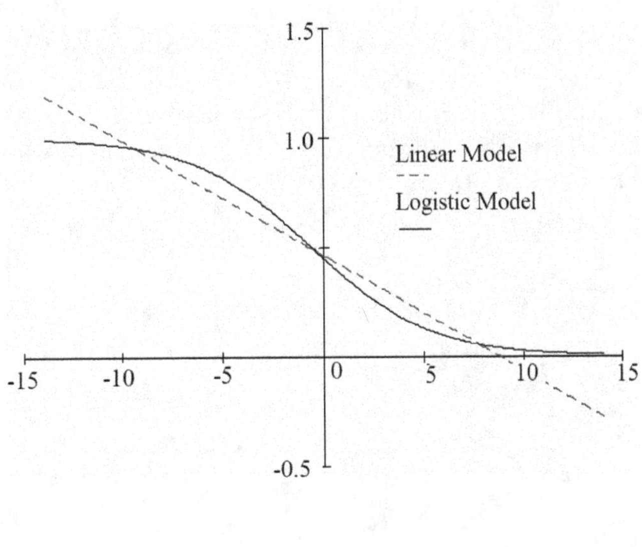

Part (b) The marginal effects for the two functions are plotted below. The marginal effect for the linear probability function is constant but that of the logistic model varies across the range of the exogenous variable. The marginal effect for the linear model can be thought of as an approximate average of the marginal effect from the logistic model across the range of values taken by the exogenous variable. Note that the derivatives here are calculated as approximate values using a small increment h in the right-hand side variable.

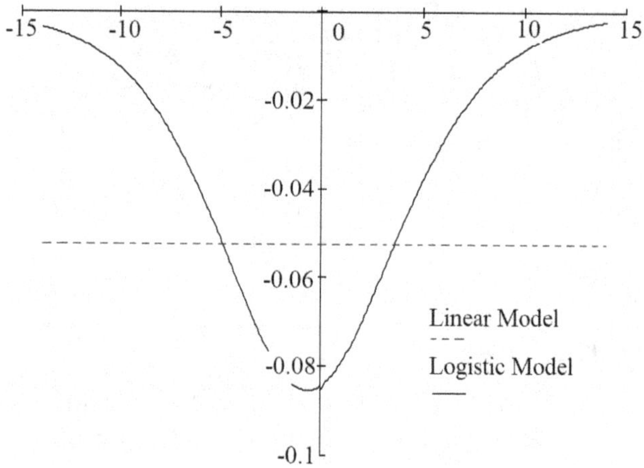

EXERCISE 7.5

The results of estimating the model using the logit, probit, and extreme value functions are given below. To assess which provides the best predictor, we look at either the McFadden R-squared or the percentage of correct predictions. All three models are very similar but the Logit model is marginally better. It has a higher McFadden R-squared than the other two models and the percentage of correct predictions is just higher, at 68%, than the other two models, for which it is 67%.

```
Logit Estimates

Newton-Raphson Method
Dependent Variable CADB
Sample Size 1358
Iterations 5
Variable          Coefficient          Std Err          T-Ratio

Constant            -0.142114          0.059470        -2.389679
FTSE               110.504091          8.580861        12.877972
```

```
Mean of RHS Variable      0.30746548e-3
SDev of RHS Variable      0.30757875e-3
Log Likelihood              -821.7302
Restricted LogL             -940.0549
McFadden R-Squared           0.125870
Marginal Effect             27.412968

Percentage distribution of outcomes

                      y=1              y=0

P> 0.50               0.31             0.15
                    ( 0.31)          ( 0.15)
P< 0.50               0.17             0.37
                    ( 0.17)          ( 0.37)

Figures in parentheses are deviations from 'naive' probabilities

Probit Estimates

Newton-Raphson Method
Dependent Variable CADB
Sample Size 1358
Iterations 5
Variable          Coefficient          Std Err          T-Ratio

Constant            -0.087058          0.036089        -2.412349
FTSE                63.954677          4.684762        13.651639

Mean of RHS Variable      0.30746548e-3
SDev of RHS Variable      0.30757875e-3
Log Likelihood              -824.0568
Restricted LogL             -940.0549
McFadden R-Squared           0.123395
Marginal Effect             25.456347

Percentage distribution of outcomes

                      y=1              y=0

P> 0.50               0.30             0.15
                    ( 0.30)          ( 0.15)
P< 0.50               0.17             0.37
                    ( 0.17)          ( 0.37)
```

```
Figures in parentheses are deviations from 'naive' probabilities

Extreme Value Estimates

Newton-Raphson Method
Dependent Variable CADB
Sample Size 1358
Iterations 5
Variable          Coefficient          Std Err          T-Ratio

Constant             0.313167         0.040954         7.646741
FTSE                71.758555         5.158511        13.910712

Mean of RHS Variable      0.30746548e-3
SDev of RHS Variable      0.30757875e-3
Log Likelihood              -822.7478
Restricted LogL             -940.0549
McFadden R-Squared           0.124788
Marginal Effect            1778.600760

Percentage distribution of outcomes

                  y=1             y=0

P> 0.50           0.32            0.17
                ( 0.32)         ( 0.17)
P< 0.50           0.16            0.35
                ( 0.16)         ( 0.35)

Figures in parentheses are deviations from 'naive' probabilities
```

CHAPTER 8: STOCHASTIC REGRESSORS

EXERCISE 8.1

Part (a) The changes in consumption expenditures and GDP are linked through the national income accounting identity. Consider a simple model of the form

$$\Delta c_t = \beta_1 + \beta_2 \Delta y_t + u_t$$
$$\Delta y_t = \Delta c_t + \Delta i_t + \Delta g_t$$

The reduced form equation for the change in GDP is

$$\Delta y_t = \frac{1}{1-\beta_2}\left(\beta_1 + u_t + \Delta i_t + \Delta g_t\right).$$

It follows that

$$\operatorname{cov}\left(\Delta y, u\right) = \frac{1}{1-\beta_2}\sigma_u^2,$$

and therefore

$$\operatorname{plim}\hat{\beta}_2 = \beta_2 + \frac{\operatorname{cov}\left(\Delta y, u\right)}{\operatorname{var}\left(\Delta y\right)} \neq \beta_2,$$

that is, OLS is inconsistent for this model.

Part (b) From the model presented in part (a), changes in investment spending and changes in government spending are possible instrument, in this case. Both are assumed to be exogenous in the simple Keynesian income-expenditure model. Of course, these assumptions may not be correct.

EXERCISE 8.3

Part (a) Each equation contains two endogenous variables and one of the two exogenous variables. Therefore, in both cases, we have $g-1=K-k=1$ and, by the order condition, both equations are just identified.

Part (b) The easiest way to solve for the reduced form is to write the system down in matrix form. We have

$$\begin{bmatrix} 1 & -\beta_{11} \\ -\beta_{12} & 1 \end{bmatrix}\begin{bmatrix} Y_{1t} \\ Y_{2t} \end{bmatrix} = \begin{bmatrix} \gamma_{11}X_{1t} + u_{1t} \\ \gamma_{22}X_{2t} + u_{2t} \end{bmatrix}$$

$$\begin{bmatrix} Y_{1t} \\ Y_{2t} \end{bmatrix} = \frac{1}{1-\beta_{11}\beta_{12}}\begin{bmatrix} 1 & \beta_{11} \\ \beta_{12} & 1 \end{bmatrix}\begin{bmatrix} \gamma_{11}X_{1t} + u_{1t} \\ \gamma_{22}X_{2t} + u_{2t} \end{bmatrix}$$

$$Y_{1t} = \frac{1}{\Delta}\left(\gamma_{11}X_{1t} + \beta_{11}\gamma_{22}X_{2t} + u_{1t} + \beta_{11}u_{2t}\right)$$

$$Y_{2t} = \frac{1}{\Delta}\left(\beta_{21}\gamma_{11}X_{1t} + \gamma_{22}X_{2t} + \beta_{21}u_{1t} + u_{2t}\right)$$

$$\Delta = 1 - \beta_{11}\beta_{21}$$

We, therefore, have

$$\pi_{11} = \frac{\gamma_{11}}{\Delta}; \pi_{12} = \frac{\beta_{11}\gamma_{22}}{\Delta}; \pi_{21} = \frac{\beta_{21}\gamma_{11}}{\Delta}; \pi_{22} = \frac{\gamma_{22}}{\Delta}.$$

Solving for the structural parameters yields

$$\beta_{11} = \frac{\pi_{12}}{\pi_{22}} \qquad \beta_{21} = \frac{\pi_{21}}{\pi_{11}}$$

$$\gamma_{11} = \pi_{11}\left(1 - \beta_{11}\beta_{21}\right) \qquad \gamma_{22} = \pi_{22}\left(1 - \beta_{11}\beta_{21}\right).$$

CHAPTER 9: DYNAMIC MODELS

EXERCISE 9.1

The purpose of this exercise is for students to work through an important result in dynamic modeling which generates too much tedious algebra to be put in the main text. We have the following

$$\hat{\beta} = \frac{\sum_{t=2}^{T} Y_{t-1}Y_t}{\sum_{t=2}^{T} Y_{t-1}^2} = \frac{\sum_{t=2}^{T} Y_{t-1}\left(\beta Y_{t-1} + u_t\right)}{\sum_{t=2}^{T} Y_{t-1}^2} = \beta + \frac{\sum_{t=2}^{T} Y_{t-1}u_t}{\sum_{t=2}^{T} Y_{t-1}^2}.$$

Therefore,

$$\text{plim}\,\hat{\beta} = \beta + \frac{\text{plim}(1/T)\sum_{t=2}^{T} Y_{t-1}u_t}{\text{plim}(1/T)\sum_{t=2}^{T} Y_{t-1}^2}.$$

By the assumption of stationarity and, using the result derived in the main text, we have

$$\text{plim}\frac{1}{T}\sum_{t=2}^{T} Y_{t-1}u_t = \text{cov}\left(Y_{t-1}u_t\right) = \frac{\rho}{1-\rho\beta}\sigma_u^2$$

Now let us consider $\text{plim}(1/T)\sum_{t=2}^{T} Y_{t-1}^2$. By the assumption of stationarity, this will be equal to the variance of Y. This can be derived as follows

$$\sigma_Y^2 = E\left(\beta Y_{t-1} + u_t\right)^2 = \beta^2 E\left(Y_{t-1}^2\right) + 2\beta\,\text{cov}\left(Y_{t-1}u_t\right) + \sigma_u^2.$$

Again, by stationarity we have $E\left(Y_{t-1}^2\right)=\sigma_Y^2$, and we have already derived an expression for $\operatorname{cov}\left(Y_{t-1}u_t\right)$. Substituting into our expression yields

$$\operatorname{plim}\frac{1}{T}\sum_{t=2}^{T}Y_{t-1}^2=\sigma_Y^2=\frac{1}{\left(1-\beta^2\right)}\left[\frac{2\beta\rho}{1-\rho\beta}+1\right]\sigma_u^2=\frac{1}{\left(1-\beta^2\right)}\left(\frac{1+\beta\rho}{1-\beta\rho}\right)\sigma_u^2.$$

Substituting both these probability limits into the expression for $\operatorname{plim}\hat{\beta}$ yields

$$\operatorname{plim}\hat{\beta}=\beta+\frac{\rho\left(1-\beta^2\right)}{\left(1+\beta\rho\right)}.$$

This gives us an explicit expression for the size of the inconsistency in the OLS estimator for β. For example, if $\beta=\rho=0.9$ then the inconsistency will equal 0.094. The OLS estimator converges on the value 0.994 in probability limit which, despite the fact that the underlying process is stationary, looks very much like a random walk. Since most tests for random walk processes lack power for values of β close to, but less than, one, this has potentially important implications when it comes to testing for unit roots.

EXERCISE 9.3

The purpose of this exercise is for students to be aware of the effects of altering the dynamic structure on regression estimates of parameters of interest.

Part (a) In this case, the parameters of interest are the elasticities for air travel with respect to income and price. The log-linear form of the regression equations means that the coefficients are of the form $d\ln(A)/d\ln(Y)$ or $(Y/A)dA/dY$, that is, the elasticity. In this case, we have an elasticity of air travel with respect to income of 1.75, and with respect to price of -1.26. The signs and magnitudes are reasonable from the point of view of economic theory.

Part (b) The problem with this regression is that there is significant autocorrelation as evidenced by the Durbin-Watson statistic of 0.10. From the tables, we see that the 5% lower bound for this test is 1.57 and therefore, there is significant positive autocorrelation. If the explanatory variables are positively autocorrelated, then this will lead to downward bias in the equation standard errors. We don't have any direct evidence that the explanatory variables are positively autocorrelated, but this is very common for time-series economic data and therefore downward bias in the standard errors is very likely.

A possible strategy for dealing with these problems is to respecify the equation to allow for a more plausible dynamic structure. For example, the partial adjustment model might provide more reliable statistical results. This is illustrated in part (c).

Part (c) The long-run elasticities can be calculated as

$$\eta_Y = \frac{0.3633}{1-0.8749} = 2.9041$$

$$\eta_P = \frac{0.2623}{1-0.8749} = 2.0967$$

The income elasticity is still positive but has increased in magnitude. Is an income elasticity of 2.9 plausible? Arguably yes, since the demand for air travel is likely to be a luxury good and therefore, likely to increase significantly as income rises. The price elasticity, however, is no longer consistent with any economic theory, since it is now positive and quite large in magnitude.

Part (d) Using the data in the workfile, we calculate the following Ljung-Box test statistics

	AR Coeff	Q-stat	5% crit val
Order 1	0.2312	4.1724	3.841
Order 2	-0.3087	11.7147	5.991
Order 3	-0.2972	18.7996	7.815
Order 4	-0.2246	22.9033	9.488

These confirm our conclusion from the Durbin-Watson test, that is, there is significant first-order autocorrelation in this equation. The Breusch-Godfrey test statistic for first-order autocorrelation is calculated as 4.80. This is asymptotically distributed as chi-squared with 1 degree of freedom, so we again reject the null hypothesis at the 5% level.

CHAPTER 10: TIME SERIES ANALYSIS AND ARIMA MODELING

EXERCISE 10.1

The purpose of this exercise is for students to see a more general moving average process than is discussed in the text and to see that the results stated in the text generalize to this (and other) cases.

Part (a) We have $X_t = \varepsilon_t + a_1 \varepsilon_{t-1} + a_2 \varepsilon_{t-2}$ and therefore

$$\begin{aligned} V(X_t) &= E(\varepsilon_t + a_1 \varepsilon_{t-1} + a_2 \varepsilon_{t-2})^2 \\ &= E(\varepsilon_t^2) + a_1^2 E(\varepsilon_{t-1}^2) + a_2^2 E(\varepsilon_{t-2}^2), \end{aligned}$$

since all cross-product terms involving $E(\varepsilon_t \varepsilon_{t-j}) = 0; j \neq 0$. Thus, we have

$$V(X_t) = (1 + a_1^2 + a_2^2)\sigma_\varepsilon^2$$

The first-order autocovariance is given by

$$\begin{aligned} E(X_t X_{t-1}) &= E\{(\varepsilon_t + a_1 \varepsilon_{t-1} + a_2 \varepsilon_{t-2})(\varepsilon_{t-1} + a_1 \varepsilon_{t-2} + a_2 \varepsilon_{t-3})\} \\ &= a_1 \sigma_\varepsilon^2, \end{aligned}$$

and the second-order autocovariance is given by

$$\begin{aligned} E(X_t X_{t-2}) &= E\{(\varepsilon_t + a_1 \varepsilon_{t-1} + a_2 \varepsilon_{t-2})(\varepsilon_{t-2} + a_1 \varepsilon_{t-3} + a_2 \varepsilon_{t-4})\} \\ &= a_2 \sigma_\varepsilon^2. \end{aligned}$$

In both cases, we make use of the property that $E(\varepsilon_t \varepsilon_{t-j}) = 0; j \neq 0$ to simplify these expressions.

Part (b) For all higher-order autocovariances we have

$$E(X_t X_{t-j}) = E\{(\varepsilon_t + a_1 \varepsilon_{t-1} + a_2 \varepsilon_{t-2})(\varepsilon_{t-j} + a_1 \varepsilon_{t-j-1} + a_2 \varepsilon_{t-j-2})\}.$$

Since $j \geq 3$, there are no cases in which the values of ε are contemporaneous. Hence, all the expectations will be zero and all higher-order autocovariances will be zero.

EXERCISE 10.3

The purpose of this exercise is to give students practice in identifying and estimating an ARIMA model. The data for this exercise has been generated artificially and is a realization of an ARIMA(1,1,0) stochastic process. It was generated using the EViews regression package using the following code.

```
create u 1 200
smpl 1 1
series x = 0
series u = 0
smpl 2 200
rndseed 100
series u = 0.7*u(-1)+@qnorm(rnd)
series x = 0.5+x(-1)+u
smpl 101 200
```

Part (a)

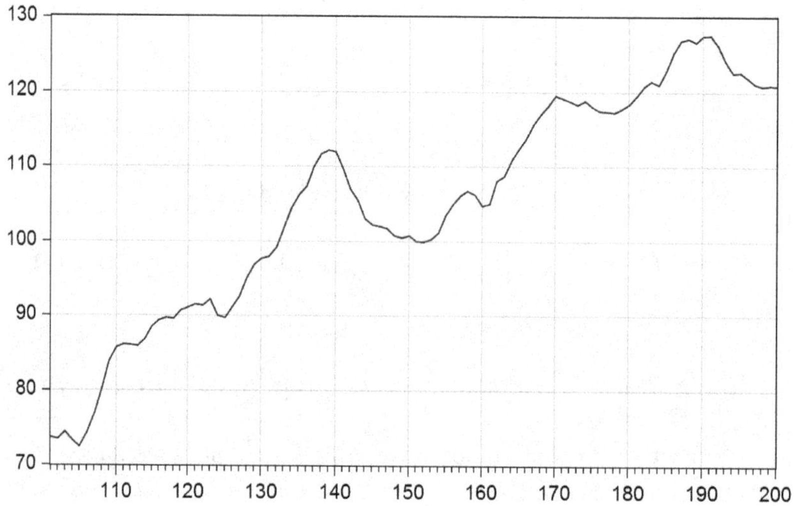

If we plot the series, then we see that it has a strong upward trend. This suggests that differencing may be appropriate, either to remove the trend or to deal with possibly non-stationarity of the process.

Part (b)

Sample: 101 200
Included observations: 100

Autocorrelation	Partial Correlation		AC	PAC	Q-Stat	Prob
		1	0.968	0.968	96.619	0.000
		2	0.931	-0.114	186.78	0.000
		3	0.891	-0.047	270.22	0.000
		4	0.847	-0.085	346.37	0.000
		5	0.799	-0.065	414.94	0.000
		6	0.753	-0.001	476.40	0.000
		7	0.709	0.019	531.47	0.000
		8	0.668	0.023	580.91	0.000
		9	0.630	0.021	625.44	0.000
		10	0.595	-0.012	665.51	0.000
		11	0.560	-0.017	701.49	0.000
		12	0.527	-0.018	733.66	0.000
		13	0.493	-0.036	762.17	0.000
		14	0.462	0.016	787.44	0.000
		15	0.435	0.053	810.12	0.000
		16	0.411	0.033	830.68	0.000
		17	0.390	0.002	849.38	0.000
		18	0.367	-0.060	866.16	0.000
		19	0.345	-0.024	881.13	0.000
		20	0.322	-0.030	894.33	0.000

This is typical of the correlogram of a non-stationary time series. The autocorrelations die down linearly to zero and there is a single partial autocorrelation outside the standard error bounds. If we examine the correlogram of the first differenced series then we obtain the following results.

Sample: 101 200
Included observations: 100

Autocorrelation	Partial Correlation		AC	PAC	Q-Stat	Prob
		1	0.626	0.626	40.368	0.000
		2	0.358	-0.055	53.718	0.000
		3	0.163	-0.064	56.527	0.000
		4	0.036	-0.044	56.667	0.000
		5	0.033	0.086	56.784	0.000
		6	0.013	-0.035	56.802	0.000
		7	-0.029	-0.061	56.896	0.000
		8	-0.065	-0.037	57.368	0.000
		9	-0.136	-0.098	59.454	0.000
		10	-0.215	-0.123	64.707	0.000
		11	-0.219	-0.014	70.203	0.000
		12	-0.156	0.046	73.031	0.000
		13	-0.219	-0.223	78.667	0.000
		14	-0.256	-0.098	86.455	0.000
		15	-0.222	0.036	92.364	0.000
		16	-0.129	0.077	94.372	0.000
		17	0.018	0.080	94.410	0.000
		18	0.118	0.054	96.130	0.000
		19	0.153	0.025	99.091	0.000
		20	0.136	-0.034	101.45	0.000

The autocorrelations now decline exponentially and there is a single partial autocorrelation outside the standard error bands. This is consistent with stationary AR(1) process.

Part (c)

Dependent Variable: D(X)
Method: ARMA Maximum Likelihood (BFGS)
Sample: 101 200
Included observations: 100
Convergence achieved after 4 iterations
Coefficient covariance computed using outer product of gradients

Variable	Coefficient	Std. Error	t-Statistic	Prob.
C	0.521460	0.266261	1.958452	0.0530
AR(1)	0.635134	0.081512	7.791917	0.0000
SIGMASQ	1.015249	0.140968	7.201962	0.0000

R-squared	0.401609	Mean dependent var	0.497398
Adjusted R-squared	0.389271	S.D. dependent var	1.309110
S.E. of regression	1.023059	Akaike info criterion	2.918176
Sum squared resid	101.5249	Schwarz criterion	2.996332
Log likelihood	-142.9088	Hannan-Quinn criter.	2.949807
F-statistic	32.55067	Durbin-Watson stat	1.904337
Prob(F-statistic)	0.000000		

Inverted AR Roots	.64

If we estimate an ARIMA(1,1,0) model then we obtain the results shown above. We now check the correlogram of the residuals to see if there is any information left that we can incorporate into our model. The correlogram, shown below, is very flat. This indicates that our model has incorporated all the systematic information in the data and there is no gain to including more autoregressive or moving average terms.

Sample: 101 200
Included observations: 100
Q-statistic probabilities adjusted for 1 ARMA term

Autocorrelation	Partial Correlation		AC	PAC	Q-Stat	Prob
		1	0.036	0.036	0.1328	
		2	-0.002	-0.004	0.1334	0.715
		3	-0.026	-0.026	0.2069	0.902
		4	-0.120	-0.119	1.7496	0.626
		5	0.019	0.027	1.7880	0.775
		6	0.018	0.015	1.8214	0.873
		7	-0.019	-0.026	1.8598	0.932
		8	0.021	0.010	1.9105	0.965
		9	-0.014	-0.009	1.9327	0.983
		10	-0.121	-0.120	3.6052	0.935

CHAPTER 11: UNIT ROOTS AND SEASONALITY

EXERCISE 11.1

The purpose of this exercise is to expand on the treatment of stationarity testing in the main text and to demonstrate that the test normally applied is not a sufficient, condition.

PROPOSITION: Consider the process $X_t = \theta_1 X_{t-1} + \theta_2 X_{t-2} + \varepsilon_t$. The conditions for stability are $-\theta_2 < 1$, $1 - \theta_1 - \theta_2 > 0$ and $1 + \theta_1 - \theta_2 > 0$. (These are standard stability conditions for a second-order linear difference equation with constant coefficients.)

PROOF: This has characteristic equation

$$\lambda^2 - \theta_1 \lambda - \theta_2 = 0,$$

and therefore, the roots can be derived as

$$\lambda_{1,2} = \frac{\theta_1 \pm \sqrt{\theta_1^2 + 4\theta_2}}{2}.$$

We, therefore, have

$$\lambda_1 + \lambda_2 = \theta_1$$
$$\lambda_1 \lambda_2 = -\theta_2$$

From the second condition, if the roots are real then $|\theta_2| > 1$ immediately establishes that the equation is unstable because at least one root must be greater than one in absolute value. If the roots are complex, then their product is equal to the square of the modulus and, again, $-\theta_2 > 1$ immediately establishes that the equation is unstable.

Now, considering only cases in which $|\theta_2| < 1$, and assuming that the roots are real, we have instability if $(1 - \lambda_1)(1 - \lambda_2) < 0$, since at least one root must be greater than one. We can write this condition as

$$(1 - \lambda_1)(1 - \lambda_2) = 1 - (\lambda_1 + \lambda_2) + \lambda_1 \lambda_2 < 0$$
$$\Leftrightarrow 1 - \theta_1 - \theta_1 < 0.$$

Hence, the equation is unstable if $1 - \theta_1 - \theta_2 > 0$. The equation is also unstable if $(\lambda_1 + 1)(\lambda_2 + 1) < 0$, since at least one root must be less than -1. We can write this condition as

$$(\lambda_1+1)(\lambda_2+1)=\lambda_1\lambda_2+(\lambda_1+\lambda_2)+1<0$$
$$\Leftrightarrow -\theta_2+\theta_1+1<0$$

Hence, the equation is also unstable if $1+\theta_1-\theta_2<0$. Thus, if we consider the plane defined by the coefficients θ_1 and θ_2, we can use these conditions to define boundaries between regions in which the values of the coefficients yield stable and unstable solutions when the roots are real.

Finally, we return to the case of complex roots. These can only occur when $\theta_2<0$ since we need $\theta_1^2+4\theta_2<0$. This condition also defines the boundary between regions of complex and real roots in the (θ_1,θ_2) plane.

The possible types of solution are illustrated in the following diagram. All stable solutions lie somewhere in the interval $-1<\theta_2<1$. The curved lines indicate the dividing lines between the regions of stable and unstable real roots, with the region of stability lying below the line $\theta_1=\theta_2-1$ and above the line $\theta_1=1-\theta_2$. Within this region, the stable solutions may have real roots, which are both less than one in absolute value, or complex roots, with a modulus less than one. The boundary between these regions is given by the curve $\theta_1=2\sqrt{\theta_2}$. Note that this curve is tangent to the line $\theta_1=\theta_2-1$ at $(-1,2)$ and to the line $\theta_1=\theta_2-1$ at $(-1,-2)$. Stable regions with real roots are given by the darker shaded regions and stable regions with complex roots are given by the lighter shaded regions.

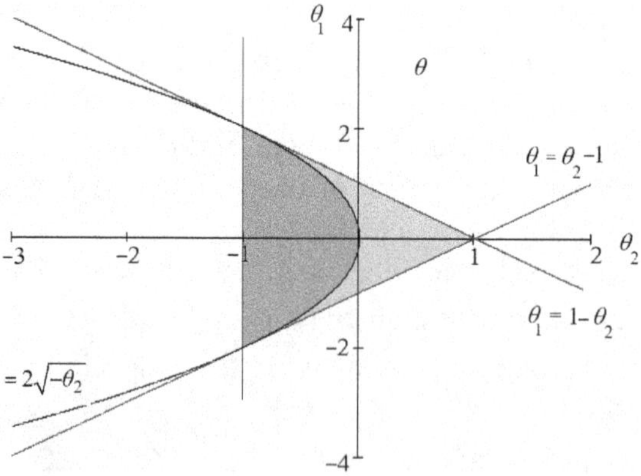

Returning to our question, we see, from the results that we have derived, that $\theta_1+\theta_1-1<0$ is only one of the stability conditions. We also need $|\theta_2|<1$ and $1+\theta_1-\theta_2>0$. It follows that the alternative in the standard Dickey-Fuller test is a necessary, but not sufficient, condition for stability in second (and higher-order) difference equations.

EXERCISE 11.3

If we plot the series EXRATE, then there is no obvious time trend.

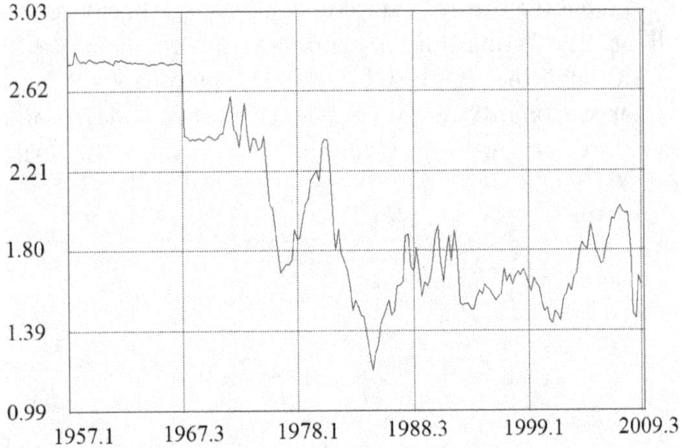

Variable: EXRATE   Scaling Factor 10^0

However, it is still possible for a process to have a unit root, even if it is not trended. If we estimate the Dickey-Fuller test equation, then we obtain the following results.

```
Ordinary Least Squares Regression Results
Sample period: 1957.3 to 2009.3
Dependent Variable D(LE)
Sample Size 209

Variable             Coefficient          Std Err         T-Ratio

C                       0.014352         0.009538        1.504721
LE(-1)                 -0.024421         0.013227       -1.846206
D(LE(-1))               0.161327         0.068559        2.353106

R-squared             0.3883 E-1     F-statistic           4.1608
SEE                 0.458720 E-1     RSS                 0.433474
Durbin-Watson             1.9351     LogL              349.069817
ARCH(1) Test              2.5638     AIC                -3.311673
Jarque-Bera              62.9765     SIC                -3.263697
```

The test statistic is therefore –1.84. If we were to use a conventional t-test then this would be significant at the 5% level (since this is a one-tailed test). However, we need to use the Dickey-Fuller critical values. In this case, we find that the 5% critical value is –2.87 and therefore, we cannot reject the null hypothesis that the series contains a unit root at the 5% level.

Although the graph did not show any obvious time trend, we will also test for stationarity around a linear trend in order to illustrate the testing procedure. The test regression in this case, takes the form

```
Ordinary Least Squares Regression Results
Sample period: 1957.3 to 2009.3
Dependent Variable D(LE)
Sample Size 209
```

Variable	Coefficient	Std Err	T-Ratio
C	0.054763	0.023099	2.370760
TREND	-0.000166	0.000086	-1.918296
LE(-1)	-0.057699	0.021764	-2.651139
D(LE(-1))	0.181232	0.068903	2.630232

R-squared	0.5578 E-1	F-statistic	4.0365
SEE	0.455765 E-1	RSS	0.425830
Durbin-Watson	1.9352	LogL	350.929012
ARCH(1) Test	1.7551	AIC	-3.319895
Jarque-Bera	46.3765	SIC	-3.255927

In this case, the test statistic is equal to –2.65 but the critical value is now –3.43. Therefore, it is still not possible to reject the null of a unit root at the 5% level.

Does this process contain more than one unit root? To test for this, we difference the series again and test the null hypothesis that the differenced series contains a unit root. This gives the following results

```
Ordinary Least Squares Regression Results
Sample period: 1957.4 to 2009.3
Dependent Variable D(DE)
Sample Size 208
```

Variable	Coefficient	Std Err	T-Ratio
C	-0.002839	0.003161	-0.898205
DE(-1)	-1.020084	0.089877	-11.349723
D(DE(-1))	0.204139	0.070332	2.902486
R-squared	0.4467	F-statistic	82.7420
SEE	0.045438	RSS	0.423240
Durbin-Watson	1.9272	LogL	349.385611
ARCH(1) Test	3.3136	AIC	-3.330631
Jarque-Bera	46.0260	SIC	-3.282493

The test statistic here is −11.35 and the 5% critical value is again −2.87. Therefore, in this case, we can reject the null hypothesis at the 5 level. We, therefore, conclude that this process contains a single unit root.

CHAPTER 12: COINTEGRATION

EXERCISE 12.1

The purpose of this exercise is to help students understand how a multi-variable stochastic system can be solved so that it is expressed in terms of a single variable. In doing so, a first-order system in two variables becomes a second-order system in a single variable.

Part (a) We have

$$X_t = 0.05 + X_{t-1} + \varepsilon_{1t}$$
$$Y_t = 0.25X_t + 0.75Y_{t-1} + \varepsilon_{2t}.$$

Using the lag operator, we can write this as

$$X_t(1-L) = 0.05 + \varepsilon_{1t}$$
$$Y_t(1-0.75L) = 0.25X_t + \varepsilon_{2t},$$

and therefore

$$(1-0.75L)Y_t = 0.25 \times \frac{0.05+\varepsilon_{1t}}{1-L} + \varepsilon_{2t}$$
$$(1-L)(1-0.75L)Y_t = 0.0125 + 0.25\varepsilon_{1t} + (1-L)\varepsilon_{2t}.$$

Note, the presence of the unit root in Y. We now have an ARIMA(1,1,1) model in the Y variable.

Part (b) We can derive an error correction form for this model by first substituting for the current value of the X variable in the second equation to obtain

$$Y_t = 0.25(0.05 + X_{t-1} + \varepsilon_{1t}) + 0.75Y_{t-1} + \varepsilon_{2t}.$$

Now, subtract Y_{t-1} from both sides and multiply out to obtain.

$$\Delta Y_t = 0.0125 + 0.25X_{t-1} + 0.25\varepsilon_{1t} - 0.25Y_{t-1} + \varepsilon_{2t}$$
$$= 0.0125 - 0.25(Y_{t-1} - X_{t-1}) + 0.25\varepsilon_{1t} + \varepsilon_{2t}$$

This is an error correction equation. Note that it is not the only way of writing an error correction equation. An exactly equivalent alternative is given by

$$\Delta Y_t = 0.25\Delta X_t - 0.25(Y_{t-1} - X_{t-1}) + \varepsilon_{2t}$$

In both cases the error correction model contains a mixture of differenced and levels terms. Let us take the first version, given that the Y series contains a single unit root, it follows that ΔY_t is $I(0)$. On the right-hand side we have $0.25\varepsilon_{1t}$ and ε_{2t} which are both $I(0)$ by assumption. Therefore, $(Y_{t-1} - X_{t-1})$ must also be $I(0)$ for the equality between the RHS and the LHS to hold. It follows that there is a cointegrating relationship.

Part (c) Given the equation we have derived, the cointegrating parameter is equal to one and the speed of adjustment is equal to 0.25. That is, Y adjusts such that one-quarter of any deviation from equilibrium is eliminated in each time period.

EXERCISE 12.3

The purpose of this exercise is for students to practice implementing unit root tests and to use the information acquired to specify a regression model.

Part (a) Using the data set provided, we can carry out augmented Dickey-Fuller unit root tests for both the Treasury Bill Rate and the Government Bond Yield. The equations take the form

$$\Delta X_t = \gamma_0 + \gamma_1 X_{t-1} + \gamma_2 \Delta X_{t-1} + \varepsilon_t$$

The test statistic is $\tau = \hat{\gamma}_1 / \mathrm{SE}(\hat{\gamma}_1)$. The test did not include a trend in the test equation since neither series is trended. For the Treasury Bill rate, we obtain a test statistic of -1.99 and, for the bond yield, we obtain a test statistic of -2.65. The 5% critical value from the MacKinnon response surfaces is -2.89 so neither statistic is significant at this level. However, that for the bond yield is not too far from the 5% critical value, and this would be significant at the 10% level, where the critical value is -2.58.

Part (b) Next, we regress the bond yield on the Treasury Bill rate and obtain the following results.

```
Ordinary Least Squares Regression Results
Sample period: 1980.2 to 2001.4
Dependent Variable R
Sample Size 87

Variable          Coefficient          Std Err        T-Ratio

C                    0.352717         0.420809       0.838187
TBR                  1.122105         0.044478      25.228217

R-squared              0.8822      F-statistic       636.4630
SEE                  1.317364      RSS             147.513082
Durbin-Watson          0.5434      LogL           -146.416033
ARCH(1) Test          20.1744      AIC               3.411863
Jarque-Bera            1.1832      SIC               3.468550
```

We would expect a cointegrating parameter close to one, and this is consistent with the results. However, is there evidence here that this is a cointegrating relationship and not a spurious regression? To assess this, we can use the Engle-Granger test, that is, perform an ADF test on the equation residuals. The results obtained are as follows.

```
Ordinary Least Squares Regression Results
Sample period: 1980.4 to 2001.4
Dependent Variable D(RES)
Sample Size 85

Variable          Coefficient            Std Err         T-Ratio

C                    0.005935           0.092766        0.063978
RES(-1)             -0.341956           0.076501       -4.469924
D(RES(-1))           0.289033           0.102169        2.828966

R-squared              0.2099     F-statistic            10.8922
SEE                  0.854957     RSS                  59.938005
Durbin-Watson          2.0324     LogL               -105.762805
ARCH(1) Test       0.5990 E-2     AIC                   2.559125
Jarque-Bera           19.1717     SIC                   2.645336
```

The test statistic is, therefore, –4.47. The 5% critical value from MacKinnon's response surface is –3.41. We can, therefore, conclude that there is evidence that this is a genuine cointegrating relationship rather than a spurious regression.

Part (c) Finally, we estimate an error correction model for the government bond yield and obtain the following results.

```
Ordinary Least Squares Regression Results
Sample period: 1980.3 to 2001.4
Dependent Variable D(R)
Sample Size 86

Variable          Coefficient            Std Err         T-Ratio

C                   -0.452007           0.097280       -4.646435
D(TBR)               0.250152           0.035772        6.992848
R(-1)               -0.374256           0.024494      -15.279456
TBR(-1)              0.474524           0.029120       16.295425

R-squared              0.8032     F-statistic           111.5433
SEE                  0.294323     RSS                   7.103340
Durbin-Watson          0.8160     LogL                -14.796080
ARCH(1) Test          18.9630     AIC                   0.437118
Jarque-Bera            9.0223     SIC                   0.551274
```

The Ericsson and MacKinnon cointegration test statistic is given by the t-ratio for the lagged government bond yield, that is, –15.27. The 5% critical value for this test is –3.25, so we again reject the null of no cointegration, confirming the earlier result from the Engle-Granger test. As a final check, we can use an F-test for the joint significance of the two lagged levels variables in this equation. The test statistic was equal to 146.75 and the 5% critical value is 5.91. There is, therefore, some reasonably strong evidence that this is a genuine cointegrating relationship.

CHAPTER 13: VECTOR AUTOREGRESSIONS

EXERCISE 13.1

The purpose of this exercise is for students to practice writing higher-order systems in companion form.

Consider the equation $X_t = a_1X_{t-1} + a_2X_{t-2} + a_3X_{t-3} + u_t$. We define $Z_{1t} = X_{t-1}$ and $Z_{2t} = Z_{1t-1} = X_{t-2}$. This means that we can write the original equation as

$$\begin{pmatrix} X_t \\ Z_{1t} \\ Z_{2t} \end{pmatrix} = \begin{pmatrix} a_1 & a_2 & a_3 \\ 1 & 0 & 0 \\ 0 & 1 & 0 \end{pmatrix} \begin{pmatrix} X_{t-1} \\ Z_{1t-1} \\ Z_{2t-1} \end{pmatrix} + \begin{pmatrix} u_t \\ 0 \\ 0 \end{pmatrix}$$

This is the companion form of the system. The eigenvalues of the transition matrix are defined as

$$\begin{vmatrix} a_1 - \lambda & a_2 & a_3 \\ 1 & -\lambda & 0 \\ 0 & 1 & -\lambda \end{vmatrix} = 0$$

$$(a_1 - \lambda)\begin{vmatrix} -\lambda & 0 \\ 0 & -\lambda \end{vmatrix} - a_2\begin{vmatrix} 1 & 0 \\ 0 & -\lambda \end{vmatrix} + a_3\begin{vmatrix} 1 & -\lambda \\ 0 & 1 \end{vmatrix} = 0$$

$$-\lambda^3 + a_1\lambda^2 + a_2\lambda + a_3 = 0.$$

The characteristic equation for the original equation takes the form

$$\lambda^3 - a_1\lambda^2 - a_2\lambda - a_3 = 0.$$

Therefore, the solutions for λ are the same. These are just alternative ways of solving for the roots of the system.

EXERCISE 13.3

The purpose of this exercise is for students to gain some hands-on experience in estimating and interpreting a vector autoregression. The example chosen is deliberately simple so that they are not distracted by excessive complications or a VAR which is difficult to interpret.

Part (a) Since this is annual data, the lag length has been set at one. The estimates of the VAR are as follows

Vector Autoregression Estimates
Sample (adjusted): 1868 1911
Included observations: 44 after adjustments
Standard errors in ( ) & t-statistics in [ ]

	Q	P
Q(-1)	0.016044	-0.345955
	(0.24305)	(0.37303)
	[0.06601]	[-0.92741]
P(-1)	0.426491	-0.681005
	(0.17088)	(0.26227)
	[2.49581]	[-2.59657]
C	2.896545	15.27055
	(4.60813)	(7.07256)
	[0.62857]	[2.15913]
R-squared	0.349012	0.239147
Adj. R-squared	0.317256	0.202032
Sum sq. resids	24391.71	57457.55
S.E. equation	24.39098	37.43533
F-statistic	10.99058	6.443429
Log likelihood	-201.4251	-220.2748
Akaike AIC	9.292050	10.14885
Schwarz SC	9.413699	10.27050
Mean dependent	6.284318	7.794545
S.D. dependent	29.51891	41.90722

Determinant resid covariance (dof adj.)	276800.3
Determinant resid covariance	240341.6

Note that, in both equations, the lagged price variable is significant, but the lagged quantity variable is not. In VAR modeling, however, we retain all lags, whether they are significant or not.

Part (b) For the impulse responses, we choose an horizon length of five years. This is sufficient for the dynamics of the system to work themselves out and the model to reach an equilibrium solution following a shock. We present the results in graphical form because this is the easiest way to interpret impulse responses.

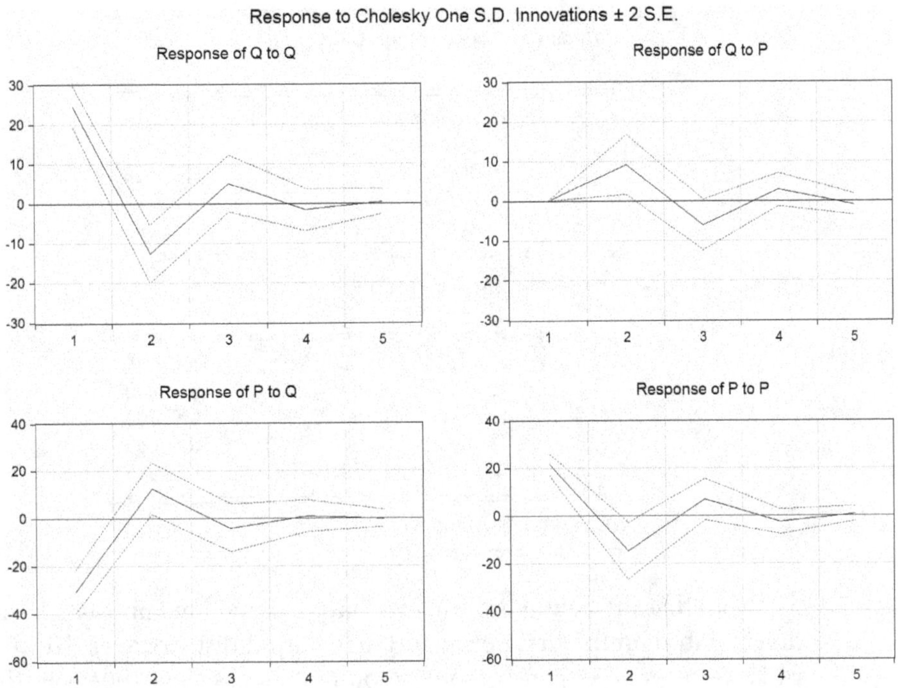

The solid lines here indicate the central estimate of the impulse response, while the broken lines show a 95% confidence interval. Note that we have chosen a causal ordering here by entering the variables in a particular order. This causal ordering means that price changes do not affect quantity contemporaneously, but changes in quantity do have an immediate effect on price. What we see is that a shock to quantity reduces price in the short run. The dynamics here are consistent with the cobweb model which is often applied to agricultural markets.

Part (c) The variance decomposition is presented in table form because this is the easiest way to read and interpret this output. The results are shown below

Variance Decomposition of Q:

Period	S.E.	Q	P
1	24.39098	100.0000	0.000000
2	28.97875	89.92213	10.07787
3	30.05066	86.48434	13.51566
4	30.22700	85.77679	14.22321
5	30.24736	85.67920	14.32080

Variance Decomposition of P:

Period	S.E.	Q	P
1	37.43533	66.79949	33.20051
2	42.08197	61.54190	38.45810
3	42.82446	60.32700	39.67300
4	42.91092	60.13941	39.86059
5	42.91762	60.12135	39.87865

Cholesky Ordering: Q P

Note that the causal ordering we have assumed means that all the variation in output in the first period is due to output disturbances. However, the variance of price depends more on output shocks than shocks to the price variable itself. There is some movement in these ratios over time, with price shocks becoming more important for both quantity and price itself. Note that the results here depend on the causal ordering we have assumed and would change if we adopted a different ordering. It is, therefore, important that we choose a causal ordering for a reason rather than just randomly. If we do not have a good reason for a particular causal ordering, then alternatives should be investigated and the sensitivity of the results to the final choice should be assessed.

INDEX

S

www.ingramcontent.com/pod-product-compliance
Lightning Source LLC
LaVergne TN
LVHW060055220926
848484LV00006BA/1281